Organic Fertilizers in Horticulture

Guest Editors
Francesco De Mastro
Department of Soil, Plant and Food Sciences
University of Bari Aldo Moro
Bari
Italy

Gennaro Brunetti
Department of Soil, Plant and Food Sciences
University of Bari Aldo Moro
Bari
Italy

Karam Farrag
Central Laboratory for Environmental Quality Monitoring (CLEQM)
National Water Research Center (NWRC)
Cairo
Egypt

Huadong Zang
College of Agronomy and Biotechnology
China Agricultural University
Beijing
China

Editorial Office
MDPI AG
Grosspeteranlage 5
4052 Basel, Switzerland

This is a reprint of the Special Issue, published open access by the journal *Horticulturae* (ISSN 2311-7524), freely accessible at: https://www.mdpi.com/journal/horticulturae/special_issues/02GKCVF27U.

For citation purposes, cite each article independently as indicated on the article page online and as indicated below:

Lastname, A.A.; Lastname, B.B. Article Title. *Journal Name* **Year**, *Volume Number*, Page Range.

ISBN 978-3-7258-3875-2 (Hbk)
ISBN 978-3-7258-3876-9 (PDF)
https://doi.org/10.3390/books978-3-7258-3876-9

Contents

horticulturae

Editorial

Sustainable Horticulture: Advancements and Challenges in Organic Fertilizer Applications

Francesco De Mastro [1,*], Gennaro Brunetti [1], Karam Farrag [2] and Huadong Zang [3]

1 Department of Soil, Plant and Food Sciences, University of Bari Aldo Moro, Via Amendola 165/A, 70126 Bari, Italy; gennaro.brunetti@uniba.it
2 Central Laboratory for Environmental Quality Monitoring (CLEQM), National Water Research Center (NWRC), Cairo 13621, Egypt; karam_farrag@hotmail.com
3 College of Agronomy and Biotechnology, China Agricultural University, Beijing 100193, China; zanghuadong@cau.edu.cn
* Correspondence: francesco.demastro@uniba.it

1. Introduction

Modern horticulture is increasingly focused on enhancing crop quality while prioritizing environmental sustainability. Beyond traditional quality attributes, consumers are demanding fruits and vegetables with higher nutritional value and minimal pesticide residues. This shift necessitates sustainable management practices that optimize plant–environment interactions and reduce reliance on external inputs. While intensive horticulture still depends heavily on synthetic fertilizers to ensure rapid nutrient replenishment and high yields [1,2], their prolonged and excessive use has led to significant environmental challenges, including soil degradation, microbial imbalance, erosion, acidification, and groundwater pollution [3]. Transitioning to organic horticulture offers a promising solution, as organic fertilizers have demonstrated the potential to improve yields, enhance soil health, and promote sustainable agricultural practices [4].

Organic fertilizers provide a balanced mix of macronutrients and essential micronutrients such as iron and zinc. They facilitate the gradual release of nutrients into the soil, stimulating microbial growth and activity, which, in turn, enhances nutrient and water availability. These processes contribute to improved soil structure, supporting healthy root development in vegetable crops [5]. Additionally, organic fertilizers help reduce soil acidity, mitigate heavy metal contamination [6], suppress pests and diseases, and minimize nutrient leaching. Their slow decomposition ensures a steady supply of nutrients over an extended period, making them a sustainable alternative to synthetic inputs [7]. Innovations in horticulture, such as the integration of cover crops and diversified crop rotations with organic fertilizers, further enhance soil properties, microbial diversity, and nutrient availability, thereby reducing the environmental impact of chemical fertilizers.

This integrated approach not only enhances crop productivity but also mitigates the environmental impact of chemical fertilizers. The primary challenge today lies in developing innovative organic fertilizers capable of transforming intensive vegetable cultivation practices. The scientific community is increasingly focused on exploring a diverse range of products and waste-derived materials. Recent advancements include the creation of bioformulations that integrate organic matter with mineral particles, nanomaterials, and plant-growth-promoting microorganisms. These "smart" fertilizers are designed to improve nutrient use efficiency and facilitate the transition to more sustainable agricultural systems [8]. Among these innovations, algae-based formulations have garnered significant

Received: 20 February 2025
Revised: 4 March 2025
Accepted: 6 March 2025
Published: 11 March 2025

Citation: De Mastro, F.; Brunetti, G.; Farrag, K.; Zang, H. Sustainable Horticulture: Advancements and Challenges in Organic Fertilizer Applications. *Horticulturae* **2025**, *11*, 307. https://doi.org/10.3390/horticulturae11030307

attention due to their environmental friendliness and cost-effectiveness. Their application has demonstrated productivity increases in various horticultural crops, including tomatoes [9]

However, several knowledge gaps persist regarding the use of new organic fertilizers. A major challenge is the variability in nutrient content across different organic materials, which complicates the precise fulfillment of the specific nutritional needs of vegetable crops. Additionally, the slow-release nature of these fertilizers may not align with the immediate nutrient demands of certain horticultural crops, potentially leading to reduced yields. There is also a critical need for long-term, field-scale studies conducted under diverse climatic conditions to thoroughly evaluate the effects of organic inputs. Furthermore, the improper management of high-nitrogen organic fertilizers can result in increased nitrous oxide emissions, a potent greenhouse gas, under specific conditions. To address these challenges, further research is essential to optimize organic fertilizer formulations and application methods. This will ensure that these fertilizers meet crop nutrient requirements effectively while minimizing negative environmental impacts. By overcoming these barriers, the horticultural sector can fully realize the potential of organic fertilizers in promoting sustainable and productive agricultural systems.

2. The Importance of Organic Fertilizers in Horticulture

In light of these considerations, the use of organic fertilizers in horticulture is a timely and important topic in the context of environmental sustainability, and it has been explored in different ways in this Special Issue (SI). Therefore, the goal of this SI is to highlight innovative research on organic fertilizers in horticulture to further disseminate foundational knowledge in this field and to promote its practical application on a larger scale. This issue presents a series of papers covering a range of topics related to the use of different matrices as fertilizers for horticultural crops. In this editorial, we briefly describe ten research articles (contributions 1–10). Readers will be able to find information about the potential use of unconventional solid matrices, such as cereal flours and by-products from the cold pressing of fruits and seeds (contribution 1–3), sewage sludge and by-products from insect farming (contribution 4,5), biostimulants (contribution 6,7), and liquid matrices like vermicompost extracts (contribution 8,9), as well as companion cropping systems (contribution 10) to improve crop yields and soil quality.

3. Unconventional Solid Matrices

Parecido et al. (contribution 1) assessed the impact of pure and mixed castor meal with ground hydrothermalized phonolite rock (CM+HP mixture) on nutrient supply, particularly nitrogen (N) and potassium (K), as well as on the optimization of sweet potato yield and quality. The field experiment, which involved treatments with and without synthetic fertilizers combined with varying percentages of organic fertilizers, revealed that the CM+HP mixture helped maintain adequate N and K levels in the plant leaves. Additionally, organic fertilizers increased both the number of reserve roots per plant and the sweetness of those roots, while synthetic fertilizers enhanced the average weight of the reserve roots. Castor meal, when combined with synthetic fertilizers, also contributed to improved soil health. The combined application of synthetic fertilizers with 2.4 Mg ha^{-1} of castor meal or 4.5 Mg ha^{-1} of the CM+HP mixture yielded the greatest benefits for storage root production, with an average increase of 128% in marketable storage root yield and nutrient removal compared to the use of organic fertilizers alone.

Butterfield et al. (contribution 2) also explored the potential use of corn gluten meal (CGM) and soybean meal (SBM) as organic fertilizers. The aim of this study was to evaluate the feasibility of integrating weed and nitrogen management by examining the effects of varying seed meal application rates within plastic-mulched film planting holes on weed density, soil nitrogen availability, and crop yield in tomato (*Solanum lycopersicum*) and broccoli (*Brassica oleracea*). Increasing seed meal rates led to a reduction in weed density, regardless of the type used. However, while soil nitrogen availability increased with the application rate, ammonium mineralized from seed meals applied at the highest rates was likely phytotoxic to both weeds and crops. In terms of crop yield, the seed meal did not have a positive effect. For tomatoes, yields were reduced by 39% to 64% in 2018 compared to the no-weed control. These findings suggest that using the seed meal in plastic-mulched planting holes for weed and nitrogen management is not a viable option, as the application rates required for effective weed suppression also proved to be toxic to the crops.

As an alternative to seed flour, by-products from the oil extraction process of oilseeds can also serve as organic fertilizers. This is exemplified by the experimental trial conducted by Tong et al. (contribution 3), which aimed to uncover the mechanisms underlying the effects of various fertilization treatments—such as chemical fertilizer alone, chemical fertilizer combined with cake fertilizer, chemical fertilizer combined with manure fertilizer, and a combination of all three—on soil fertility, soil enzyme activities, and pecan fruit quality. The results showed that combined fertilization could enhance both the yield and quality of pecan nuts. Among the treatments, the combination of chemical fertilizer with both organic fertilizers yielded the most promising results, with the pecan kernel oil content and unsaturated fatty acid levels reaching 72% and 98%, respectively. While the combined fertilization treatments had no significant impact on soil trace elements, they did significantly increase the activities of available phosphorus, total nitrogen, soil organic matter, and S-ACP (soil acid phosphatase).

4. Sewage Sludge and Insect Farming By-Products

Another valuable by-product that can be repurposed as an organic fertilizer is sewage sludge. Globally, nearly 75–100 million tons of dry matter (DM) are produced annually, with projections estimating an increase to 127.5 million tons by 2030 [10]. Its reuse in agriculture presents a sustainable solution that reduces waste while enhancing soil fertility and horticultural productivity. AL-Huqail et al. (contribution 4) investigated the effects of sewage sludge (DM) amendments on the growth, yield, and biochemical properties of marigolds (*Tagetes erecta* L. var. *Pusa Basanti Gainda*). Their findings revealed a significant improvement in plant growth, yield, and biochemical attributes as the DM concentration increased from 0% to 10%. The optimal treatment was 10% DM, which resulted in the highest flower yield (318 g per plant). However, the bioaccumulation factor (BAF) values (>1) calculated by the authors indicated that marigold plants absorbed substantial amounts of six heavy metals, following the order Cd < Cr < Cu < Zn < Mn < Fe. Additionally, predictive models based on multiple linear regression (MLR) effectively estimated heavy metal uptake in marigold plants.

Over the past few years, byproducts derived from rearing Black Soldier Fly Larvae (BSFL) have emerged as some of the most extensively studied organic fertilizers. Their growing popularity worldwide stems from their significant potential to enhance environmental sustainability and support the circular economy. BSFL excreta contribute to improving soil health, enhancing plant immunity, and ultimately boosting crop quality. Romano et al. (contribution 5) examined the use of BSFL frass as a nutrient supplement in an aquaponic system for cultivating sweet potato seedlings. In their study, BSFL eggs were placed on two different substrates: spoiled fish feed formulated for catfish (Rangen; 32% protein) and a mix of fruit (orange peels, banana peels, apple cores, and strawberries) and vegetables (sweet potato and peas). The findings suggest that the initial substrate used to produce BSFL frass did not significantly impact sweet potato slip production. Despite variations in mineral composition among the frass types, neither water quality nor slip production and sugar content were affected. This indicates that a broad range of substrates could be effectively utilized to produce BSFL frass as a fertilizer in aquaponic systems.

5. Biostimulants

In recent years, considerable efforts have been made to develop new fertilizers and fertilization systems for organic horticulture that reduce chemical fertilizer inputs. The goal is to enhance nutrient uptake, promote vegetable growth and development, and improve quality, productivity, and environmental sustainability. For this purpose, biostimulant products have emerged on the market. These products are derived from plant extracts, algae, fungi, bacteria, or animal hydrolysates and contain components such as oligosaccharides, vitamins, humic substances (e.g., mixtures of humic and fulvic acids), micronutrients, and protein hydrolysates [11]. The use of biostimulants to promote plant growth has been extensively studied. For instance, Maková et al. (contribution 6) evaluated the combined effects of experimental (PGPB) and commercial (G) microbial biostimulants, along with a humic substance product (A), in combination with a mineral nitrogen fertilizer (N), on soil microbial communities and strawberry yield over a two-year field trial. The results demonstrated that enzymatic activity (FDA hydrolysis and phosphatase activity) was positively influenced by the N+G, N+G+A, and N+PGPB+A treatments. Additionally, plant biostimulants increased basal- and substrate-induced respiration but did not significantly affect the culturable bacteria population. On the other hand, although the N+PGPB and N+PGPB+A treatments negatively affected the number of culturable fungi, the latter resulted in a 95% higher strawberry yield in the second year compared to the control. Therefore, combining microbial biostimulants with humic substances may offer an effective solution to enhance the production of vegetables, such as strawberries.

Also, in the study by Ollio et al. (contribution 7), the effects of biofertilizers and inorganic fertilizers on a vegetable crop were compared, focusing on their influence on soil microbial abundance, microbial community structure, functional genetic diversity, broccoli yield, and greenhouse gas emissions. In contrast to the previous study, the results revealed that reduced fertilization, along with the application of both biofertilizer products, had no significant impact on soil nutrients, microbial populations, microbial activity, or yield when compared to conventional inorganic fertilization.

6. Liquid Matrices

It is well established that recent regulations have restricted the use of mineral and synthetic fertilizers to mitigate the environmental impact of horticultural systems [12]. As an alternative, fertilizers derived from composted solid organic waste and vermicomposting processes have gained attention. While the beneficial effects of solid vermicompost on soil and plant health are well documented, research on the potential of aqueous extracts and percolates derived from vermicompost remains limited. One such product, known as vermiliquer, is a nutrient-rich liquid that percolates through worm beds containing vermicomposted waste, bedding materials, and worm populations. Vermiliquer has been reported to be abundant in essential plant nutrients, enhancing plant growth and mineral uptake. In addition to dissolved organic and inorganic materials, it contains complex microbiota, plant growth regulators, and humic acids [13]. Furthermore, this organic fertilizer has demonstrated a broad range of benefits, including improving plant resistance to abiotic and biotic stresses, such as controlling insect pests and diseases, alleviating soil salinity, and mitigating drought stress [14]. In this context, Kosem et al. (contribution 8) investigated the effects of liquid vermicompost applications (25%, 50%, 75%, and 100%) on the agronomic traits, phenolic composition, and essential oil content of basil plants subjected to drought stress. Their study revealed that vermicompost applications significantly influenced nearly all measured parameters, except for leaf length in well-watered plants. Regarding essential oil compounds, estragole was identified as the dominant component (85–90%), with the highest levels observed in the 25% vermicompost + water stress, water stress, and control groups. Among the main phenolic compounds, caffeic acid levels declined under drought stress but increased with vermicompost treatments. Meanwhile, rosmarinic acid content rose under water stress conditions, reaching its peak at 25% vermicompost application. Overall, vermicompost applications at 25% and 50% enhanced the phenolic compound content in basil plants, regardless of the irrigation conditions, suggesting its potential as a natural stress mitigator.

For vegetable cultivation, aqueous extracts of vermicompost derived from horticultural crop residues can also serve as a viable option for hydroponic fertilization. For this purpose, Salas-Sanjuán et al. (contribution 9) assessed the effectiveness of both aerated and non-aerated aqueous extracts when used as recirculating nutrient solutions in a hydroponic nutrient film technique (NFT) system for lettuce. Their performance was compared to a conventional nutrient solution containing mineral or synthetic fertilizers. The results demonstrated that lettuce plants fertilized with the organic-aerated nutrient solution achieved not only good yields but also enhanced quality, with nitrate ($N\text{-}NO_3^-$) levels in edible leaves reduced by sixfold compared to those treated with mineral fertilizers. These findings highlight the potential of aqueous extracts as a sustainable nutrient source, contributing to circular agriculture and improving efficiency in intensive production systems.

7. Companion Cropping

To reduce reliance on chemical fertilizers, in parallel with the study of innovative organic fertilizers, researchers are increasingly focused on developing environmentally friendly agroecological strategies. In this context, agronomic management practices such as companion cropping can play a crucial role in maximizing yields while preserving soil health and horticultural biodiversity [15]. In the study by Moran-Chamorro et al. (contribution 10), interactions between fruit trees and bean plants were analyzed under three fertilization levels. The combination of cape gooseberry and blackberry yielded particularly positive results, with increased leaf production and reduced pest incidence, highlighting the benefits of companion planting. Additionally, the UF system (*Physalis peruviana* and *Phaseolus vulgaris*) exhibited the greatest plant height, while the TF system (tomato and bean) demonstrated the best stem perimeter development. These findings also suggest a growing trend toward integrating chemical and organic fertilizers, which the authors identified as a promising approach to reducing costs while enhancing crop growth.

8. Conclusions and Future Directions

The research presented in this SI underscores the potential of organic fertilizers to enhance crop yields, improve vegetable quality, and promote soil health while reducing environmental impacts. However, challenges such as nutrient variability, slow-release dynamics, and potential greenhouse gas emissions require further investigation. Future research should focus on optimizing organic fertilizer formulations, developing precision application methods, and conducting long-term field studies under diverse climatic conditions. Standardized guidelines for organic fertilizer use will be essential to maximize benefits and minimize risks. By integrating innovative organic amendments, biostimulants, and agroecological practices, modern horticulture can achieve a balance between productivity and sustainability, addressing global food security and environmental challenges.

Author Contributions: F.D.M., G.B., K.F. and H.Z. wrote the editorial. All authors have read and agreed to the published version of the manuscript.

Acknowledgments: As guest editors of the Special Issue "Organic Fertilizers in Horticulture", we would like to extend our sincere thanks to all the authors who contributed their valuable articles to the success of this Special Issue.

Conflicts of Interest: The authors declare no conflicts of interest.

List of Contributions:

1. Parecido, R.; Soratto, R.; Fernandes, A.; Blanes, M.; Fidelis, L.; Gitari, H.; Dutra, S. Castor Meal and Ground Hydrothermalized Phonolite Optimize Sweet Potato Nutrition, Yield, and Quality. *Horticulturae* **2024**, *10*, 775. https://doi.org/10.3390/horticulturae10080775.
2. Butterfield, A.; Wortman, S. Exploring the Feasibility of Integrating Weed and Nitrogen Management with Seed Meals in Organic Vegetables. *Horticulturae* **2024**, *10*, 75. https://doi.org/10.3390/horticulturae10010075.
3. Tong, Y.; Wang, Z.; Gong, D.; Huang, C.; Ma, X.; Ma, X.; Yuan, F.; Fu, S.; Feng, C. Enhancing Soil Fertility and Elevating Pecan Fruit Quality through Combined Chemical and Organic Fertilization Practices. *Horticulturae* **2024**, *10*, 25. https://doi.org/10.3390/horticulturae10010025.
4. AL-Huqail, A.; Kumar, P.; Abou Fayssal, S.; Adelodun, B.; Širić, I.; Goala, M.; Choi, K.; Taher, M.; El-Kholy, A.; Eid, E. Sustainable Use of Sewage Sludge for Marigold (*Tagetes erecta* L.) Cultivation: Experimental and Predictive Modeling Studies on Heavy Metal Accumulation. *Horticulturae* **2023**, *9*, 447. https://doi.org/10.3390/horticulturae9040447.
5. Romano, N.; Webster, C.; Datta, S.; Pande, G.; Fischer, H.; Sinha, A.; Huskey, G.; Rawles, S.; Francis, S. Black Soldier Fly (*Hermetia illucens*) Frass on Sweet-Potato (*Ipomoea batatas*) Slip Production with Aquaponics. *Horticulturae* **2023**, *9*, 1088. https://doi.org/10.3390/horticulturae9101088.
6. Maková, J.; Artimová, R.; Javoreková, S.; Adamec, S.; Paulen, O.; Andrejiová, A.; Ducsay, L.; Medo, J. Effect of Application of Nitrogen Fertilizer, Microbial and Humic Substance-Based Biostimulants on Soil Microbiological Properties During Strawberry (*Fragaria* × *ananassa* Duch.) Cultivation. *Horticulturae* **2025**, *11*, 119. https://doi.org/10.3390/horticulturae11020119.
7. Ollio, I.; Santás-Miguel, V.; Gómez, D.; Lloret, E.; Sánchez-Navarro, V.; Martínez-Martínez, S.; Egea-Gilabert, C.; Fernández, J.; Calviño, D.; Zornoza, R. Effect of Biofertilizers on Broccoli Yield and Soil Quality Indicators. *Horticulturae* **2024**, *10*, 42. https://doi.org/10.3390/horticulturae10010042.
8. Kosem, H.; Kocak, M.; Kaysim, M.; Celikcan, F.; Kulak, M. Liquid Leachate Produced from Vermicompost Effects on Some Agronomic Attributes and Secondary Metabolites of Sweet Basil (*Ocimum basilicum* L.) Exposed to Severe Water Stress Conditions. *Horticulturae* **2022**, *8*, 1190. https://doi.org/10.3390/horticulturae8121190.
9. Salas-Sanjuán, M.; Ruíz-Zubiate, J.; Valenzuela, J.; Campos, A. Nutrient Solution from Aqueous Extracts as an Alternative to Fertigation in Hydroponic. *Horticulturae* **2023**, *9*, 1281. https://doi.org/10.3390/horticulturae9121281.
10. Moran-Chamorro, O.; Andrade-Díaz, D.; Chirivi-Salomon, J.; Velasquez-Vasconez, P. Effect of Fertilization in Companion Cropping Systems of Andean Fruit Trees in the Municipality of Ipiales. *Horticulturae* **2024**, *10*, 1107. https://doi.org/10.3390/horticulturae10101107.

References

1. Tarantino, E.; Disciglio, G.; Frabboni, L.; Libutti, A.; Gatta, G.; Gagliaridi, A.; Tarantino, A. Effects of Biostimulant Application on Quali-Quantitative Characteristics of Cauliflower, Pepper and Fennel Crops under Organic and Conventional Fertilization. *World Acad. Sci. Eng. Technol. Int. J. Biol. Biomol. Agric. Food Biotechnol. Eng.* **2015**, *9*, 734–738. [CrossRef]
2. Bergstrand, K.-J. Organic Fertilizers in Greenhouse Production Systems—A Review. *Sci. Hortic.* **2022**, *295*, 110855. [CrossRef]
3. Gai, X.; Liu, H.; Zhai, L.; Tan, G.; Liu, J.; Ren, T.; Wang, H. Vegetable Yields and Soil Biochemical Properties as Influenced by Fertilization in Southern China. *Appl. Soil Ecol.* **2016**, *107*, 170–181. [CrossRef]
4. Khan, M.T.; Aleinikovienė, J.; Butkevičienė, L.-M. Innovative Organic Fertilizers and Cover Crops: Perspectives for Sustainable Agriculture in the Era of Climate Change and Organic Agriculture. *Agronomy* **2024**, *14*, 2871. [CrossRef]

5. Singh, T.B.; Naeem, M.; Ansari, A.; Gill, S. Role of Organic Fertilizers in Improving Soil Fertility. In *Contaminants in Agriculture*; Naeem, M., Ansari, A., Gill, S., Eds.; Springer: Cham, Switzerland, 2020. [CrossRef]
6. Lin, W.; Lin, M.; Zhou, H.; Wu, H.; Li, Z.; Lin, W. The Effects of Chemical and Organic Fertilizer Usage on Rhizosphere Soil in Tea Orchards. *PLoS ONE* **2019**, *14*, e0217018. [CrossRef] [PubMed]
7. Shaji, H.; Chandran, V.; Mathew, L. Organic Fertilizers as a Route to Controlled Release of Nutrients. In *Controlled Release Fertilizers for Sustainable Agriculture*; Lewu, F.B., Volova, T., Thomas, S., Rakhimol, K.R., Eds.; Academic Press: London, UK, 2021; pp. 231–245. [CrossRef]
8. Mora, M.L.; Calabi-Floody, M.; Rumpel, C. Closing Biogeochemical Cycles and Meeting Plant Requirements by Smart Fertilizers and Innovative Organic Amendments. *Agronomy* **2021**, *11*, 1158. [CrossRef]
9. Yao, Y.; Wang, X.; Chen, B.; Zhang, M.; Ma, J. Seaweed Extract Improved Yields, Leaf Photosynthesis, Ripening Time, and Net Returns of Tomato (*Solanum lycopersicum* Mill.). *ACS Omega* **2020**, *5*, 4242–4249. [CrossRef] [PubMed]
10. Nakić, D.; Posavčić, H.; Licht, K.; Vouk, D. Application of Novel Biochar Derived from Experimental Sewage Sludge Gasification as an Adsorbent for Heavy Metals Removal. *Sustainability* **2025**, *17*, 997. [CrossRef]
11. Zhang, X.; Yin, J.; Ma, Y.; Peng, Y.; Fenton, O.; Wang, W.; Zhang, W.; Chen, Q. Unlocking the Potential of Biostimulants Derived from Organic Waste and By-Product Sources: Improving Plant Growth and Tolerance to Abiotic Stresses in Agriculture. *Environ. Technol. Innov.* **2024**, *34*, 103571. [CrossRef]
12. Vico, A.; Sáez, J.A.; Pérez-Murcia, M.D.; Martinez-Tomé, J.; Andreu-Rodríguez, J.; Agulló, E.; Bustamante, M.A.; Sanz-Cobena, A.; Moral, R. Production of Spinach in Intensive Mediterranean Horticultural Systems Can Be Sustained by Organic-Based Fertilizers without Yield Penalties and with Low Environmental Impacts. *Agric. Syst.* **2020**, *178*, 102765. [CrossRef]
13. Zhai, Z.; Ehret, D.L.; Forge, T.; Helmer, T.; Lin, W.; Dorais, M.; Papadopoulas, A.P. Organic Fertilizers for Greenhouse Tomatoes: Productivity and Substrate Microbiology. *HortScience* **2009**, *44*, 800–809. [CrossRef]
14. Rehman, S.U.; De Castro, F.; Aprile, A.; Benedetti, M.; Fanizzi, F.P. Vermicompost: Enhancing Plant Growth and Combating Abiotic and Biotic Stress. *Agronomy* **2023**, *13*, 1134. [CrossRef]
15. Sanjerehei, M.M. *Mixed Cropping with Companion Crops*; Tafakkor Talaei: Yazd, Iran, 2023; p. 62, ISBN 978-622-8159-01-0.

horticulturae

MDPI

Article

Liquid Leachate Produced from Vermicompost Effects on Some Agronomic Attributes and Secondary Metabolites of Sweet Basil (*Ocimum basilicum* L.) Exposed to Severe Water Stress Conditions

Hatice Kosem [1], Mehmet Zeki Kocak [2], Mustafa Guven Kaysim [3], Ferdi Celikcan [1] and Muhittin Kulak [2,*]

1 Department of Organic Farming Management, College of Applied Science, Igdir University, Igdir 76000, Türkiye
2 Department of Herbal and Animal Production, Vocational School of Technical Sciences, Igdir University, Igdir 76000, Türkiye
3 Department of Field Crops, Faculty of Agriculture, Igdir University, Igdir 76000, Türkiye
* Correspondence: muhyttynx@gmail.com or muhittin.kulak@igdir.edu.tr

Abstract: Water stress is one of the most critical threats to the growth and productivity of plants and is one of the most studied topics in agricultural sciences. In order to enhance the tolerance of plants to water stress conditions, synthetic fertilizers have been widely used in the field. However, due to their toxic effects, recent reports have focused on organic options. In this study, the effects of liquid vermicompost applications (25, 50, 75, and 100%) on the agronomic attributes, phenolic compounds, and essential oil compounds of basil plants exposed to drought stress conditions were investigated. Accordingly, water stress critically reduced the factors of plant height, plant fresh weight, root fresh weight, leaf length, and leaf diameter. On the other hand, vermicompost applications significantly affected all of the parameters considered, except the leaf length of well-watered basil plants. However, a two-way ANOVA analysis revealed that the interactions of water stress and vermicompost were significant on root length and root fresh weight. Regarding the essential oil compounds, the contents of humulene, anethol, eucalyptol, estragole, bisabolene, germacrene, and caryophyllene were quantified. Estragole was determined as a major component by 85–90%. The results revealed that the highest estragole content was determined in the 25% vermicompost + water stress, water stress, and control groups. Of the major phenolic compounds, caffeic acid decreased as a result of water stress conditions but increased with vermicompost treatments. The rosmarinic acid content increased during water stress conditions, attaining the highest content at 25% via the vermicompost and water stress interaction. In general, the 25% and 50% vermicompost applications increased the content of phenolic compounds in plants under either well-watered or stress conditions.

Keywords: waste management; organic amendment; phenolics; terpenoids; abiotic stress

Citation: Kosem, H.; Kocak, M.Z.; Kaysim, M.G.; Celikcan, F.; Kulak, M. Liquid Leachate Produced from Vermicompost Effects on Some Agronomic Attributes and Secondary Metabolites of Sweet Basil (*Ocimum basilicum* L.) Exposed to Severe Water Stress Conditions. *Horticulturae* **2022**, *8*, 1190. https://doi.org/10.3390/horticulturae8121190

Academic Editors: Francesco De Mastro, Gennaro Brunetti, Karam Farrag and Huadong Zang

Received: 22 October 2022
Accepted: 1 December 2022
Published: 13 December 2022

1. Introduction

Plants are exposed to both biotic and abiotic environmental factors throughout their lives; however, relevant environmental factors cause stress to the plant if they exceed the plant's ability to cope with the level of stress. Although the effects of the relevant stress factors depend on the plant species or the severity, type, and duration of the stress, these factors can often delay the growth and development of plants, reduce their productivity, and may ultimately cause their death [1,2].

Of the vital abiotic stress factors that exist, drought stress is one of the major problems for crops. For that reason, drought stress is one of the most widely investigated abiotic stress factors in agricultural sciences. Its deleterious effects have been reported for numerous plant species in general [3,4] and for *Ocimum* sp. in particular [5–8]. The adverse effects manifest as stunted growth and performance, which, in turn, result in a critical loss of crop productivity. As reported in a numerous reviews and research reports [9–12], the slowing down of plant growth is crucially linked to damage to the photosynthesis system, which subsequently affects assimilate production and its allocation to plant tissues. Corresponding to the perturbations occurring in the plant metabolism process, critical shifts from primer to secondary metabolism processes occur due to a carbon surplus. The relevant shifts are deemed to be the non-enzymatic adaptive strategies of the plants [13–15]. Secondary metabolites are some of the crucial non-enzymatic compounds exerting protective roles against either biotic or abiotic stress elicitors [16,17]. Considering the secondary metabolites (phenolics and terpenoids), *Ocimum basilicum* is a reputed species characterized by a high content of rosmarinic, caffeic, and chicoric acids, as well as methyl chavicol (estragol), eugenol, linalool, methyl cinnamate, and camphor [18,19].

In order to ensure global food security and to make the plants more compatible with water constraints, synthetic or semi-synthetic fertilizers have been employed in the research, and, subsequently, a higher yield of crops has been produced. However, the high levels of chemical inputs have caused the contamination of the soil, water, and air. Due to the residues produced by the relevant agricultural chemicals, adverse effects on human and animal health have also been reported in the literature [20]. These negative effects have led researchers to use organic fertilizers that are compatible with the natural environment and do not pose a toxic threat to other living organisms. One of the organic fertilizers commonly used is vermicompost. Vermicompost is not only an important compost and bio-control factor, but also an effective means of solid waste management. These organic amendments are also crucial for the sustainability of agricultural activities and subsequently food security due to their contributions to the physical, chemical, and biological properties of soil [21].

Vermicompost is an organic fertilizer obtained from food processed in the digestive system of certain waste-eating worm species (*Eisenia fetida, Eisenia andrei, Dendrobaena veneta, Lumbricus rubellus, Perionyx excavatus*) [22]. Vermicompost increases the water holding capacity of soil, increases the plant's resistance by competing with the beneficial bacteria in its structure, is non-toxic, regulates the soil's pH level, and positively affects certain parameters, such as plant fresh and dry weights and yield [23,24]. Considering its effects on basil plants, several reports revealed the affirmative effects of vermicompost on the vegetative growth of basil [25–31]. Enhanced plant performances have been attributed to the regulation of photosynthesis, antioxidant enzyme activity, and secondary metabolites [30,32–35]. Although organic fertilizers, such as vermicompost, are known to have a positive effect on plant growth, the action mechanisms of vermicompost concerned with the physiology and biochemistry of plants are still not fully elucidated in the literature. Similar to the case of secondary metabolism, the research mostly focuses on the essential oil yield and composition of plants under well-watered conditions. Considering the phenolic acids and flavonoids and their alterations against stress conditions, Celikcan et al. [30] reported that the vermicompost-enhanced crop productivity of the plants under well-watered conditions and the relevant amendments might not be effective in coping with water stress conditions. However, the critical changes occurring in phenolics or terpenoid compounds were noted alongside the treatments, but the changes in the secondary metabolites were not manifested or translated into the enhanced tolerance of basil against water stress conditions.

According to our research and knowledge, the effects of a liquid vermicompost fertilizer application to both the agronomic properties and secondary metabolites of basil plants have not yet been studied in the literature. For that reason, the current study aims to investigate the potential uses of vermicompost effluent (leakage) for basil plants grown under water stress conditions. Moreover, organic fertilizers are compatible with the soil structure and plant and provide significant contributions due to their high nutrient and organic matter contents, in general. Corresponding to the uses of organic fertilizers, the nutrient status and organic matter content of the soil lost over time might be maintained/buffered as a result. In the present study, the experimental soils were enriched with vermicompost prior to being subjected to water stress conditions. The hypothesis of the study addresses the enrichment of the soil. Due to the compounds with molecular structure analogues similar to the hormones, enzymes, elements, and bacterial flora available in vermicompost, we hypothesized that the enrichment of the soil with vermicompost would result in significant changes in the agronomic attributes of the basil plant and in secondary metabolite composition. Due to the contribution of vermicompost to the root system of the plants, we further hypothesized that the vermicompost-mediated development in the root systems of the basil plants would provide a greater tolerance to water stress conditions.

2. Materials and Methods

2.1. Experimental Site, Plant Materials, Submitting Water Stress, and Harvest Time

The experiment was conducted at the research greenhouses of the Agricultural Research and Application Centre, Igdir University, Türkiye. The study was performed as a factorial experiment using a completely randomized design with three replicates. Sweet basil (*Ocimum basilicum* L.) seeds were purchased from Simagro Agro & Seed Company (Konya, Türkiye). Of the medicinal and aromatic plant taxa, basil plants are one of the preferred species due to their chemical composition characterized by a high-essential-oil and phenolic content. For that reason, we used basil plants for the current study. In this regard, the seeds were initially surface-sterilized using 1% (*v*/*v*) hypochlorite for 2–3 min, and then the seeds were rinsed with distilled water to remove the residue of the disinfectant. After re-drying the seeds to their original moisture content using a tissue paper at room temperature, the seeds were sown in 2 L plastic pots containing peat and grown in greenhouses for a 14/10 h photoperiod, 26–30 °C/ day and 16–20 °C/ night; relative humidity: 60%. From germination to the final harvest, the irrigation of the control plants was based on the field capacity of the experimental soil. Regarding the estimation of soil water content/pot water capacity, the experimental soils were firstly fully saturated and then the pots were weighed. Subsequently, the soil samples were dried at 105 °C until a constant weight was obtained. The differences between the weight of fully saturated and dried soil samples were quantified, which were considered as the water weight required for the pot water capacity. Regarding the irrigation levels, the pot weights were estimated every second day and then transpiration-mediated water losses were buffered with re-watering to obtain the soil water capacity [30]. Once the basil seedlings grew 6–8 true leaves, the seedlings underwent severe water stress by water-holding for eleven days. The seedlings were susceptible to the water-holding stage after 11 days; for this reason, the experiments were terminated and the relevant samplings were performed following an 11-day drought period. The stress period was based on the wilting point of the plants. Concerned with the vermicompost treatments and their interactions with the stress, prior to submitting them to water stress conditions, the experimental soils were firstly enriched with 25%, 50%, 75%, and 100% concentrations of vermicompost once a week for four weeks. At the end of this period, the plants were exposed to drought stress for 11 days. All the measurements were performed with three replicates and each replicate corresponded to ten plants. The experimental design of the study is presented in Table 1.

Table 1. Experimental design of the study.

Acronym	Vermicompost Treatments	Irrigation Level
Control	Leachate amended	Well-watered plants
Water Stress (WS) *	Non-leachate amended	Severe water stressed plants
25% LVC **	Leachate amended (25% LVC/75% distilled water, *v*/*v*)	Well-watered plants
25% LVC + WS	Leachate amended (25% LVC/75% distilled water, *v*/*v*) ***	Severe water stressed plants
50% LVC	Leachate amended (50% LVC/50% distilled water, *v*/*v*)	Well-watered plants
50% LVC + WS	Leachate amended (50% LVC/50% distilled water, *v*/*v*)	Severe water stressed plants
75% LVC	Leachate amended (75% LVC/25% distilled water, *v*/*v*)	Well-watered plants
75% LVC + WS	Leachate amended (75% LVC/25% distilled water, *v*/*v*)	Severe water stressed plants
100% LVC	Leachate amended (100% LVC/0% distilled water, *v*/*v*)	Well-watered plants
100% LVC + WS	Leachate amended (100% LVC/0% distilled water, *v*/*v*)	Severe water stressed plants

* WS: water stress; ** LVC: liquid leachate obtained from vermicompost, *** *v*/*v*: Volume/volume.

2.2. Vermicompost Preparation and Physicochemical Properties of Liquid Leachate of Vermicompost

The vermicompost was produced, as we previously reported [30], in dark conditions with (20 ± 2 °C) and 75% humidity. The sources of the vermicompost were cow manure and *Eisenia fetida*. The analysis of the physicochemical composition of relevant liquid leachate was conducted at the Soil, Fertilizer and Water Resources Central Research Institute (Ministry of Agriculture and Forestry, Türkiye). The analysis revealed that the composition was organic matter content (0.44%); pH (6.98); EC (18.12 dS/m); total nitrogen (N) (0.14%); total potassium (K) (0.30%); total copper (trace level; Tr); total phosphorus (P) (0.05%); total calcium (Ca) (0.01%); total magnesium (Mg) (0.01%); total iron (Fe) (Tr); total manganese (Mn) (Tr); and total zinc (Zn) (Tr).

2.3. Agronomic Traits

Agronomic traits such as plant height, plant fresh weight, root length, root fresh weight, leaf fresh weight, leaf length, and leaf width, were assayed with ten plants for each replicate with a total of thirty basil plants corresponding to the three replicates.

2.4. Solid-Phase Micro-Extraction (SPME) of Essential Oils and GC-MS Conditions

For the extraction of essential oils, the method optimized by [30] was assayed. Briefly, 0.5 g of dried and powdered basil leaves were mixed with 10 mL of double distillate water, and then the relevant mixture was stirred at 45 °C for 30 min. The stirring was followed by the trapping of essential oil using an SPME holder (Supelco 57330-U) needle for a 7-minute period. Then, the trapped volatiles on the SPME holder needle were injected into the GC-MS and left for 4 min on the relevant septum. The analysis for the identification of the essential oil components lasted for 33 min. Each analysis was performed with three replicates. Considering the identification and relevant analysis of the essential oil components, Thermo GC-MS Trace Ultra (USA) was used. Regarding the GC-MS conditions, a DB-5MS column (30 m × 0.25 mm × 0.25 μm) was used and the flow rate of the carrier gas of helium was set as 1.0 mL/min. The oven temperature was kept at 40 °C for 1 min and then increased from 40 to 120 °C at a rate of 5 °C/min and maintained for 2 min. The temperature was then increased to 240 °C with a rate of 10 °C/min and maintained for 3 min. The injection part temperature was set to 240 °C. The mass spectrometer was operated in EI mode at 70 eV. The split ratio was set as 20:1. Mass range: 45–450 m/z; scan speed (amu/s): 1000. The components were identified in comparison to NIST08, Willey7n.1, and HPCH1607 libraries reference compounds.

2.5. Extraction and Quantification of Phenolics Using LC–MS/MS

The harvested basil leaves were, firstly, dried and powdered. Then, shaker-aided and sequential extraction were performed at 120 rpm at room temperature for 24 h. In that context, 3 g of basil leaves were extracted using 50 mL of methanol. The extraction was repeated three times with the same plant materials to collect all residues following the extraction. The filtrates of each extraction were then vacuo-dried using a rotary evaporator (Heidolph 94200, Bioblock Scientific, Germany). The extracts were then preserved at +4 °C until chromatographic analysis was performed. For the quantification of the phenolic acid and flavonoids, ultrahigh performance liquid chromatography (Shimadzu Nexera, Kyoto, Japan) coupled with a tandem mass spectrometer (LCMS8040 model) was used. Considering the conditions of LS-MS/MS, similar modified and optimized conditions of [36,37] were applied. This was performed as the reversed-phase UHPLC was equipped with a SIL-30AC model autosampler, a CTO-10ASvp model column oven, LC-30CE model binary pumps, and a DGU20A3R model degasser. Different analytical columns, *viz.*, RP-C18 Inertsil ODS-4 (100 mm × 2.1 mm, 2 μm) and 120 EC-C18 models (150 mm × 2.1 mm, 2.7 μm), were used and the column temperature was set to 40 °C. Methanol and acetonitrile were used as the mobile phases, while ammonium formate, ammonium acetate, acetic acid, and formic acid were used as the mobile-phase additives. The gradient elutions were 20% B (35–45 min), 100% B (25–35 min), and 20–100% B (0–25 min). The flow rate was set to

0.5 mL/min and the injection volume was 5 μL. An ionization source (ESI) was used to perform spectrometric detection. ESI was operated in positive-ionization mode for vanillin, daidzin, piceid, coumarin, and hesperidin, while ESI was operated in negative for other standards. MS conditions: drying gas (N2) flow: 15 L/min; nebulizing gas (N2) flow: 3 L/min; interface temperature: 350 °C; heat block temperature: 400 °C; and DL temperature: 250 °C [36,37].

2.6. Experimental Design and Statistical Analysis

The experimental design corresponded to a factorial model in a completely randomized block, with treatments being irrigated/non-irrigated with liquid leachate, and drought-stressed/non-drought-stressed plants. For each measurement, three replications were used, and each replicate corresponded to ten plants. The experimental data were analyzed via two-way ANOVA. The relevant variances were related to major treatments (liquid leachate and water stress conditions) and their interactions. The means were separated using Duncan's multiple range test at a 5% probability level (p< 0.05) (SPSS 22). Additionally, heat map clustering was conducted in order to visualize and associate the parameters (ClustVis online). Principal component and correlation analyses were performed using JAMOVI and GraphPad Prism, and a network plot analysis was performed using PAST Software.

3. Results

3.1. Agronomic Attributes

Water stress significantly affected the agronomic attributes of the sweet basil, as estimated from the shorter plant height and leaf length, taller root length, lighter plant FW and leaf FW, heavier root FW, and smaller leaf width (p = 0.000) (Table 2), as was the case commonly observed for sweet basil under a restricted water supply [6,8,38,39]. However, independent from water stress conditions, applications of liquid leachate significantly affected the relevant attributes of the basil, except the root FW and leaf length, as observed from the values corresponding to the taller plant height and root length, heavier plant FW and leaf FW, as well as wider leaf width. Additionally, the impacts of the leachate were concentration-dependent, suggesting an increase from 25 to 50% for the plant height, plant FW, leaf FW, and leaf width, and decrease from 75 to 100%. Furthermore, the root length increased by the increasing concentration of leachate. Considering the interactions of water stress and vermicompost, only the root length and root FW were observed to be significant (p = 0.000). Under water stress conditions, root FW substantially decreased with the leachate amendments, whilst root length increased with the treatments, in general (Table 2).

3.2. Heat Map Clustering, Correlation, Principal Component, and Network Plot Analyses of the Agronomic Attributes Corresponding to the Treatments

In addition to the two-way ANOVA analysis we performed, the relevant data of the agronomic attributes were subjected to an array of statistical analyses in order to reduce the dimension, and correlate, visualize, and clarify the experimental results corresponding to the treatments. Such analyses are quite common in research with a high number of variables. Firstly, we constructed a heat map. According to the clustering presented in the heat map, water stress and vermicompost treatments were clearly sorted into two distinct clusters. The first cluster included "*control* and *all vermicompost treatments*", corresponding to the *well-watered groups*. On the other hand, the second cluster included "*stress* and *its interaction with vermicompost treatments*", corresponding to the *stress-submitted groups*. The results suggest that irrigation status is a critical predictor with respect to the agronomic attributes. Of the estimated attributes, the under-ground components of the plant, *viz.*, root length and root FW, were clearly separated from the above-ground components of the plants corresponding to the treatments (Figure 1). Additionally, the correlation analysis clearly revealed the negative coefficients between under- and above-ground parts, but a significant correlation was only noted between leaf length and root length (r = −0.84) (Figure 2).

Table 2. Effects of liquid leachate obtained from vermicompost (25, 50, 75, and 100%) on some agronomic attributes of sweet basil (*O. basilicum* L.) under drought stress conditions.

Treatments	Plant Height (cm)	Plant FW (g)	Root Length (cm)	Root FW (g)	Leaf FW (g)	Leaf Length (cm)	Leaf Width (cm)
Control	13.480 ± 1.000 cde	3.900 ± 0.229 d	14.460 ± 0.841 c	0.353 ± 0.045 fg	1.096 ± 0.110 c	4.020 ± 0.453 b	1.610 ± 0.079 bcd
WS *	8.333 ± 0.666 f	2.683 ± 0.480 e	24.847 ± 1.264 a	1.024 ± 0.132 bc	0.520 ± 0.076 d	2.820 ± 0.072 c	1.320 ± 0.092 d
25% LVC **	16.700 ± 1.410 b	4.767 ± 0.737 abc	14.933 ± 0.306 c	0.747 ± 0.095 cde	1.550 ± 0.132 b	4.203 ± 0.300 ab	1.870 ± 0.066 ab
25% LVC+ WS	12.517 ± 0.797 de	4.203 ± 0.211 cd	22.277 ± 0.751 b	0.653 ± 0.115 de	1.253 ± 0.105 c	2.808 ± 0.357 c	1.338 ± 0.078 d
50% LVC	18.933 ± 1.504 a	5.227 ± 0.261 a	21.350 ± 1.103 b	0.910 ± 0.168 bcd	2.033 ± 0.260 a	4.230 ± 0.305 ab	2.080 ± 0.203 a
50% LVC+ WS	15.333 ± 0.950 bc	4.457 ± 0.172 bcd	20.550 ± 0.853 b	0.790 ± 0.236 b–e	1.543 ± 0.081 b	2.967 ± 0.153 c	1.787 ± 0.220 ab
75% LVC	15.200 ± 0.900 bc	5.060 ± 0.333 ab	14.767 ± 0.751 c	0.597 ± 0.015 ef	1.990 ± 0.105 a	4.660 ± 0.295 a	1.867 ± 0.090 ab
75% LVC+ WS	11.333 ± 1.258 e	4.120 ± 0.209 cd	22.083 ± 1.551 b	1.034 ± 0.070 b	1.597 ± 0.257 b	2.877 ± 0.125 c	1.679 ± 0.427 bc
100% LVC	14.200 ± 2.138 cd	3.867 ± 0.252 d	15.703 ± 0.754 c	0.260 ± 0.036 g	1.093 ± 0.110 c	3.947 ± 0.311 b	1.940 ± 0.052 ab
100% LVC+ WS	13.450 ± 0.606 cde	2.573 ± 0.459 e	26.363 ± 1.061 a	1.343 ± 0.316 a	0.557 ± 0.067 d	2.633 ± 0.153 c	1.383 ± 0.104 cd
p-value	LVC: 0.000 WS: 0.000 LVCxWS: 0.054	LVC: 0.000 WS: 0.000 LVCxWS: 0.433	LVC: 0.000 WS: 0.000 LVCxWS: 0.000	LVC: 0.283 WS: 0.000 LVCxWS: 0.000	LVC: 0.000 WS: 0.000 VCxWS: 0.485	LVC: 0.070 WS: 0.000 LVCxWS: 0.411	LVC:0.003 WS: 0.000 LVCxWS: 0.321

* WS: water stress; ** LVC: liquid leachate obtained from vermicompost. Different letters indicate significant difference according to a Duncan's multiple range test ($p < 0.05$).

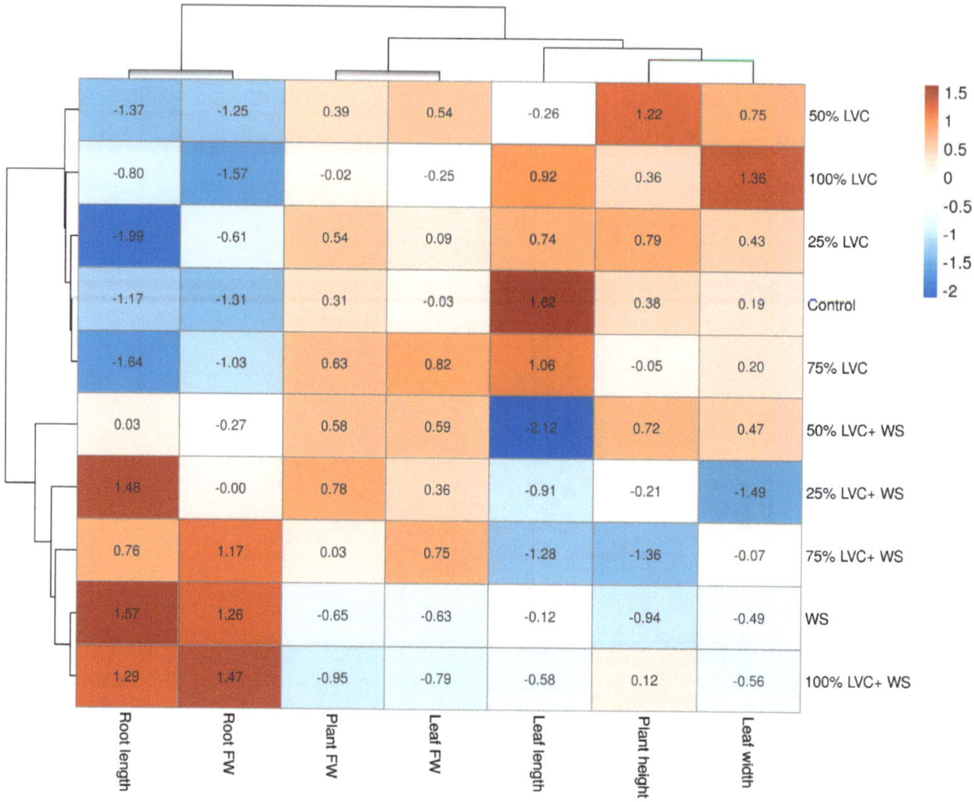

Figure 1. Heat map clustering of agronomic attributes corresponding to the treatments.

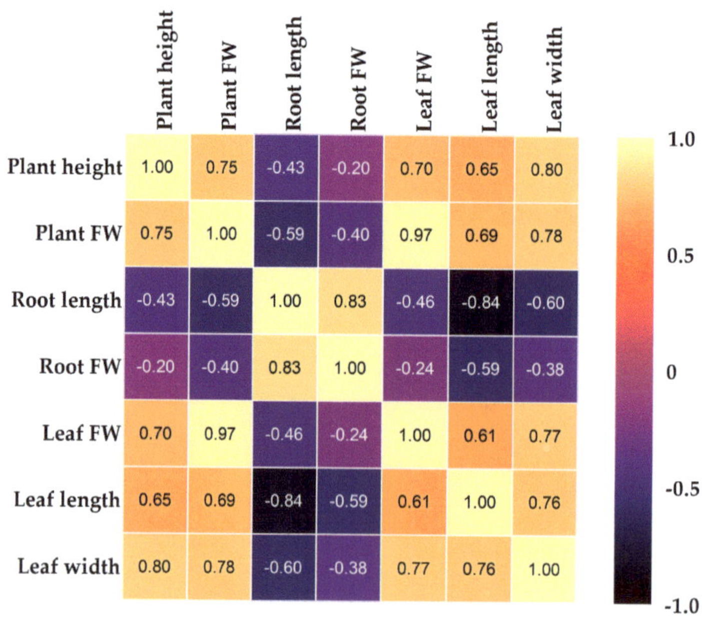

Figure 2. Correlation analysis of agronomic attributes corresponding to the treatments.

Furthermore, the agronomic attributes and treatments were discriminated on a biplot pair (Figure 3). Accordingly, two principal components with eigenvalues >1.0 accounted for 86.96% of the variability of the original data. Such a high explained variance suggests that principal component analysis can be a significant predictor in the assessment of relevant dependent variables corresponding to the treatments performed. The first principal component, PC_1 (*eigenvalue*: 4.788), accounting for 68.40% of the total variation, exhibited significant positive correlations with the plant height, plant FW, leaf FW, leaf length, and leaf width. On the other hand, in the second principal component, PC_2 (*eigenvalue*: 1.298), accounting for 18.56% of the total variation, root length and root FW had higher eigenvectors.

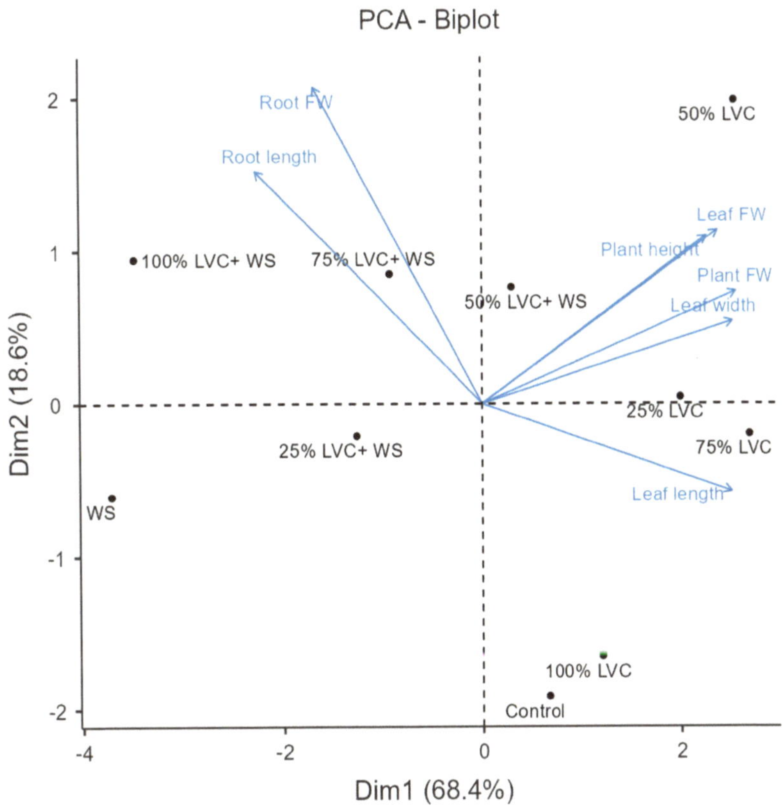

Figure 3. Principal component analysis of agronomic attributes corresponding to the treatments.

We finally performed a network plot analysis to reveal the link between individual vermicompost concentrations and water stress based on their performance on agronomic attributes (Figure 4). The plot consists of nodes via lines, and the depth of the line reveals the relation among the experimental groups. The thinner/lighter line presents the weaker relation whilst the thicker line shows the strong relations with each other. According to the network plot analysis, WS and WS + 100% LVC were clearly separated from the other treatments. WS + 100% LVC exhibited a similar performance to WS. Based on the network and thickness of the lines, it was clear that, based on the responses of the aforementioned agronomic attributes, the vermicompost applications without stress conditions were closely associated with each other, whereas the vermicompost-stress-interacting groups were also scattered close to each other.

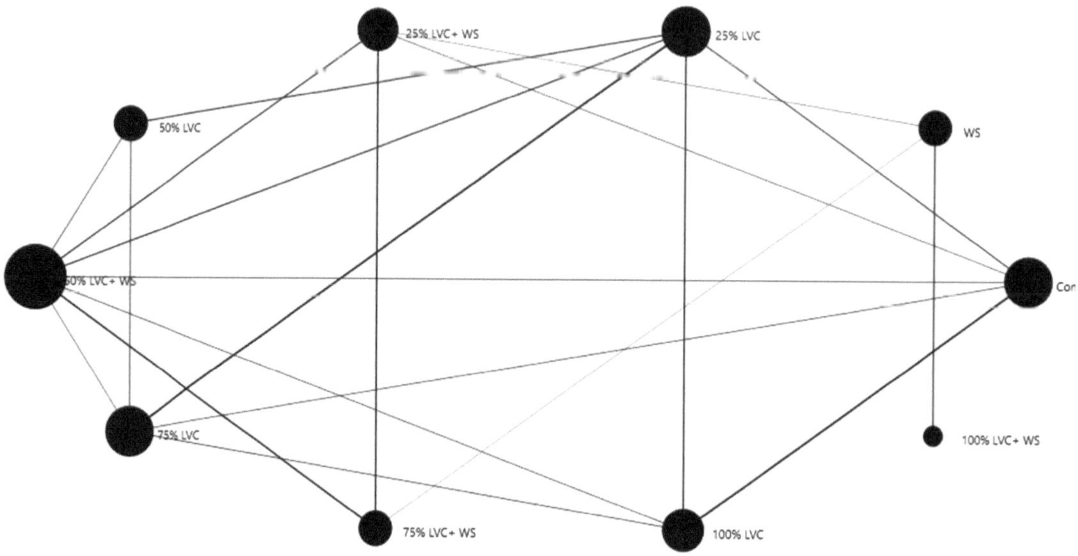

Figure 4. Network plot analysis of agronomic attributes corresponding to the treatments.

3.3. Essential Oil Compounds

The essential oil compounds identified in the basil leaves are presented in Table 3, following their elution orders on the HP-5 column. Of the identified compounds, estragole (methyl chavicol), eucalyptol, anethole, caryophyllene, humulene, germacrene D, and bisabolene were the predominant compounds. According to the statistical analysis, it can be observed that water stress, vermicompost, and their interactions significantly affect the percentage of the compounds ($p < 0.05$). In comparison to the control, water stress did not significantly affect the percentage of estragole. Regarding vermicompost treatments, neither 25% LVC nor 50% LVC critically affected the percentage value, but neither 75% LVC nor 100% LVC significantly decreased the percentage of estragole in well-watered basil plants. With respect to the vermicompost and water stress interactions, the highest percentage of estragole was noted at 25% LVC + WS, and the percentage decreased by higher concentrations of vermicompost and water stress interactions.

Water stress conditions reduced the percentage of eucalyptol, but 25% LVC increased the percentage value. However, higher concentrations of LVC significantly decreased the percentage of the compound in well-watered basil plants. In plants suffering from water stress conditions, vermicompost treatments, except 100% LVC, did not result in critical changes in the percentage of the compound present in comparison to the well-watered plants. Of the minor compound identified in the study, all increased with water stress conditions as well as interactions of 25% LVC + WS.

Table 3. Effects of liquid leachate obtained from vermicompost (25, 50, 75, and 100%) on essential oil compounds in basil leaves under water stress conditions (%).

Treatments	Eucalyptol	Estragole	Anethole	Caryophyllene	Humulene	Germacrene D	Bisabolene
Control	2.37 ± 0.08 b	89.12 ± 2.10 b	0.32 ± 0.06 d	0.51 ± 0.08 e	0.23 ± 0.04 e	0.20 ± 0.04 g	0.33 ± 0.05 e
WS *	1.75 ± 0.18 ef	87.77 ± 0.44 bc	0.44 ± 0.06 c	0.64 ± 0.06 de	0.26 ± 0.03 e	0.29 ± 0.01 f	0.54 ± 0.06 de
25% LVC **	2.04 ± 0.05c d	88.33 ± 0.72 b	0.31 ± 0.03 d	0.69 ± 0.05 d	0.53 ± 0.07 c	0.69 ± 0.01 c	0.94 ± 0.22 ab
25% LVC +WS	2.20 ± 0.12 bc	91.28 ± 0.85 a	0.65 ± 0.05 b	0.91 ± 0.05 bc	0.68 ± 0.01 b	0.60 ± 0.02 d	0.92 ± 0.02 abc
50% LVC	1.86 ± 0.04 de	87.58 ± 0.65 bc	0.41 ± 0.01 c	0.86 ± 0.04 c	0.40 ± 0.01 d	0.52 ± 0.04 e	0.96 ± 0.05 ab
50% LVC +WS	2.11 ± 0.15 c	89.06 ± 0.16 b	0.77 ± 0.04 a	1.03 ± 0.05 ab	0.83 ± 0.04 a	0.59 ± 0.02 d	0.76 ± 0.10 bcd
75% LVC	1.83 ± 0.06 def	86.29 ± 0.59 cd	0.46 ± 0.01 c	0.91 ± 0.02 bc	0.38 ± 0.02 cd	0.82 ± 0.03 b	1.00 ± 0.14 ab
75% LVC +WS	2.37 ± 0.05 b	87.49 ± 0.46 bc	0.64 ± 0.04 b	0.94 ± 0.03 bc	0.73 ± 0.04 b	0.61 ± 0.02 d	0.69 ± 0.14 cd
100% LVC	1.63 ± 0.08 f	84.42 ± 0.74 d	0.56 ± 0.04 b	1.11 ± 0.15 a	0.35 ± 0.03 d	0.90 ± 0.01 a	1.03 ± 0.05 a
100% LVC +WS	2.96 ± 0.05 a	85.20 ± 0.32 d	0.57 ± 0.04 b	0.89 ± 0.05 bc	0.58 ± 0.02 c	0.62 ± 0.02 d	0.82 ± 0.06 abc
p-value	LVC: <0.004 WS: <0.001 LVCxWS: <0.001	LVC: <0.001 WS: <0.017 LVCxWS: 0.035	LVC: <0.001 WS: <0.001 LVCxWS: <0.001	LVC: <0.001 WS: <0.046 LVCxWS: <0.001	LVC: <0.001 WS: <0.001 LVCxWS: <0.001	LVC: <0.001 WS: <0.001 LVCxWS: <0.001	LVC: <0.001 WS: <0.032 LVCxWS: <0.001

* WS: water stress; ** LVC: liquid leachate obtained from vermicompost. Different letters indicate significant difference according to a Duncan's multiple range test ($p < 0.05$).

3.4. Heat Map Clustering, Correlation, Principal Component, and Network Plot Analyses of the Essential Oil Compound Corresponding to the Treatments

Heat map clustering revealed two clusters of essential oil compounds corresponding to the treatments performed (Figure 5). Considering the treatments, as in the case of agronomic traits, groups that were well-watered and submitted to stress were clearly separated into distinct clusters. Of the identified compounds, caryophyllene, germacrene D, and bisabolene were grouped into the first cluster, while eucalyptol, estragole, anethole, and humulene were scattered into the second cluster. Estragole was dominant and its values peaked at a solo water stress (WS) level and 25% LVC + WS treatments. Interestingly, the predominant compounds, estragole and eucalyptol, were not correlated with any compounds ($p > 0.05$) according to the correlation analysis (Figure 6). In addition, essential oil compounds and treatments were scattered on a biplot pair via PCA (Figure 7). Accordingly, two principal components with eigenvalues > 1.0 accounted for 75.89% of the variability of the original data. The first principal component, PC_1 (*eigenvalue*: 3.33), accounting for 47.57% of the total variation, exhibited significant positive correlations with caryophyllene, germacrene D, and bisabolene. The second principal component, PC_2 (*eigenvalue*: 1.298), accounting for 28.32% of the total variation, was related to eucalyptol, estragole, anethole, and humulene. Regarding the network plot analysis (Figure 8), the linkage of individual vermicompost concentrations and water stress with each other based on their effect on essential compounds (eucalyptol, estragole, anethole, caryophyllene, humulene, germacrene D, and bisabolene) was established. According to the network plot analysis, it was clear that, based on the responses of the aforementioned essential oil compounds, the vermicompost applications without being subjected to stress were closely associated with each other, whereas the vermicompost-stress-interacted groups were also scattered close to each other. Furthermore, the control and stress groups presented a strong association in correspondence with vermicompost and its stress interactions.

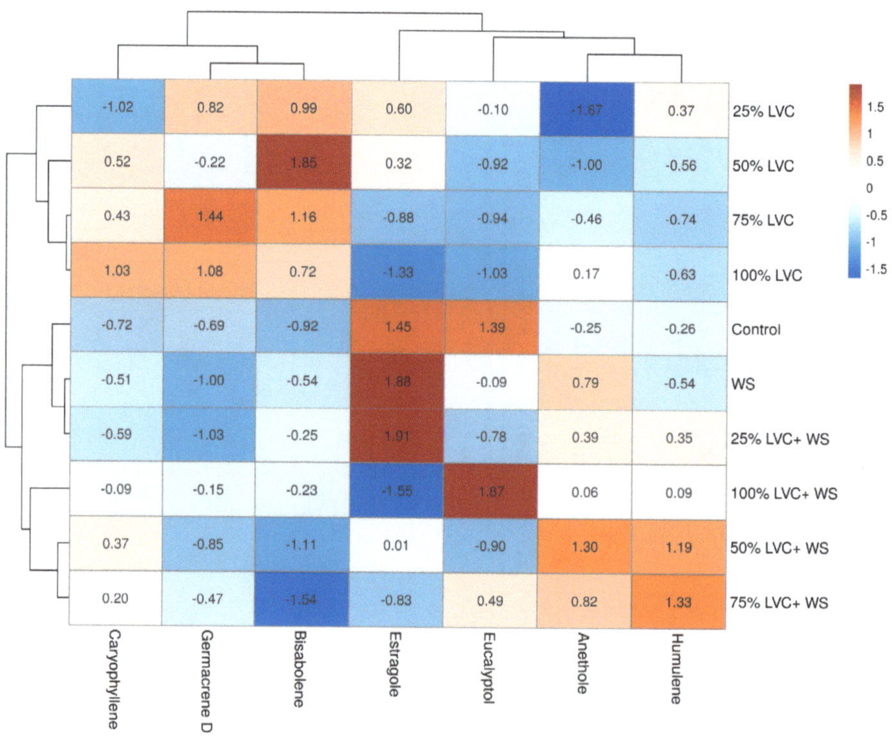

Figure 5. Heat map clustering of essential oil compounds corresponding to the treatments.

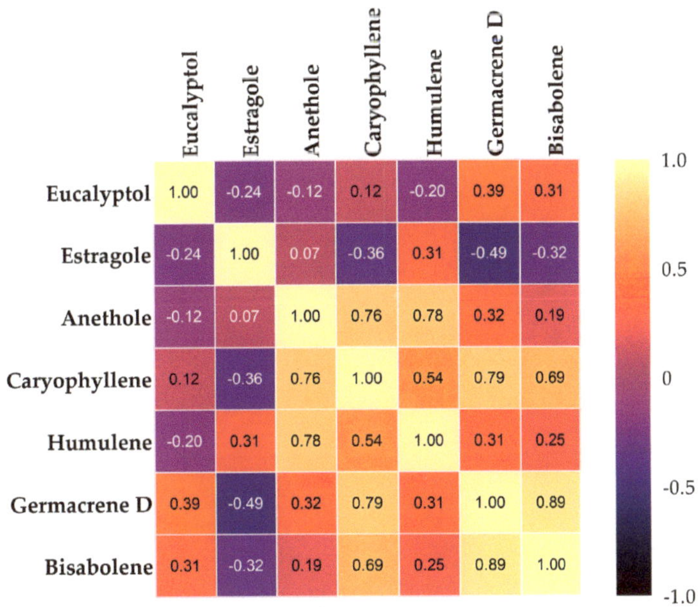

Figure 6. Correlation analysis of essential oil compounds corresponding to the treatments.

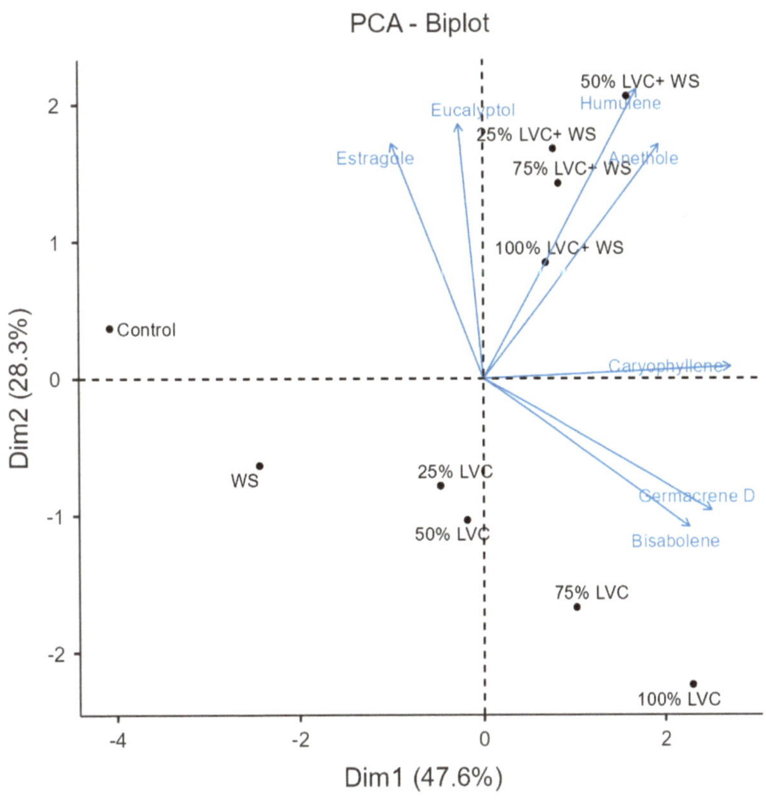

Figure 7. Principal component analysis of essential oil compounds corresponding to the treatments.

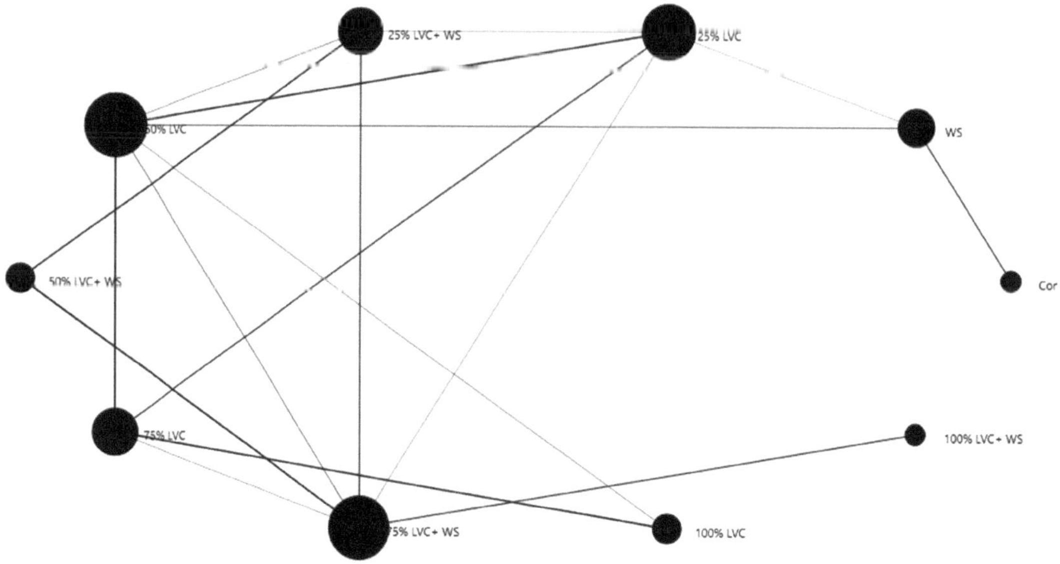

Figure 8. Network plot analysis of essential oil compounds corresponding to the treatments.

3.5. Phenolic Acids and Flavonoids

According to the LC–MS/MS analysis of the phenolics, twenty-six phenolic compounds were identified and quantified in the leaf samples (Table 4). Of the identified compounds, ascorbic acid was not significantly affected by either main-treatment water stress (p = 0.654); vermicompost (p = 0.373); or the interaction (p = 0.071). Shikimic acid content was critically affected by water stress (p = 0.005) and vermicompost (p = 0.007), but the interactive effects of the treatments were not significant (p = 0.105) (Table 5). The content of the compound was increased by approximately two-fold by water stress conditions. In comparison to the control, 50% LVC increased the content by 25.19%, but other vermicompost treatments did not exhibit substantial effects on the shikimic acid content in well-watered basil plants. As previously noted, although the interactive effects were not significant, the interaction decreased the shikimic acid content in comparison to the solo water stress treatment.

Of the major compounds of the basil plant, caffeic acid was significantly affected by water stress (p = 0.037), vermicompost (p = 0.000), and the interaction of the treatments (p = 0.005). For instance, water stress critically reduced the content. While 25 and 50% of vermicompost treatments increased the content, critical decreases were noted with the increasing concentrations of vermicompost. Considering the interaction, it was determined that 25, 50, and 75% vermicompost concentrations and stress interactions increased the quantity of the related compound. Interestingly, 100% vermicompost concentrations have been noted to significantly inhibit caffeic acid biosynthesis, regardless of its solo uses or its interaction with stress. Similar to the case of caffeic acid, rosmarinic acid content was also significantly responsive to the treatments (water stress; p = 0.000; vermicompost; p = 0.000; interaction of water stress and vermicompost; p = 0.000).

Quercimeritrin was significantly affected by the main treatments and their interactions (p = 0.000). Water stress increased the content by approximately two-fold. However, solo treatments of vermicompost reduced the content in comparison to the control. As in the case of interaction, 75% LVC + WS treatments peaked the content of the compound. Of the identified compounds, water stress critically increased the content (p = 0.021), whereas vermicompost treatments were not significant predictors for the content (p = 0.071). However, the interaction of the treatments was significant (p = 0.016).

Table 4. Effects of liquid leachate obtained from vermicompost (25, 50, 75, and 100%) on phenolic acids and flavonoids in basil leaves under water stress conditions (ng/uL).

Compounds	Control	25% LVC **	50% LVC	75% LVC	10% LVC	WS *	25% LVC + WS	50% LVC + WS	75% LVC + WS	100% LVC + WS
Ascorbic acid	105.61 ± 4.26	104.58 ± 5.38	107.06 ± 3.06	104.79 ± 5.13	114.23 ± 9.18	115.05 ± 10.61	107.03 ± 2.68	112.13 ± 7.60	103.86 ± 0.98	103.13 ± 2.77
Shikimic acid	451.56 ± 76.13	461.57 ± 76.95	565.73 ± 103.00	353.15 ± 47.74	356.80 ± 40.00	816.95 ± 333.28	756.45 ± 150.81	561.68 ± 83.68	442.51 ± 97.31	391.64 ± 77.30
Gallic acid	0.00 ± 0.00	0.00 ± 0.00	60.30 ± 19.71	53.40 ± 11.34	8.62 ± 0.84	12.73 ± 1.42	0.99 ± 0.20	3.74 ± 2.30	7.34 ± 0.32	7.88 ± 2.36
Protocatechuic acid	0.00 ± 0.00	0.34 ± 0.59	0.00 ± 0.00	0.00 ± 0.00	0.00 ± 0.00	11.06 ± 0.92	0.00 ± 0.00	0.00 ± 0.00	0.00 ± 0.00	0.00 ± 0.30
Chlorogenic acid	1.91 ± 0.53	10.40 ± 5.18	6.62 ± 0.74	1.32 ± 0.14	1.02 ± 0.16	5.00 ± 0.72	1.32 ± 0.25	1.20 ± 0.39	1.13 ± 2.53	5.91 ± 0.18
4-Hydroxy-benzaldehyde	0.00 ± 0.00	0.00 ± 0.00	0.96 ± 0.85	0.00 ± 0.00	0.00 ± 0.00	0.33 ± 0.30	0.65 ± 0.57	0.06 ± 0.11	0.00 ± 0.00	0.62 ± 0.54
Caffeic acid	29.65 ± 3.47	87.91 ± 6.59	67.15 ± 20.07	14.39 ± 0.00	0.00 ± 0.82	8.12 ± 2.30	62.79 ± 12.62	73.92 ± 7.85	20.53 ± 0.00	0.00 ± 1.56
Syringic acid	125.91 ± 8.97	129.82 ± 9.16	127.10 ± 3.15	143.32 ± 1.21	154.89 ± 7.16	125.34 ± 7.88	117.19 ± 2.03	128.50 ± 1.72	136.23 ± 6.73	127.63 ± 2.70
P-coumaric acid	0.82 ± 0.76	0.00 ± 0.01	0.00 ± 0.00	0.00 ± 0.00	0.00 ± 0.00	0.86 ± 0.76	0.00 ± 0.00	0.00 ± 0.00	0.00 ± 0.0 0	0.00 ± 0.00
Polydatine	0.92 ± 0.27	0.07 ± 0.12	0.03 ± 0.06	0.00 ± 0.00	0.00 ± 0.00	0.16 ± 0.27	0.26 ± 0.31	0.00 ± 0.00	0.00 ± 0.00	1.16 ± 2.01
Trans-ferulic acid	103.48 ± 54.08	1722.96 ± 176.75	1614.05 ± 242.85	93.27 ± 8.62	68.85 ± 20.43	716.28 ± 70.83	2447.94 ± 475.02	151.43 ± 8.06	506.89 ± 98.17	34.65 ± 8.59
Quercimeritrin	87.16 ± 39.27	9.90 ± 1.41	48.59 ± 4.05	58.18 ± 7.49	41.38 ± 2.51	238.27 ± 14.54	245.87 ± 69.37	27.00 ± 5.29	497.84 ± 153.88	102.78 ± 8.98
Cynarin	24.54 ± 3.30	23.70 ± 1.49	24.87 ± 1.19	23.38 ± 1.68	22.82 ± 0.80	22.48 ± 1.52	23.20 ± 0.73	23.48 ± 1.88	23.89 ± 1.29	25.89 ± 1.33
Hyperocide	24.17 ± 11.03	5.16 ± 1.52	10.05 ± 7.35	27.74 ± 1.22	28.16 ± 7.99	295.74 ± 15.51	113.92 ± 4.30	32.17 ± 3.01	205.73 ± 5.84	34.52 ± 10.24
Quercetin-3-glucoside	10.73 ± 5.09	5.22 ± 2.88	8.66 ± 2.50	2.89 ± 0.41	9.31 ± 2.73	80.59 ± 12.89	39.86 ± 3.85	10.85 ± 1.26	53.93 ± 18.63	10.55 ± 3.19
Rutin	229.90 ± 12.22	231.50 ± 11.82	258.32 ± 13.01	218.27 ± 7.17	243.24 ± 10.60	354.55 ± 102.51y	311.79 ± 70.89	223.30 ± 1.48	240.65 ± 5.30	235.38 ± 4.72
Isoquercitrin	11.01 ± 5.21	6.38 ± 1.69	8.05 ± 2.47	2.65 ± 0.45	10.29 ± 0.68	73.27 ± 17.35	38.05 ± 2.37	10.61 ± 0.71	56.19 ± 19.92	10.25 ± 3.41
Resveratrol	5.32 ± 1.14	7.32 ± 2.36	4.33 ± 2.34	9.15 ± 2.35	8.22 ± 2.14	5.32 ± 1.00	10.94 ± 1.26	7.39 ± 0.99	7.93 ± 0.70	10.42 ± 1.19
Naringin	313.50 ± 18.35	296.67 ± 12.64	283.66 ± 5.77	313.50 ± 5.41	299.47 ± 14.59	294.55 ± 5.98	325.34 ± 9.34	297.84 ± 11.80	304.97 ± 15.94	305.51 ± 11.09
Rosmarinic acid	682.88 ± 285.372	5950.13 ± 110.82	11251.46 ± 1230.71	352.23 ± 75.79	4690.59 ± 276.48	13904.23 ± 2.460.64	29879.68 ± 600.89	11319.82 ± 1149.15	1798.03 ± 446.64	5192.40 ± 329.97
Neohesperidin	8.68 ± 1.51	2.98 ± 0.83	6.20 ± 0.95	0.00 ± 0.00	17.39 ± 4.04	13.34 ± 1.02	31.12 ± 1.83	23.06 ± 2.58	48.92 ± 10.52	5.36 ± 0.77
Ellagic acid	34.67 ± 8.81	0.00 ± 0.00	0.00 ± 0.00	0.00 ± 0. 00	0.00 ± 0.00	12.82 ± 2.33	0.00 ± 0.00	0.00 ± 0.00	0.00 ± 0.00	35.63 ± 8.11
Naringenin	64.78 ± 6.840484	72.23 ± 11.57	38.51 ± 17.38	20.28 ± 1.71	18.05 ± 2.98	6.41 ± 0.93	18.19 ± 1.21	17.65 ± 6.71	14.67 ± 1.27	22.38 ± 9.79
Silibinin	10.53 ± 0.60	10.59 ± 0.58	10.62 ± 0.55	11.06 ± 1.06	10.63 ± 0.32	10.50 ± 0.66	10.50 ± 0.62	10.74 ± 0.46	10.63 ± 0.37	10.80 ± 0.38
3-Hydroxyflavone	21.20 ± 1.83	20.50 ± 1.72	22.48 ± 2.15	19.69 ± 1.29	21.57 ± 3.00	23.55 ± 2.12	22.31 ± 3.56	20.79 ± 1.22	19.73 ± 3.27	21.63 ± 2.07
Diosgenin	5.15 ± 2.91	2.29 ± 1.05	3.49 ± 0.39	4.21 ± 0.86	4.10 ± 0.62	4.17 ± 0.31	2.23 ± 0.72	2.08 ± 0.27	1.60 ± 0.72	6.40 ± 1.80

* WS: water stress; ** LVC: liquid leachate obtained from vermicompost. Different letters indicate significant difference according to a Duncan's multiple range test ($p < 0.05$).

Table 5. Variance analysis of phenolic compounds corresponding to the treatments.

Compounds	Water Stress	Vermicompost	Water Stress × Vermicompost
Ascorbic acid	0.654 ns	0.373 ns	0.071 ns
Shikimic acid	0.005	0.007	0.105 ns
Gallic acid	0.000	0.000	0.000
Protocatechuic acid	0.000	0.000	0.000
Chlorogenic acid	0.063 ns	0.006	0.000
4-Hydroxybenzaldehyde	0.326 ns	0.226 ns	0.012
Caffeic acid	0.037	0.000	0.005
Syringic acid	0.000	0.000	0.000
P-coumaric acid	0.946 ns	0.001	1.000 ns
Polydatine	0.645 ns	0.383 ns	0.199 ns
Trans-ferulic acid	0.454 ns	0.000	0.000
Quercimeritrin	0.000	0.000	0.000
Cynarin	0.906 ns	0.844 ns	0.126 ns
Hyperocide	0.000	0.000	0.000
Quercetin-3-glucoside	0.000	0.000	0.000
Rutin	0.021	0.071 ns	0.016
Isoquercitrin	0.000	0.000	0.000
Resveratrol	0.021	0.001	0.095 ns
Naringin	0.335 ns	0.061 ns	0.020
Rosmarinic acid	0.000	0.000	0.000
Neohesperidin	0.000	0.000	0.000
Ellagic acid	0.065 ns	0.000	0.000
Naringenin	0.000	0.000	0.000
Silibinin	0.820 ns	0.872 ns	0.918 ns
3-Hydroxyflavone	0.558 ns	0.404 ns	0.599 ns
Diosgenin	0.229 ns	0.001	0.028

ns: non-significant.

3.6. Heat Map Clustering, Correlation, Principal Component and Network Plot Analyses of the Phenolics and Flavonoids Corresponding to the Treatments

According to the heat map clustering (Figure 9), it can be observed that two distinct clusters were obtained in relation to the treatments. The first cluster was composed of 75% LVC, control, and 75% LVC + WS, while the other treatments were grouped under the second cluster. However, the stress or non-stress groups were not well-discriminated. With respect to the phenolic compounds, the first cluster included ascorbic acid, syringic acid, shikimic acid, rutin, and naringin. Those compounds attained their highest content values in treatments of 75% LVC. The major compound of basil plants, rosmarinic acid, was grouped in the second cluster. Contrary to the compounds in the first cluster, the lowest content of rosmarinic acid was recorded in treatments of 75% LVC. According to the correlation analysis (Figure 10) of major compounds (caffeic and rosmarinic acids), caffeic acid was only significantly correlated with trans-ferulic acid ($r = 0.667$; $p < 0.05$) and diosgenin ($r = -0.638$; $p < 0.05$). Similar to the case of the major essential oil compound (estragole), the major phenolic compound (rosmarinic acid) was not significantly correlated with any phenolic compounds ($p > 0.05$). In addition, we performed PCA analysis to scatter the phenolic compounds and treatments on a biplot pair (Figure 11). Accordingly, two principal components (PC_1 *eigenvalue*: 8.93 (89.31%) and PC_2 *eigenvalue*: 0.92 (9.20%)) accounted for 98.51% of the variability of the original data. According to the loading factors, shikimic acid, naringin, and rosmarinic acid were clearly separated from the other compounds on the biplot pair. Similar to the cases concerned with the relations of the experimental groups and their performance phenolic compounds, the network plot analysis revealed the only relations among 25% LVC, 75% LVC, 100% LVC, and 50% LVC + WS (Figure 12).

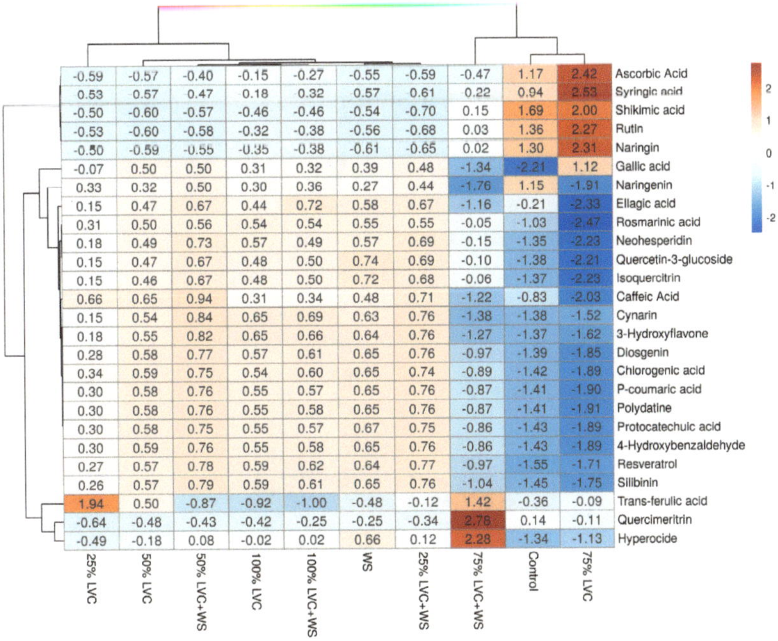

Figure 9. Heat map clustering of phenolics and flavonoids corresponding to the treatments.

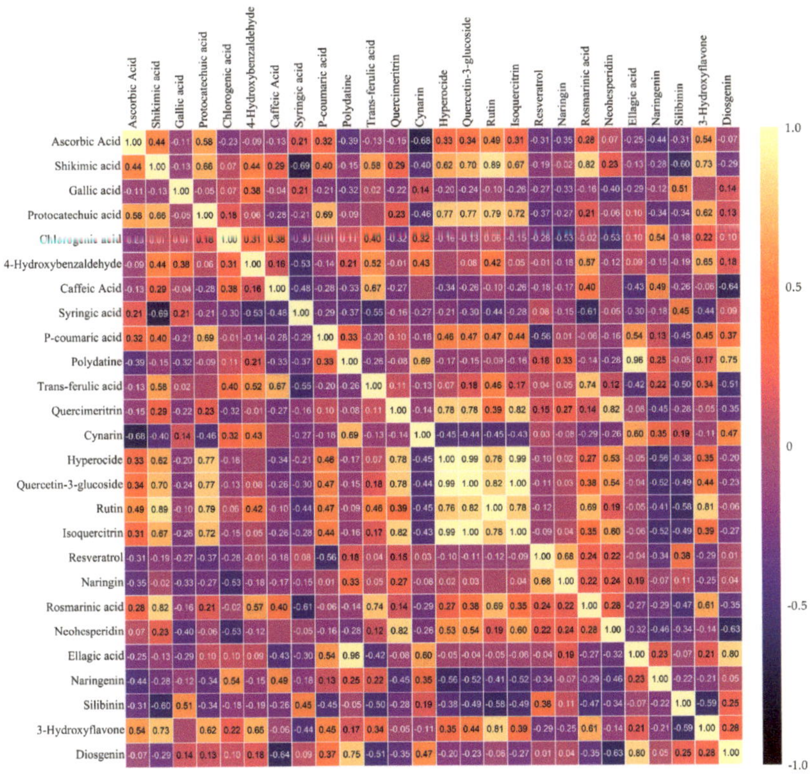

Figure 10. Correlation analysis of phenolics and flavonoids corresponding to the treatments.

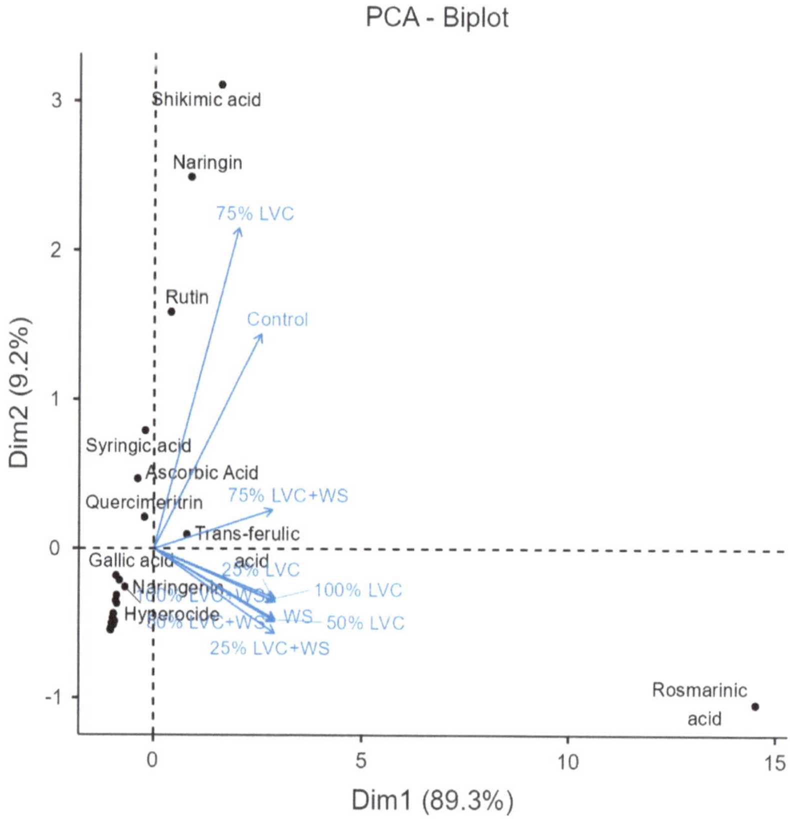

Figure 11. Principal component analysis of phenolics and flavonoids corresponding to the treatments.

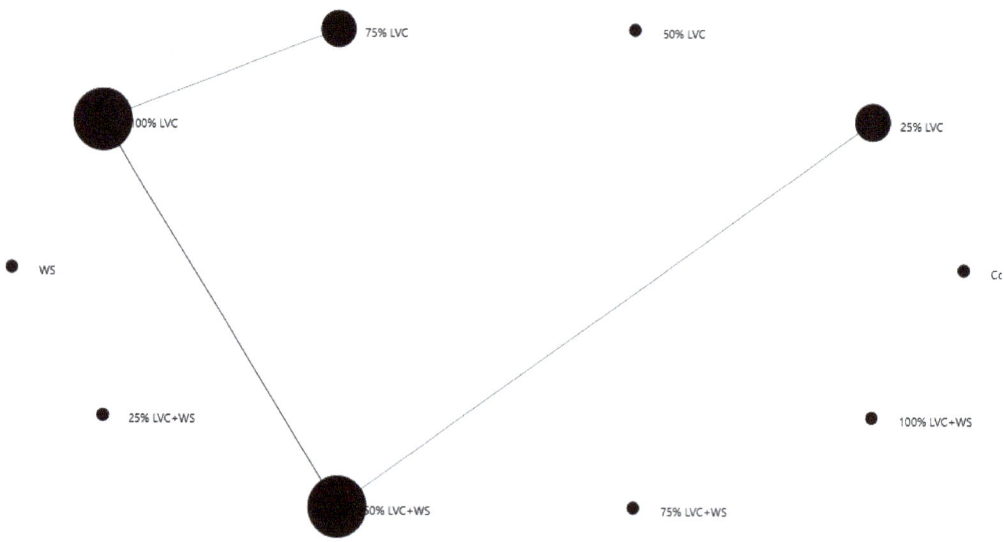

Figure 12. Network plot analysis of phenolics and flavonoids corresponding to the treatments.

4. Discussion

Water stress is one of the most important environmental problems that causes drastic problems in both developed and developing countries. After conducting a basic search on SCOPUS with criteria, including "drought stress OR water stress" on 20 July 2022, 253.552 documents obtained from SCOPUS were recorded. Considering the devastating effects of water stress, the direct effects of the stress were manifested as retarded plant growth and performance as well as a loss of crop productivity, in particular [40]. For that reason, intensive research has been conducted in order to cope with drought stress and to understand the action mechanism of droughts on crop and non-crop plant species. Just as the effects of limited water supply are not the same for all plant species, the duration, severity, frequency, and period of occurrence of water stress cause critical effects on each plant species [4,41]. These results, in a way, limit our ability to make general postulations about the mechanisms of drought. For these reasons, this deserves further investigations in the relevant fields. In an attempt to combat water stress conditions, due to fact that breeding studies are a process requiring many years and a lot of expertise, short-term solutions, such as chemical inputs in agricultural fields, have been suggested. However, the long-term and excessive chemical fertilizer input causes serious negative effects on plants and the ecosystem [42]. For these reasons, the demand for organic-sourced fertilizers, which produce fewer negative effects on the natural environment, has recently increased [43–45]. Although organic fertilizers can be obtained from quite varying sources, vermicompost amendments are the most commonly and recently used organic fertilizers [30,46–49]. It should be emphasized that the related studies mostly focus on the development and productivity of plants. Although this varies according to the concentration and application time, the positive effect of vermicompost on the growth and development of plants has been clearly demonstrated in the research. In our previous study [30], we tested the potential effects of a solid form of vermicompost on basil plants subjected to water stress conditions. Being very similar to the present study, we monitored the changes in the agronomic traits and secondary metabolites of basil under water stress conditions. As an effective approach to waste management, in this study, the liquid leachate obtained from vermicompost was assayed for its potential effects in basil against water stress conditions. Based on the analysis regarding the physico-chemical composition of liquid leachate, contrary to the solid form of vermicompost, leakage was observed to be poor in relation to the organic content and other elements. However, the microbial composition of the vermicompost was not analyzed in our study. As the effects of vermicompost are not only dependent on the organic and element content, but also on hormone-like compounds and the microbial composition [50–52], we hypothesized that liquid leachate might also be a critical discriminative factor and predictor in buffering the adverse impacts of basil plants submitted to water stress conditions.

Water stress caused critical damage to the agronomic traits of basil. These results are consistent with the previous reports on basil plants suffering from water stress conditions [7,8,30]. However, the root systems of the plants might be positively affected by the decline in water levels in the soil [30]. The significant increases in both the root length and root FW were noted and those parameters were positively correlated ($r = 0.83$; $p < 0.05$). On the other hand, as expected, positive results were obtained for the agronomic properties of the basil plant with liquid vermicompost applications independent of stress conditions. Similar to the stress-suffered basil plants, the vermicompost amendment affirmatively influenced the under-ground components of the basil plants. Pant et al. [53] also reported that plant and root growth as well as overall crop productivity were achieved with the vermicompost. These results might suggest that the root system of a plant could be an important distinguishing and predictive factor. However, basil leaves are significant in view of industrial demand. For that reason, vermicompost treatments might not be efficient to combat stress in basil plants, but these applications may be more realistic and effective for plants evaluated for their underground parts. In general, the augmented vegetative growth attributes of plants exposed to vermicompost were explained by the enrichment of the growth media in terms of both nutrients and organic matter [30,53,54]. Additionally previous reports revealed that the foliar application of vermicompost leakage enhanced the photosynthesis efficiency in either control plants or those submitted to stress conditions [55]. Although both organic and nutritional element contents were low according to the vermicompost fertilizer analysis, the trial soil might be enriched with the enzymes or hormone-like substances of the relevant fertilizer. Depending on the enrichment, additional microorganism, enzyme, and hormone inputs might be added to the soil structure which in turn might directly contribute to the vegetative growth of the plants [51,52]. As clearly reported by [56,57], the compounds with molecular structure analogues similar to auxin and cytokinin were available in the compost. In this study, we did not measure the phytohormones or plant growth-stimulating compounds. However, we can suggest their potential and plausible roles in plant responses. Furthermore, the plant growth might be linked to the absorption of elements through the plasmatic membrane H+-ATPase-aided activation of macro- or micro-nutrient uptakes [50].

Basil plants are reputed medicinal and aromatic plants due to their secondary metabolites (terpenoids and phenolic compounds). The alterations in the patterns of their metabolites as a response to water stress conditions have been reported in numerous studies [8,30,58–60]. However, the interactions of liquid vermicompost and water stress have not been investigated hitherto. It has been widely reported that the excess carbon that is not used in growth and development processes (primary metabolism) in plant systems is used in secondary metabolism. The shift of carbon surplus from primary to secondary metabolism is one of the critical defense strategies used by plants against stress conditions. In addition to the enzymatic antioxidant system, plants have also developed a non-enzymatic defense system with the construction of secondary metabolites [41,61–64]. According to the current results, however, the major compound, estragole percentage, was not critically affected by water stress, in comparison to the control, but the percentage peaked at the interaction of 25% LVC and water stress. On the other hand, eucalyptol percentage was significantly decreased in water stress-submitted plants, in comparison to the control, but the percentage of the compound reached the highest value at the interaction of 100% LVC and water stress. The decline in the content of both compounds was also previously reported in basil plants exposed to water stress conditions [30].

Phenolic compounds are one of the remarkable groups of metabolites acknowledged for their critical roles in reducing the oxidative stress, being reported in numerous studies [65–67]. However, the studies conducted with respect to the plasticity of phenolic compounds as a response to vermicompost and interaction of vermicompost with water stress are quite limited. Of those reports, Celikcan et al. [30] assayed the solid form of vermicompost for basil plants against water stress conditions. The major phenolic compounds of basil plant are caffeic, rosmarinic, and chicoric acids [68]. In the present study, the contents of caffeic and rosmarinic acids were quantified and significant decreases were noted in rosmarinic acid content in relation to water stress conditions, in accordance with the results obtained from previous studies [69–71]. Additionally, water stress critically reduced the caffeic acid content. In addition, it has been noted that regardless of stress conditions, 100% vermicompost concentrations used alone or independent of their interaction with stress significantly inhibited caffeic acid biosynthesis. However, it was observed that lower concentrations of liquid vermicompost applications increased the quantity of the related compound interacting with stress conditions. However, solid vermicompost forms, drought stress, and their interactions significantly increased the caffeic acid content of the basil plant [30]. These differences might be explained by the physico-chemical composition of vermicompost and the type as well as duration of stress factors, since other factors, *viz.*, basil cultivars, growing media, or stress-timing, were the same. Previous reports have revealed that organic amendment positively affected the quantity of total phenolic content [72]. In this study, we profiled the quantities of phenolic acids and flavonoids instead of total phenolic content of the basil plants as a response of vermicompost and its interaction with stress. However, the enrichment of the growing media with organic fertilizer critically affected the phenolic acids [73].

5. Conclusions

Water stress, as expected, critically resulted in reductions in agronomic attributes, such as plant height, plant fresh weight, root fresh weight, leaf length, and leaf diameter. Despite the water stress conditions, enriching the growth media with liquid leakage obtained from vermicompost crucially affected the agronomic attributes of well-watered basil plants. In particular, the highest values with respect to the above-ground parts were observed at a 50% concentration, whilst the highest values of under-ground parts were recorded at a 100% concentration of leakage. Considering the interactions of water stress and vermicompost, however, the interaction only had significant effects on the root length and root fresh weight. Regarding the major essential oil compound (estragole), the highest estragole content was determined in the 25% vermicompost + water stress, water stress, and control groups. Of the major phenolic compounds, caffeic acid decreased as a result of water stress but increased with the vermicompost treatments. The rosmarinic acid content increased as a result of water stress, reaching the highest content at 25% vermicompost and water stress interaction. In general, 25% and 50% vermicompost applications increased the content of phenolic compounds in plants under either well-watered or stress conditions. To the best of our knowledge, the present study is one of the first studies of its kind to analyze essential oil, phenolic acid, and flavonoids present in basil plants submitted to water stress conditions.

Author Contributions: Conceptualization, M.K.; methodology, M.K. and M.Z.K.; software, M.K. and M.G.K.; validation, M.K.; formal analysis, M.K.; investigation, H.K., M.Z.K., F.C. and M.G.K.; data curation, H.K., M.Z.K. and F.C.; writing—original draft preparation, M.K.; writing—review and editing, M.K.; visualization, M.K. and M.G.K.; supervision, M.K.; project administration, M.K. and M.Z.K.; funding acquisition, M.K. All authors have read and agreed to the published version of the manuscript.

Funding: The study was financially supported by the project coordination unit of Igdir University (Türkiye) with project number: TBY1220Y18. In this regard, we would like to send our deep thanks to Igdir University.

Data Availability Statement: The data used to support the findings were all included in this study.

Acknowledgments: The present study was derived from the Master's thesis of H. Kosem (supervised by M. Kulak).

Conflicts of Interest: The authors declare that they have no known competing financial interest or personal relationship that could have appeared to influence the work reported in this paper.

References

1. Agarwal, P.K.; Jha, B. Transcription Factors in Plants and ABA Dependent and Independent Abiotic Stress Signalling. *Biol. Plant.* **2010**, *54*, 201–212. [CrossRef]
2. Gill, S.S.; Tuteja, N. Polyamines and Abiotic Stress Tolerance in Plants. *Plant Signal. Behav.* **2010**, *5*, 26–33. [CrossRef]
3. Petropoulos, S.A.; Daferera, D.; Polissiou, M.G.; Passam, H.C. The Effect of Water Deficit Stress on the Growth, Yield and Composition of Essential Oils of Parsley. *Sci. Hortic.* **2008**, *115*, 393–397. [CrossRef]
4. Sun, Y.; Wang, C.; Chen, H.Y.; Ruan, H. Response of Plants to Water Stress: A Meta-Analysis. *Front. Plant Sci.* **2020**, *11*, 978. [CrossRef]
5. Heidari, M.; Golpayegani, A. Effects of Water Stress and Inoculation with Plant Growth Promoting Rhizobacteria (PGPR) on Antioxidant Status and Photosynthetic Pigments in Basil (*Ocimum basilicum* L.). *J. Saudi Soc. Agric. Sci.* **2012**, *11*, 57–61. [CrossRef]
6. Kalamartzis, I.; Dordas, C.; Georgiou, P.; Menexes, G. The Use of Appropriate Cultivar of Basil (*Ocimum basilicum*) Can Increase Water Use Efficiency under Water Stress. *Agronomy* **2020**, *10*, 70. [CrossRef]
7. Kalamartzis, I.; Menexes, G.; Georgiou, P.; Dordas, C. Effect of Water Stress on the Physiological Characteristics of Five Basil (*Ocimum basilicum* L.) Cultivars. *Agronomy* **2020**, *10*, 1029. [CrossRef]
8. Kulak, M.; Jorrín-Novo, J.V.; Romero-Rodriguez, M.C.; Yildirim, E.D.; Gul, F.; Karaman, S. Seed Priming with Salicylic Acid on Plant Growth and Essential Oil Composition in Basil (*Ocimum basilicum* L.) Plants Grown under Water Stress Conditions. *Ind. Crops Prod.* **2021**, *161*, 113235. [CrossRef]

9. Reddy, A.R.; Chaitanya, K.V.; Vivekanandan, M. Drought-Induced Responses of Photosynthesis and Antioxidant Metabolism in Higher Plants. *J. Plant Physiol.* **2004**, *161*, 1189–1202. [CrossRef]
10. Rao, D.E.; Chaitanya, K.V. Photosynthesis and Antioxidative Defense Mechanisms in Deciphering Drought Stress Tolerance of Crop Plants. *Biol. Plant.* **2016**, *60*, 201–218. [CrossRef]
11. Zargar, S.M.; Gupta, N.; Nazir, M.; Mahajan, R.; Malik, F.A.; Sofi, N.R.; Salgotra, R.K. Impact of Drought on Photosynthesis: Molecular Perspective. *Plant Gene* **2017**, *11*, 154–159. [CrossRef]
12. Ding, L.; Lu, Z.; Gao, L.; Guo, S.; Shen, Q. Is Nitrogen a Key Determinant of Water Transport and Photosynthesis in Higher Plants upon Drought Stress? *Front. Plant Sci.* **2018**, *9*, 1143. [CrossRef]
13. Mazid, M.; Khan, T.A.; Mohammad, F. Role of Secondary Metabolites in Defense Mechanisms of Plants. *Biol. Med.* **2011**, *3*, 232–249.
14. Zaynab, M.; Fatima, M.; Abbas, S.; Shari, Y.; Umair, M.; Zafar, M.H.; Bahadar, K. Role of Secondary Metabolites in Plant Defense against Pathogens. *Microb. Pathog.* **2018**, *124*, 198–202. [CrossRef]
15. Erb, M.; Kliebenstein, D.J. Plant Secondary Metabolites as Defenses, Regulators, and Primary Metabolites: The Blurred Functional Trichotomy. *Plant Physiol.* **2020**, *184*, 39–52. [CrossRef]
16. Aguirre-Becerra, H.; Vazquez-Hernandez, M.C.; Saenz de la O, D.; Alvarado-Mariana, A.; Guevara-Gonzalez, R.G.; Garcia-Trejo, J.F.; Feregrino-Perez, A.A. Role of Stress and Defense in Plant Secondary Metabolites Production. In *Bioactive Natural Products for Pharmaceutical Applications*; Springer: Cham, Switzerland, 2021; pp. 151–195.
17. Gulmez, C.; Kulak, M. New Insights to Enhance the Desired Anti-Diabetic Compounds in Medicinal and Aromatic Plants Exposed to Abiotic Stress Factors. In *Biotechnology of Anti-Diabetic Medicinal Plants*; Gantait, S., Verma, S.K., Sharangi, A.B., Eds.; Springer: Singapore, 2021; pp. 285–306.
18. Karaman, S.; Kirecci, O.A.; Ilcim, A. Influence of Polyamines (Spermine, Spermidine and Putrescine) on the Essential Oil Composition of Basil (*Ocimum basilicum* L.). *J. Essent. Oil Res.* **2008**, *20*, 288–292. [CrossRef]
19. Kahveci, H.; Bilginer, N.; Diraz-Yildirim, E.; Kulak, M.; Yazar, E.; Kocacinar, F.; Karaman, S. Priming with Salicylic Acid, β-Carotene and Tryptophan Modulates Growth, Phenolics and Essential Oil Components of *Ocimum basilicum* L. Grown under Salinity. *Sci. Hortic.* **2021**, *281*, 109964. [CrossRef]
20. Kumar, R.; Kumar, R.; Prakash, O. The Impact of Chemical Fertilizers on Our Environment and Ecosystem. In *Research Trends in Environmental Sciences*, 2nd ed.; Sharma, P., Ed.; Akinik Publications: Khasmir, India, 2019; pp. 69–86.
21. Büyük, G.; Kırpık, M.; Çelik, A.; Kuş, M.; Akça, E.; Zeydanlı, U. Adıyaman İlinin Hayvansal atik Kaynaklı Vermikompost üRetim Potansiyeli. *Adyütayam* **2017**, *5*, 31–37. (In Turkish)
22. Domínguez, J.; Edwards, C.A. Relationships between Composting and Vermicomposting. In *Vermiculture Technology: Earthworms, Organic Waste and Environmental Management*, 2nd ed.; Sherman, Ed.; CRC Press.: Boca Raton, Florida, 2011; pp. 1–14.
23. Yıldız, M.; Gürkan, M.O.; Turgut, C.; Kaya, Ü.; Ünal, G. *Tarımsal Savaşımda Kullanılan Pestisitlerin Yol açtığı çevre Sorunları. VI. Türkiye Ziraat Mühendisliği Teknik Kongresi*; TMMOB Ziraat Mühendisleri Odası: Ankara, Türkiye, 2005. (In Turkish)
24. Demir, H.; Polat, E.; Sönmez, İ. Ülkemiz Için Yeni Bir Organik Gübre: Solucan Gübresi. *Tarım Aktüel* **2010**, *14*, 54–60. (In Turkish)
25. Befrozfar, M.R.; Habibi, D.; Asgharzadeh, A.; Sadeghi-Shoae, M.; Tookallo, M.R. Vermicompost, Plant Growth Promoting Bacteria and Humic Acid Can Affect the Growth and Essence of Basil (*Ocimum basilicum* L.). *Ann. Biol. Res.* **2013**, *4*, 8–12.
26. Esmaielpour, B.; Rahmanian, M.; Heidarpour, O.; Shahriari, M.H. Effect of Vermicompost and Spent Mushroom Compost on the Nutrient and Essential Oil Composition of Basil (*Ocimum basilicum* L.). *J. Essent. Oil-Bear. Plants* **2017**, *20*, 1283–1292. [CrossRef]
27. Morelli, F.; Ferarrese, L.; Munhoz, C.L.; Alberton, O. Antimicrobial Activity of Essential Oil and Growth of *Ocimum basilicum* L. Inoculated with Mycorrhiza and Humic Substances Applied to Soil. *Genet. Mol. Res.* **2017**, *16*, 16039710. [CrossRef]
28. Chelariu, E.L.; Ghiorghe, C.; Turcu, N. Vermicompost Influence on Production of *Ocimum basilicum* L. Seedlings. *Lucr. Ştiinţifice USAMV Iaşi Ser. Hortic.* **2018**, *61*, 203–206.
29. Heidarpour, O.; Esmaielpour, B.; Soltani, A.A.; Khorramdel, S. Effect of Vermicompost on Essential Oil Composition of (*Satureja hortensis* L.) under Water Stress Condition. *J. Essent. Oil-Bear. Plants* **2019**, *22*, 484–492. [CrossRef]
30. Celikcan, F.; Kocak, M.Z.; Kulak, M. Vermicompost Applications on Growth, Nutrition Uptake and Secondary Metabolites of *Ocimum basilicum* L. under Water Stress: A Comprehensive Analysis. *Ind. Crops Prod.* **2021**, *171*, 113973. [CrossRef]
31. Reddy, A.R.; Teja, K.S.; Varghese, R.P.; Deepthi, M.; Sastry, K.P. Influence of Different Doses of Vermicompost and NPK on Growth of Herb *Ocimum tenuiflorum* var. CIM-Ayu. *Res. J. Pharm. Technol.* **2018**, *11*, 1713–1717. [CrossRef]
32. Rahmanian, M.; Esmaielpour, B.; Hadian, J.; Shahriari, M.H.; Fatemi, H. The Effect of Organic Fertilizers on Morphological Traits, Essential Oil Content and Components of Basil (*Ocimum basilicum* L.). *J. Agric. Sci. Sustain. Prod.* **2017**, *27*, 103–118.
33. Ayastuy, M.E.; Muscolino, C.; Fernandez, J.A.; Belladonna, D.; Rodriguez, R.A.; Caro, L.; Hernandez, L.F. Effect of Organic Substrate and Aqueous Extract of Vermicompost on Nursery Basil Growth. *Acta Hortic.* **2020**, 71–76. [CrossRef]
34. Reyes Araujo, D.Y.; Mora Herrera, M.E.; Lugo, J. Estabilización Por Vermicomposteo de Lodos Residuales Aplicados en la Productividad de Albahaca (*Ocimum basilicum* L.). *Rev. Int. De Contam. Ambient.* **2020**, *36*, 371–381. [CrossRef]
35. Rezaei-Chiyaneh, E.; Amani Machiani, M.; Javanmard, A.; Mahdavikia, H.; Maggi, F.; Morshedloo, M.R. Vermicompost Application in Different Intercropping Patterns Improves the Mineral Nutrient Uptake and Essential Oil Compositions of Sweet Basil (*Ocimum basilicum* L.). *J. Soil Sci. Plant Nutr.* **2021**, *21*, 450–466. [CrossRef]

36. Aras, A.; Turkan, F.; Yildiko, U.; Atalar, M.N.; Kılıç, O.; Alma, M.H.; Bursal, E. Biochemical Constituent, Enzyme Inhibitory Activity, and Molecular Docking Analysis of an Endemic Plant Species, *Thymus migricus*. *Chem Pap.* **2021**, *75*, 1133–1146. [CrossRef]
37. Yilmaz, M.A. Simultaneous Quantitative Screening of 53 Phytochemicals in 33 Species of Medicinal and Aromatic Plants: A Detailed, Robust and Comprehensive LC–MS/MS Method Validation. *Ind. Crop Prod.* **2020**, *149*, 112347. [CrossRef]
38. Farzaneh, A.; Ghani, A.; Azizi, M. The Effect of Water Stress on Morphological Characteristic and Essential Oil Content of Improved Sweet Basil (*Ocimum basilicum* L.). *J. Plant. Prod.* **2010**, *17*, 103–111.
39. Radácsi, P.; Inotai, K.; Sárosi, S.; Czövek, P.; Bernath, J.; Nemeth, E. Effect of Water Supply on the Physiological Characteristic and Production of Basil (*Ocimum basilicum* L.). *Eur. J. Hortic. Sci.* **2010**, *75*, 193.
40. Eriyagama, N.; Smakhtin, V.Y.; Gamage, N. *Mapping Drought Patterns and Impacts: A Global Perspective*; International Water Management Institute (IWMI): Colombo, Sri Lank, 2009.
41. Kulak, M.; Ozkan, A.; Bindak, R. A Bibliometric Analysis of the Essential Oil-Bearing Plants Exposed to the Water Stress: How Long Way We Have Come and How Much Further? *Sci. Hortic.* **2019**, *246*, 418–436. [CrossRef]
42. Kandpal, G. Review on Impact of Chemical Fertilizers on Environment. *Int. J. Mod. Agric.* **2021**, *10*, 758–763.
43. Alfarisy, M.Y.; Yassi, A.; Mustari, K. Increasing Productivity and Biomass of Corn Plants toward Grant Organic Fertilizer and Liquid Organic Fertilizer. *ENDLESS Int. J. Future Stud.* **2021**, *4*, 236–248. [CrossRef]
44. Bhunia, S.; Bhowmik, A.; Mallick, R.; Debsarcar, A.; Mukherjee, J. Application of Recycled Slaughterhouse Wastes as an Organic Fertilizer for Successive Cultivations of Bell Pepper and Amaranth. *Sci. Hortic.* **2021**, *280*, 109927. [CrossRef]
45. Leno, N.; Sudharmaidevi, C.R. Physicochemical and Nutrient Release Characteristics of a Thermochemical Organic Fertilizer Produced from Degradable Solid Waste and Its Effect on Productivity of Banana. *Commun. Soil Sci. Plant Anal.* **2021**, *52*, 2562–2577. [CrossRef]
46. Ding, Z.; Kheir, A.M.; Ali, O.A.; Hafez, E.M.; El Shamey, E.A.; Zhou, Z.; Seleiman, M.F. A Vermicompost and Deep Tillage System to Improve Saline-Sodic Soil Quality and Wheat Productivity. *J. Environ. Manag.* **2021**, *277*, 111388. [CrossRef]
47. Feizabadi, A.; Noormohammadi, G.; Fatehi, F. Changes in Growth, Physiology, and Fatty Acid Profile of Rapeseed Cultivars Treated with Vermicompost under Drought Stress. *J. Soil Sci. Plant Nutr.* **2021**, *21*, 200–208. [CrossRef]
48. Greco, C.; Comparetti, A.; Fascella, G.; Febo, P.; La Placa, G.; Saiano, F.; Laudicina, V.A. Effects of Vermicompost, Compost and Digestate as Commercial Alternative Peat-Based Substrates on Qualitative Parameters of *Salvia officinalis*. *Agronomy* **2021**, *11*, 98. [CrossRef]
49. Hafez, E.M.; Omara, A.E.D.; Alhumaydhi, F.A.; El-Esawi, M.A. Minimizing Hazard Impacts of Soil Salinity and Water Stress on Wheat Plants by Soil Application of Vermicompost and Biochar. *Physiol. Plant.* **2021**, *172*, 587–602. [CrossRef]
50. Canellas, L.P.; Olivares, F.L.; Okorokova-Façanha, A.L.; Façanha, A.R. Humic Acids Isolated from Earthworm Compost Enhance Root Elongation, Lateral Root Emergence, and Plasma Membrane H+-ATPase Activity in Maize Roots. *Plant Physiol.* **2002**, *130*, 1951–1957. [CrossRef]
51. Suthar, S. Evidence of Plant Hormone Like Substances in Vermiwash: An Ecologically Safe Option of Synthetic Chemicals for Sustainable Farming. *Ecol. Eng.* **2010**, *36*, 1089–1092. [CrossRef]
52. Theunissen, J.; Ndakidemi, P.A.; Laubscher, C.P. Potential of Vermicompost Produced from Plant Waste on the Growth and Nutrient Status in Vegetable Production. *Int. J. Phys. Sci.* **2010**, *5*, 1964–1973.
53. Pant, A.P.; Radovich, T.J.; Hue, N.V.; Talcott, S.T.; Krenek, K.A. Vermicompost Extracts Influence Growth, Mineral Nutrients, Phytonutrients and Antioxidant Activity in Pak Choi (*Brassica rapa* cv. Bonsai, Chinensis Group) Grown under Vermicompost and Chemical Fertiliser. *J. Sci. Food Agric.* **2009**, *89*, 2383–2392. [CrossRef]
54. Singh, R.; Agarwal, S.K. Growth and Yield of Wheat (*Triticum aestivum*) as Influenced by Levels of Farmyard Manure and Nitrogen. *Indian J. Agron.* **2001**, *46*, 462–467.
55. Guzmán-Albores, J.M.; Montes-Molina, J.A.; Castañón-González, J.H.; Abud Archila, M.; Gutiérrez-Miceli, F.A.; Ruiz-Valdiviezo, V.M. Effect of Different Vermicompost Doses and Water Stress Conditions on Plant Growth and Biochemical Profile in Medicinal Plant. *Moringa Oleifera Lam. J. Environ. Biol.* **2020**, *41*, 240–246. [CrossRef]
56. Bidabadi, S.S.; Dehghanipoodeh, S.; Wright, G.C. Vermicompost Leachate Reduces Some Negative Effects of Salt Stress in Pomegranate. *Int. J. Recycl. Org. Waste Agric.* **2017**, *6*, 255–263. [CrossRef]
57. Arthur, G.D.; Jäger, A.K.; Van Staden, J. The Release of Cytokinin-Like Compounds from Gingko Biloba Leaf Material during Composting. *Environ. Exp. Bot.* **2001**, *45*, 55–61. [CrossRef] [PubMed]
58. Martinez, I.G.; Sosa, F.C.; Saavedra, A.L.; Hernandez, M.S. Extraction of Auxin-Like Substances from Compost. *Crop Res. Hisar* **2002**, *24*, 323–327.
59. Khalid, K.A. Influence of Water Stress on Growth, Essential Oil, and Chemical Composition of Herbs (*Ocimum* sp.). *Int. Agrophysics* **2006**, *20*, 289–296.
60. Mandoulakani, B.A.; Eyvazpour, E.; Ghadimzadeh, M. The Effect of Drought Stress on the Expression of Key Genes Involved in the Biosynthesis of Phenylpropanoids and Essential Oil Components in Basil (*Ocimum basilicum* L.). *Phytochemistry* **2017**, *139*, 1–7. [CrossRef] [PubMed]
61. Mota, I.; Sánchez-Sánchez, J.; Pedro, L.G.; Sousa, M.J. Composition Variation of the Essential Oil from *Ocimum basilicum* L. cv. Genovese Gigante in Response to Glomus Intraradices and Mild Water Stress at Different Stages of Growth. *Biochem. Syst. Ecol.* **2020**, *90*, 104021. [CrossRef]

62. Akula, R.; Ravishankar, G.A. Influence of Abiotic Stress Signals on Secondary Metabolites in Plants. *Plant Signal. Behav.* **2011**, *6*, 1720–1731. [CrossRef]
63. Shitan, N. Secondary Metabolites in Plants: Transport and Self-Tolerance Mechanisms. *Biosci. Biotechnol. Biochem.* **2016**, *80*, 1283–1293. [CrossRef]
64. Ashraf, M.A.; Iqbal, M.; Rasheed, R.; Hussain, I.; Riaz, M.; Arif, M.S. Environmental Stress and Secondary Metabolites in Plants: An Overview. In *Plant Metabolites and Regulation under Environmental Stress*; Parvaiz, A., Mohammad, A.A., Vijay, P.S., Durgesh, K.T., Pravej, A., Mohammed, N.A., Eds.; Academic Press: London, UK, 2018; pp. 153–167.
65. Szabó, K.; Zubay, P.; Németh-Zámboriné, É. What Shapes Our Knowledge of the Relationship between Water Deficiency Stress and Plant Volatiles? *Acta Physiol. Plant.* **2020**, *42*, 130. [CrossRef]
66. Sánchez, E.; Soto, J.M.; García, P.C.; López-Lefebre, L.R.; Rivero, R.M.; Ruiz, J.M.; Romero, L. Phenolic Compounds and Oxidative Metabolism in Green Bean Plants under Nitrogen Toxicity. *Funct. Plant Biol.* **2000**, *27*, 973–978. [CrossRef]
67. Michalak, A. Phenolic Compounds and Their Antioxidant Activity in Plants Growing under Heavy Metal Stress. *Pol. J. Environ. Stud.* **2006**, *15*, 523–530.
68. Chalker-Scott, L.; Fuchigami, L.H. The Role of Phenolic Compounds in Plant Stress Responses. In *Low Temperature Stress Physiology in Crops*; CRC Press: Boca Raton, FL, USA, 2018; pp. 67–80.
69. Bahcesular, B.; Yildirim, E.D.; Karaçocuk, M.; Kulak, M.; Karaman, S. Seed Priming with Melatonin Effects on Growth, Essential Oil Compounds and Antioxidant Activity of Basil (*Ocimum basilicum* L.) under Salinity Stress. *Ind. Crops Prod.* **2020**, *146*, 112165. [CrossRef]
70. Fletcher, R.S.; Slimmon, T.; McAuley, C.Y.; Kott, L.S. Heat Stress Reduces the Accumulation of Rosmarinic Acid and the Total Antioxidant Capacity in Spearmint (*Mentha spicata* L.). *J. Sci. Food Agric.* **2005**, *85*, 2429–2436. [CrossRef]
71. Moradbeygi, H.; Jamei, R.; Heidari, R.; Darvishzadeh, R. Fe_2O_3 Nanoparticles Induced Biochemical Responses and Expression of Genes Involved in Rosmarinic Acid Biosynthesis Pathway in Moldavian Balm under Salinity Stress. *Physiol. Plant.* **2020**, *169*, 555–570. [CrossRef] [PubMed]
72. Keshavarz Mirzamohammadi, H.; Modarres-Sanavy, S.A.M.; Sefidkon, F.; MokhtassiBidgoli, A.; Mirjalili, M.H. Irrigation and Fertilizer Treatments Affecting Rosmarinic Acid Accumulation, Total Phenolic Content, Antioxidant Potential and Correlation between Them in Peppermint (*Mentha piperita* L.). *Irrig. Sci.* **2021**, *39*, 671–683. [CrossRef]
73. Sereme, A.; Dabire, C.; Koala, M.; Somda, M.K.; Traore, A.S. Influence of Organic and Mineral Fertilizers on the Antioxidants and Total Phenolic Compounds Level in Tomato (*Solanum lycopersicum*) var. Mongal F1. *J. Exp. Biol. Agric. Sci.* **2016**, *4*, 414–420. [CrossRef]

horticulturae

MDPI

Article

Sustainable Use of Sewage Sludge for Marigold (*Tagetes erecta* L.) Cultivation: Experimental and Predictive Modeling Studies on Heavy Metal Accumulation

Arwa A. AL-Huqail [1], Pankaj Kumar [2,*], Sami Abou Fayssal [3,4], Bashir Adelodun [5,6], Ivan Širić [7], Madhumita Goala [8], Kyung Sook Choi [6], Mostafa A. Taher [9,10], Aziza S. El-Kholy [11] and Ebrahem M. Eid [11,12,*]

1 Department of Biology, College of Science, Princess Nourah bint Abdulrahman University, P.O. Box 84428, Riyadh 11671, Saudi Arabia
2 Agro-Ecology and Pollution Research Laboratory, Department of Zoology and Environmental Science, Gurukula Kangri (Deemed to Be University), Haridwar 249404, India
3 Department of Agronomy, Faculty of Agronomy, University of Forestry, 10 Kliment Ohridski Blvd., 1797 Sofia, Bulgaria
4 Department of Plant Production, Faculty of Agriculture, Lebanese University, Beirut 1302, Lebanon
5 Department of Agricultural and Biosystems Engineering, University of Ilorin, PMB 1515, Ilorin 240003, Nigeria
6 Department of Agricultural Civil Engineering, Kyungpook National University, Daegu 41566, Republic of Korea
7 University of Zagreb, Faculty of Agriculture, Svetosimunska 25, 10000 Zagreb, Croatia
8 Department of Environment Science, Graphic Era (Deemed to Be University), Dehradun 248002, India
9 Biology Department, Faculty of Science and Arts, King Khalid University, Mohail Assir 61321, Saudi Arabia
10 Botany Department, Faculty of Science, Aswan University, Aswan 81528, Egypt
11 Botany Department, Faculty of Science, Kafrelsheikh University, Kafr El-Sheikh 33516, Egypt
12 Biology Department, College of Science, King Khalid University, Abha 61321, Saudi Arabia
* Correspondence: rs.pankajkumar@gkv.ac.in (P.K.); ebrahem.eid@sci.kfs.edu.eg (E.M.E.)

Citation: AL-Huqail, A.A.; Kumar, P.; Abou Fayssal, S.; Adelodun, B.; Širić, I.; Goala, M.; Choi, K.S.; Taher, M.A.; El-Kholy, A.S.; Eid, E.M. Sustainable Use of Sewage Sludge for Marigold (*Tagetes erecta* L.) Cultivation: Experimental and Predictive Modeling Studies on Heavy Metal Accumulation. *Horticulturae* **2023**, *9*, 447. https://doi.org/10.3390/horticulturae9040447

Academic Editors: Francesco De Mastro, Gennaro Brunetti, Karam Farrag and Huadong Zang

Received: 12 March 2023
Revised: 24 March 2023
Accepted: 28 March 2023
Published: 29 March 2023

Abstract: The present study aimed to investigate the impact of sewage sludge (SS) amendment on the growth, yield, and biochemical attributes of the marigold (*Tagetes erecta* L. var. Pusa Basanti Gainda) crop. For this purpose, marigold flowers were cultivated using three different treatments of SS, i.e., 0% (control with no SS), 5%, and 10%. Multiple linear regression (MLR) modeling was performed to develop prediction models for the impact of soil properties on heavy metals uptake by marigold plants. The results showed that the growth, yield, and biochemical attributes of marigold plants significantly ($p < 0.05$) increased with an increase in SS dose from 0 to 10%. The most feasible SS treatment was found to be 10%, which achieved a maximum flower yield of 318.42 g/plant. On the other hand, the bioaccumulation factor (BAF) values (>1) showed that the marigold plant was capable of uptaking significant contents of six heavy metals in the order of Cd < Cr < Cu < Zn < Mn < Fe. The MLR-based predictive models were capable of precisely predicting the contents of most heavy metal uptake by marigold plants as indicated by the coefficient of determination ($R^2 > 0.73$), model efficiency (ME > 0.49), root mean square error (RMSE < 3.25), and analysis of variance (ANOVA; $p < 0.05$) results. Overall, this study presented a novel approach to floriculture by sustainable management of SS while reducing public health and environmental impacts.

Keywords: biochemical components; floriculture; mathematical modeling; multiple linear regression; waste management

1. Introduction

Sewage sludge (SS), a byproduct of the wastewater treatment process, is produced in huge quantities in urban areas [1]. At the global scale, nearly 75–100 million tons of SS is produced annually, which is estimated to hit 127.5 million tons by 2030 [2]. Being rich in organic and inorganic chemicals, SS can contaminate soil and water environments [3]. Therefore, SS management has appeared as one of the emerging environmental issues, especially for developing countries. Improper disposal of SS creates a wide range of environmental problems such as soil and water pollution, greenhouse gas emission, and the spread of diseases, among others [4]. This leads to the need for proper management of SS, as it is essential for protecting public health, the environment, and water resources.

In recent times, land application of SS has become an increasing trend, which includes the use of SS on agricultural lands for improving their fertility [5]. SS can also be managed via several other methods such as anaerobic digestion, composting, incineration, and disposal in landfills [6]. At sewage treatment facilities, sludge is pre-digested under anaerobic conditions to produce fuel gas (methane). The digested SS is dried and then used as a soil amendment for agricultural purposes [7]. Composting of SS stabilizes the organic and other nutrient elements, thereby making it easier to manage [8]. Although incineration removes organic debris, it can also release toxins such as dioxins, furans, pesticides, and flame retardants into the atmosphere [3,5]. Similarly, SS-landfill is also one of the widely used methods, but it may cause severe groundwater pollution, as indicated by recent studies [9].

Out of all the available methods, the controlled agricultural use of pre-digested SS is considered the most viable option to sustainably increase crop production and boost soil health [10]. Because of its high nutrient content, SS is an excellent fertilizer for a wide range of crops, including fruits, vegetables, and grains [11,12]. It also aids to increase soil fertility and structure by supplying several macronutrients such as nitrogen (N), phosphorous (P), and potassium (K), as well as micronutrients such as zinc (Zn), iron (Fe), and copper (Cu) [13]. Furthermore, SS can aid in the retention of moisture in the soil and the enhancement of the ability to absorb and retain nutrients. The European Union (EU) Council Directive regulates the agricultural reuse of SS, especially on its use for certain types of edible crops and land applications, thus ensuring the limit for potentially harmful contaminants [14]. However, SS reuse for the production of non-edible crops is one of the most viable options that ensure both soil and consumer safety [15]. Several non-edible crops such as flowers (marigold), energy (jatropha), building (bamboo), fiber (*Sesbania* spp.), pharmaceutical (borage), and biopolymer (rubber) can be cultivated using controlled doses of SS [16–18].

Floriculture, or simply flower farming, is the production of flowers and other ornamental plants used for decoration, temple offerings, cultural and ritual uses, and the extraction of essential oils, medicines, and foods. In India, numerous types of flowers, including marigold (*Tagetes* spp.), Hibiscus (*Hibiscus rosa-sinensis*), Pansy (*Viola tricolor*), Lotus (*Nelumbo nucifera*), Dahlia (*Dahlia pinnata*), Jasmine (*Jasminum officinale*), etc., are widely cultivated and have high market demand [19]. Among them, however, the marigold flower holds first rank in terms of the annual production of 1754 thousand metric tons [20], signifying its high market demand. Marigold (*Tagetes erecta* L.) flowers are mostly used for non-edible purposes; therefore, there is very less chance of heavy metals and other contaminants being consumed by human beings along with the flowers [21,22]. Thus, marigold cultivation can be an effective solution to SS management, and it can provide a productive and sustainable way to use this material. However, there has been little research on the effect of the SS amendment on heavy metal accumulation in marigold. Previous research has primarily examined the effects of SS amendment on crop yields and soil health. However, the effects of the SS amendment on heavy metal mobility and accumulation in marigold plants are largely unexplored.

Therefore, the present study aimed to investigate the effect of SS amendment on the growth, yield, and biochemical attributes of the marigold crop. Furthermore, multiple linear regression-based prediction models were developed in order to evaluate the influence of SS-amended soil properties on heavy metal uptake by marigold plants.

2. Materials and Methods

2.1. Material Collection

For the present study, SS was collected from a sewage treatment plant (STP) located in Saliyar, Roorkee, India (29°54′05.8″ N and 77°51′53.8″ E). This STP helps in treating the municipal wastewater generated from Roorkee city. SS samples were collected in 10 kg polyvinylchloride (PVC) plastic bags and transported to the experimental site at Kulheri village, Saharanpur, Uttar Pradesh, India (29°52′51.4″ N and 77°16′17.7″ E). Experiments were conducted in a polyhouse (5 × 3 × 10 m; width × height × length) constructed using polycarbonate sheets. Before the collected SS sample was used in experiments and analysis, the sample was sun-dried and stored at room temperature. On the other hand, arable soil (AS) of loamy texture was collected from agricultural land (up to 25 cm depth) near the experimental site. Moreover, authentic and healthy seeds of marigold (*Tagetes erecta* L. var. Pusa Basanti Gainda) were procured from Bhawani Seeds and Bio-Tech (Mathura, Uttar Pradesh, India).

2.2. Experimental Design

Marigold (*T. erecta*) cultivation experiments were conducted using different doses of SS mixed with AS, including 0% (absolute AS for control), 5%, and 10% (*w*/*w*) from November 2021 to March 2022. For this purpose, plastic pots of 30 kg capacity were taken and filled with 25 kg of an appropriate dose of AS mixed with SS. The marigold seeds were germinated on trays (Boxseat Ventures Pvt. Ltd., Delhi, India) in the same soil taken from plastic pots under dark conditions and frequent water spraying. Once seedlings reached 5 cm height, they were transferred to the pots after 20 days. A total of 30 pots were used for marigold cultivation (n = 10 replicated pots for each treatment), and one healthy seedling was carefully transplanted in each pot. The plants were raised carefully under greenhouse conditions for 140 days. The average temperature and humidity of the polyhouse were maintained at 25 °C and 55% with full-day sunlight. The flowering was observed from mid-February to mid-March 2022, and picking was performed once flowers reached maximum diameter (every fourth day) and marketable grade (according to size and color). The plants were irrigated equally with a borewell water supply using a hand sprayer while defective leaves and branches were immediately removed to avoid any pest or pathogen attack.

2.3. Chemical Analyses

In the current study, SS and AS collected from sampling sites were analyzed for selected physicochemical properties such as pH and electrical conductivity (EC: dS/m) using a microprocessor-based digital meter (ESICO 1611, Parwanoo, India), organic matter (OM: g/kg) using the Walkley and Black method [23], total nitrogen (TN: g/kg) using Kjeldahl's method [24], total phosphorus (TP: g/kg) using digestion–distillation and a spectrophotometer method (Cary 60, Agilent Technologies, Santa Clara, CA, USA), and total potassium (K: g/kg) using a flame-photometer method (1382, ESICO, Parwanoo, India) as previously adopted by Kumar et al. [25]. Moreover, AS and SS were also characterized for six heavy metals, including Cd, Cr, Cu, Fe, Mn, and Zn using atomic absorption spectroscopy (Analyst 800, PerkinElmer, Waltham, MA, USA). For this purpose, 1 g of oven-dried AS or SS sample was dissolved in double distilled water and then digested in a di-acid mixture having a 1:3 ratio of HNO_3 and $HClO_4$ on an electric hot plate (150° C for 1.5 h). For heavy metal analysis in the marigold plant, 1 g of oven-dried plant tissues, i.e., root, shoot, and flowers were separately taken and digested in the same way. Further, the sample was filtered through Whatman filter paper (number 41), and the total contents were adjusted to 50 mL by the addition of 3% HNO_3 solution for final determination using AAS (recovery percentage > 98%). Heavy metal standards of 0 to 50 mg/L were prepared and AAS was calibrated accordingly. The slit width of the instrument was adjusted to 0.5 nm, and an air/acetylene gas mixture was used for combustion [26]. The current of hollow cathode lamps for each metal was adjusted as per the manufacturer's guidelines and analytical results. All chemicals and reagents used in this study were of analytical grade and procured from Merck (India), and results were validated following standard operating procedures (SOPs) and qualitative assurance (QC) [27].

2.4. Plant and Biochemical Assays

Marigold plants were evaluated for selected growth, yield, and biochemical characteristics in order to understand the impact of SS application at different rates. Firstly, the selected growth and yield attributes such as plant height (cm), the number of branches, root length (cm), first bud formation (days), flowering period (days), flower stack length (cm), flower diameter (cm), flower yield (number per plant), total flower yield (g), and the average weight of flower (g) were manually estimated using calibrated scale and weighing balance (SP500, Samson, Mohali, India). Further, the total chlorophyll content of marigold plant leaves was determined following 80% acetone extraction and spectroscopy, as previously adopted by Shah et al. [28]. The contents of catalase (U/mL) and peroxidase (µmol/mg) were spectrophotometrically estimated at 240 nm [29] and 470 nm [30]. Similarly, the contents of β-carotene were determined by using acetone as an extraction agent and taking the absorbance at 450 nm, as previously described by de Carvalho et al. [31]. The contents of total phenols were estimated by using Folin and Ciocalteu's phenol reagents and taking absorbance at 725 nm while ascorbic acid was determined after extraction of methanolic contents followed by spectroscopic determination at 515 nm [32]. Finally, the contents of lutein were determined after extracting the contents using methanol and then analyzed using high-performance liquid chromatography (HPLC), as adopted by Šivel et al. [33].

2.5. Data Analysis and Software

In this study, the bioaccumulation factor (BAF) tool was used to study the heavy metal accumulation potential of marigold plants. BAF is a measure of how much a heavy metal accumulates in the vegetative parts of the marigold plant in comparison to the content in the soil [34]. BAF was calculated by dividing the heavy metal content (mg/kg) in the plant (HM_{plant}) by the heavy metal content (mg/kg) in the soil (HM_{soil}), as shown in Equation (1):

$$BAF = HM_{plant}/HM_{soil} \tag{1}$$

On the other hand, the multiple linear regression (MLR) method was used for developing the predictive models to understand the influence of selected soil properties on heavy metals uptake by the marigold plant parts (root, shoot, and flower). For this purpose, soil properties such as pH, OM, and total heavy metal content were used as independent variables. The following Equation (2) was used for the prediction of heavy metal uptake by the marigold plant as a dependent variable [35]:

$$y = \beta + \beta_1 \times pH_{soil} + \beta_2 \times OM_{soil} + \beta_3 \times HM_{soil} \tag{2}$$

where "y" is predicted heavy metal uptake by the marigold plant, "β" refers to the MLR intercept, while "β_1", "β_2", and "β_3" refer to interactive model coefficients for soil pH, OM, and HM, respectively. Moreover, the developed models were tested for goodness of fit and accuracy by using selected validation tools such as measured vs. predicted response values, coefficient of determination (R^2), Nash-Sutcliffe model efficiency (ME) coefficient, and root mean square error (RMSE). Thus, the following Equations (3)–(5) were used for the computation of R^2, ME, and RMSE:

$$R^2 = 1 - \frac{\sum_{i=1}^{n}\left(\hat{y_{obs}} - \overline{y_{pre}}\right)^2}{\sum_{i=1}^{n}\left(y_{obs} - \overline{y_{mean}}\right)^2} \tag{3}$$

$$ME = 1 - \frac{\sum_{i=1}^{n}\left(y_{obs} - y_{pre}\right)^2}{\sum_{i=1}^{n}\left(y_{obs} - \overline{y_{mean}}\right)^2} \tag{4}$$

$$RMSE = \sqrt{\frac{\sum_{i=1}^{n}\left(y_{obs} - y_{pre}\right)^2}{n}} \quad (5)$$

where "y_{obs}", "y_{pre}", and "y_{mean}" are experimental, predicted, and mean values of heavy metal content in different parts of marigold plants. All experiments were conducted in ten replicates, and data were presented as mean followed by standard deviation. The data were analyzed using an analysis of variance (ANOVA) test to compare experimental treatments (based on Tukey's) at a significance level of probability (p) < 0.05. The data analysis and statistical modeling were performed in Microsoft Excel (Version 2019, Microsoft Corp., Redmond, Washington, United States) and OriginPro (Version 2023, OriginLab Corp., Northampton, Massachusetts, United States) software packages.

3. Results and Discussion

3.1. Impact of Sewage Sludge on Soil Physicochemical Properties

The results of the physicochemical and heavy metal analyses of AS, SS, and experimental treatments (5% and 10%) are presented in Table 1. The analyses indicated that experimental treatments had a significantly ($p < 0.05$) lower pH (6.24–6.43) than AS. However, all studied soils showed a slightly acid-to-neutral pH, which can be suitable for marigold cultivation, since it grows perfectly on soils with a pH range of 6.2–6.5 [36]. This suggests that the experimental soils are the most suitable for the growth and development of this plant. ANOVA test outlined a significant increase ($p < 0.05$) in most parameters of experimental treatments in comparison with AS. In particular, the following physicochemical and heavy metal parameters were significantly higher ($p < 0.05$) in experimental treatments compared to the control (AS): EC (2.70–3.14 dS/m), OM (2.51–3.85 g/kg), TN (2.81–3.72 mg/L), TP (2.10–2.64 g/kg), TK (0.32–0.69 g/kg), Cu (6.50–9.23 mg/kg), Mn (10.56–12.31 mg/kg), and Zn (8.07–11.20 mg/kg). It is worth noting that the application of 10% SS to AS showed higher values in terms of the aforementioned parameters in comparison with 5% experimental soil. No significant difference ($p > 0.05$) in Fe was observed between AS and experimental treatments (17.40–21.60 mg/kg). However, 5% experimental soil treatment showed no significant difference ($p > 0.05$) with AS in terms of Cd and Cr, despite that these minerals were found significantly higher ($p < 0.05$) than AS when 10% SS was added to AS (0.45 and 5.38 mg/kg, respectively).

Table 1. Average characteristics of sewage sludge, experimental, and control arable soils used for marigold cultivation.

Properties	Arable Soil	Sewage Sludge	Experimental Treatments	
			5%	10%
pH	6.74 ± 0.04^{d}	5.98 ± 0.07^{a}	6.43 ± 0.07^{c}	6.24 ± 0.05^{b}
Electrical conductivity (EC: dS/m)	2.40 ± 0.03^{a}	6.30 ± 0.12^{d}	2.70 ± 0.06^{b}	3.14 ± 0.08^{c}
Organic matter (OM: g/kg)	1.39 ± 0.04^{a}	25.04 ± 2.80^{c}	2.51 ± 0.11^{b}	3.85 ± 0.14^{bc}
Total nitrogen (TN: g/kg)	1.70 ± 0.02^{a}	20.99 ± 1.58^{d}	2.81 ± 0.03^{b}	3.72 ± 0.07^{c}
Total phosphorus (TP: g/kg)	1.26 ± 0.05^{a}	14.75 ± 0.94^{d}	2.10 ± 0.05^{b}	2.64 ± 0.10^{c}
Total potassium (TK: g/kg)	0.10 ± 0.02^{a}	5.09 ± 0.17^{d}	0.32 ± 0.04^{b}	0.69 ± 0.09^{c}
Cadmium (Cd: mg/kg)	0.27 ± 0.05^{a}	1.98 ± 0.10^{c}	0.38 ± 0.06^{a}	0.45 ± 0.04^{b}
Chromium (Cr: mg/kg)	3.64 ± 0.20^{a}	14.62 ± 3.02^{c}	4.26 ± 0.41^{a}	5.38 ± 0.27^{b}
Copper (Cu: mg/kg)	4.10 ± 0.28^{a}	49.32 ± 5.90^{d}	6.50 ± 0.24^{b}	9.23 ± 0.56^{c}
Iron (Fe: mg/kg)	17.40 ± 2.46^{a}	41.60 ± 4.08^{b}	19.42 ± 1.87^{a}	21.60 ± 2.01^{a}
Manganese (Mn: mg/kg)	9.06 ± 0.54^{a}	32.03 ± 5.62^{d}	10.56 ± 0.34^{b}	12.31 ± 1.14^{c}
Zinc (Zn: mg/kg)	3.80 ± 0.40^{a}	84.20 ± 8.28^{d}	8.07 ± 0.16^{b}	11.20 ± 0.48^{c}

Values are mean followed by the standard deviation of ten replicates; $^{a-d}$: the same letters indicate no significant difference between treatment groups at $p < 0.05$.

An increase in the soil EC enhances the solubility of nutrients [31], thus their higher availability for the growth of cultivated marigold. The increased OM content in experimental soils would result in improved soil structure and permeability [37]. Although crucial for crop growth development, high TN, TP, and TK levels can lead to the dominance of invasive alien plants and reduced soil microbial population [38]. The latter is responsible for the promotion of crop growth and development via several simple and complex mechanisms [26,39]. Soil heavy metals were in the following decreasing order: Fe > Mn > Cu > Zn > Cr > Cd. Despite being essential microelements, higher levels of these metals can bring cytotoxic effects on soils, and phytotoxicity in growing crops, which is associated with stunt growth and chlorosis [40]. However, the current values are low and do not show any potential risk to soil health and quality, and further on the cultivated marigold.

3.2. Impact of Sewage Sludge on Growth and Yield of Marigold

Table 2 shows the impact of SS amendment on the growth, yield, and flower characteristics of cultivated marigold. Results showed that 5% and 10% SS experimental treatments significantly ($p < 0.05$) improved the growth and performance of marigold. More specifically, plant height (52.40 cm), number of branches (11.06), and root length (21.48 cm) were the highest when AS was amended with 10% SS, followed by 5% experimental soil compared to the control. Several studies reported the potential and safe use of SS in agricultural activities yielding promising yields [25,41–43]. Besides this, it is considered safer than chemical and other organic fertilizers. In order to ensure that SS is safe for use as fertilizer, it undergoes a rigorous treatment process such as screening, settling, disinfection, digestion, and drying that help remove pathogens, heavy metals, and other contaminants. Additionally, SS has a slow-release nutrient profile, which means that the nutrients are released slowly over time, reducing the risk of over-fertilization and nutrient leaching [25]. Bi et al. [44] amended greenhouse-cultivated French marigold with broiler chicken litter-based organic fertilizers. They observed increased plant growth parameters including root rate and demonstrated the positive impact of organic fertilizers on marigold cultivated under greenhouse conditions. The increased root length may be due to increased soil porosity as a result of organic matter improvement [37]. This corroborates with the findings outlined in Table 1 in this regard. The current study proves the potential role of SS in the sustainable production of marigold.

Table 2. Impact of sewage sludge amendment on growth, yield, and flower characteristics of cultivated marigold.

Properties	Arable Soil	Sewage Sludge Treatment	
		5%	10%
Plant height (cm)	39.08 ± 2.72^{a}	48.11 ± 1.24^{b}	52.40 ± 3.57^{b}
Number of branches (no.)	7.13 ± 0.47^{a}	10.41 ± 0.37^{b}	11.06 ± 1.02^{b}
Root length (cm)	17.30 ± 1.13^{a}	20.09 ± 0.95^{b}	21.48 ± 0.69^{b}
First bud formation (days)	130.25 ± 4.75^{c}	119.80 ± 2.20^{b}	110.38 ± 3.72^{a}
Flowering period (days)	45.50 ± 2.50^{a}	57.20 ± 1.80^{b}	62.46 ± 2.54^{c}
Flower stack (cm)	8.03 ± 0.04^{a}	8.93 ± 0.10^{b}	9.15 ± 0.15^{c}
Flower diameter (cm)	5.51 ± 0.09^{a}	6.02 ± 0.05^{b}	6.16 ± 0.07^{c}
Flower yield (no. per plant)	24.08 ± 0.10^{a}	26.59 ± 0.24^{b}	26.35 ± 0.20^{b}
Flower yield (g per plant)	255.28 ± 5.15^{a}	306.90 ± 1.86^{b}	318.42 ± 3.09^{c}
Average weight of flower (g)	10.60 ± 0.10^{a}	11.54 ± 0.09^{b}	12.08 ± 0.16^{c}

Values are mean followed by the standard deviation of ten replicates; $^{a–c}$: the same letters indicate no significant difference between treatment groups at $p < 0.05$.

Several factors have been reported to control flowering initiation, which include irradiance, photoperiod, stresses, temperature, and the extra levels of nutrients in the soil [45–47]. Our findings outlined a significantly ($p < 0.05$) shorter time to first bud formation in experimental treatments (119.80 and 110.38 days), and a significantly ($p < 0.05$) extended flowering period was observed (57.20 and 62.46 days) compared to the control. The earlier budding can be a result of high metabolite uptake by the shoots and roots of marigold. Such observation was previously denoted on sunflowers amended with organic fertilizers [48], whereas the extended flowering period may be a result of increased cytokinin production [49]. Flower stack (9.15 cm), flower diameter (6.16 cm), flower yield (318.42 g/plant), and average flower weight (12.08 g) were highest in the 10% experimental treatment. These flower parameters were significantly ($p < 0.05$) higher in experimental treatments compared to the control (10% SS > 5% SS > AS). Flower yield was the highest in 5% experimental treatment (26.59 flowers/plant), followed by 10% treatment (26.35 flowers/plant) and AS (24.08 flowers/plant). Increased flower diameter can be a result of increased salicylic acid production by the marigold [50]. Such observation was also denoted on rose flowers [51]. Increased yield and number of flowers per plant were also reported on French marigold plants when the latter was amended with broiler chicken litter-based organic fertilizers [44]. Therefore, the findings of the present study suggest SS is a promising organic fertilizer for improving the growth and yield parameters of cultivated marigold.

3.3. Impact of Sewage Sludge on Biochemical Characteristics of Marigold

The biochemical composition of marigold grown on control and SS-amended soils is presented in Table 3. Total chlorophyll content and peroxidase activity were significantly ($p < 0.05$) higher in the leaves of experimental treatments (2.47–2.52 mg/g; 4.87–6.20 µmol/mg) compared to those of the control. Peralta-Sánchez et al. [52] earlier reported a total chlorophyll content of 0.75 mg/g in the leaves of marigold. Such differences can be explained by strain type, soil composition, and environmental factors. The use of organic fertilization showed its influence on the increase in total chlorophyll content of various crops [26,35], whereas the peroxidase activity in marigold leaves was reported to be 85–90 U/min/g in the study of Tian et al. [53]. These authors outlined the effect of cultivar variation on the antioxidant activity found in marigold leaves. In the present study, the catalase activity in marigold leaves was significantly ($p < 0.05$) lower in 10% SS treatment compared to the control (2.15 and 2.90 U/mL, respectively), whereas 5% treatment did not show such a significant difference (2.75 U/mL). Tian et al. [53] mentioned a catalase activity of 11.5–12.5 U/min/g in many marigold cultivars. Marigold flowers did not contain any chlorophyll content nor possessed catalase or peroxidase activities in all treatments. It is well known that the presence of chlorophyll content in flowers is a very rare phenomenon in developed stages; only low contents can be found in early development stages [54]. However, the lack of catalase and peroxidase activities in the marigold flowers of the present study can result in the non-scavenging of H_2O_2 [55].

Our results also showed that β-carotene, ascorbic acid, and flavonoid contents were higher in the leaves than in flowers, whereas the contrary was observed regarding total phenols and lutein contents. β-carotene content was significantly ($p < 0.05$) higher in 10% treatment in comparison with the control (19.09 and 15.93 µg/g, respectively). However, no significant difference was observed between the 5% treatment (17.36 µg/g) and the control. On the other hand, marigold flowers of experimental treatments contained significantly ($p < 0.05$) higher β-carotene content (10.51–11.56 µg/g) compared to the control. Such increases can be explained by the increased gene expression and thus increased yellowish and orangish color of the flowers [56]. The present study reported total phenols contents of 12.10–17.07 mg/g, ascorbic acid contents of 1.60–1.96 mg/g, flavonoid contents of 55.40–64.11 mg/g, and lutein contents of 47.10–53.92 mg/g in marigold leaves, which were significantly higher ($p < 0.05$) in experimental treatments than in the control (10% > 5% > control). Similarly, marigold flowers enclosed total phenols contents of 33.48–51.03 mg/g, ascorbic acid contents of 0.52–0.74 mg/g, flavonoid contents of 36.27–45.20 mg/g, and lutein contents of 70.49–84.66 mg/g, which were significantly higher ($p < 0.05$) in experimental treatments than in the control (10% > 5% > control).

The increased total phenols in marigold leaves and flowers correspond to increased bioactivity and thus better antioxidant and antibacterial activities [57]. It was also reported that increased flavonoid content in French marigold resulted in improved cytoprotective activity [58]. Peralta-Sánchez et al. [52] mentioned total flavonoid contents of 1.1 and 3.7 mg/g DW in the leaves and flowers of Mexican marigold. The increase in ascorbic acid in marigold leaves improves the regulation of the redox state of photosynthetic electron carriers, thereby better photosynthesis mechanism [59]. Krzymińska et al. [60] denoted a dependence of phenolic compounds, ascorbic acid, and flavonoids on the growing substrate of marigold plants. Recent reports depicted the role of lutein on the yellow color intensity of marigold flowers [61]. Furthermore, this component plays an important role in the triggering of ROS generation [62]; thus, SS amendment to marigold plants is very promising.

Table 3. Impact of sewage sludge amendment on biochemical characteristics of flower and leaves of cultivated marigold.

Properties	Plant Part	Arable Soil	Sewage Sludge Treatments	
			5%	10%
Total chlorophyll (mg/g)	Leaves	2.30 ± 0.05 [a]	2.47 ± 0.02 [b]	2.52 ± 0.06 [b]
	Flowers	*na*	*na*	*na*
Catalase (U/mL)	Leaves	2.90 ± 0.10 [a]	2.75 ± 0.07 [a]	2.15 ± 0.03[b]
	Flowers	*na*	*na*	*na*
Peroxidase (μmol/mg)	Leaves	3.10 ± 0.06 [a]	4.87 ± 0.05 [b]	6.20 ± 0.18 [c]
	Flowers	*na*	*na*	*na*
β-carotene (μg/g)	Leaves	15.93 ± 0.30 [a]	17.36 ± 1.02 [a]	19.09 ± 0.95 [b]
	Flowers	10.08 ± 0.07 [a]	10.51 ± 0.10 [b]	11.56 ± 0.09 [c]
Total phenols (mg/g)	Leaves	12.10 ± 0.12 [a]	15.44 ± 0.28 [b]	17.07 ± 0.50 [c]
	Flowers	33.48 ± 2.09 [a]	48.30 ± 4.58 [b]	51.03 ± 3.75 [b]
Ascorbic acid (mg/g)	Leaves	1.60 ± 0.05 [a]	1.73 ± 0.04 [b]	1.96 ± 0.11 [bc]
	Flowers	0.52 ± 0.02 [a]	0.67 ± 0.05 [b]	0.74 ± 0.08 [b]
Flavonoids (mg/g)	Leaves	55.40 ± 3.84 [a]	62.90 ± 1.90 [b]	64.11 ± 2.75 [b]
	Flowers	36.27 ± 1.51 [a]	41.02 ± 3.02 [b]	45.20 ± 2.48 [b]
Lutein (mg/g)	Leaves	47.10 ± 2.07 [a]	51.75 ± 2.72 [b]	53.92 ± 1.01 [b]
	Flowers	70.49 ± 1.43 [a]	78.38 ± 2.54 [b]	84.66 ± 3.13 [c]

Values are mean followed by the standard deviation of ten replicates; [a–c]: the same letters indicate no significant difference between treatment groups at $p < 0.05$; *na*: not applicable.

3.4. Impact of Sewage Sludge on Heavy Metal Accumulation in Marigold

Results in Table 4 outlined a gradual decrease in heavy metal accumulation from the lower to upper parts of marigold plants (root > shoot > flower) in all treatments. This is a natural phenomenon in which root vacuoles tend to reduce the translocation of such potentially toxic elements to shoots [63]. As a result, the negative impacts on shoot cells, chloroplasts, and mitochondria are diminished and associated with a lower oxidative stress heaviness [64]. The current study reported Cd contents of 0.21–0.72 mg/kg, Cr contents of 0.90–1.52 mg/kg, Cu contents of 3.50–8.40 mg/kg, Fe contents of 12.40–34.02 mg/kg, Mn contents of 5.71–14.09 mg/kg, and Zn contents of 7.62–11.83 mg/kg in significantly higher marigold roots ($p < 0.05$) in experimental treatments than in the control (10% > 5% > control). Recently, Biswal et al. [65] revealed high translocation factors of Ni and Cd from polluted sites to the roots of two marigold species, namely, *T. patula* and *T. erecta*. These authors also found a correlation between increased flower yield and heavy metal bioaccumulation in marigold. Such findings corroborate our findings, where increased heavy metals in experimental treatments resulted in higher yields than in the control. Therefore, marigold can be considered a good bioremediating agent for soils polluted with SS, while the latter can play an interesting role in the improvement of marigold physicochemical characteristics. Cd and Cr contents were significantly ($p < 0.05$) more accumulated in marigold shoots of 10% treatment than in the control (0.19 and 1.10 mg/kg, respectively). However, marigold shoots bioaccumulated Cu contents of 2.17–5.73 mg/kg, Fe contents of 17.37–22.30 mg/kg, Mn contents of 4.02–10.67 mg/kg, and Zn contents of 5.27–9.35 mg/kg, which were significantly higher ($p < 0.05$) in experimental treatments than in the control (10% > 5% > control). Madanan et al. [66] reported higher bioaccumulation of Cd in roots and shoots of marigold than Zn. Our findings showed an irregular pattern in both the control and experimental treatments. The flowers harvested from experimental treatments showed no significant difference in Cd content in comparison with the control (0.01–0.02 mg/kg), whereas the flowers of 10% treatment bioaccumulated significantly ($p < 0.05$) had a higher Cr content than the control ones (0.22 and 0.16 mg/kg, respectively). Moreover, marigold flowers bioaccumulated Cu contents of 1.88–2.50 mg/kg, Fe contents of 3.53–5.10 mg/kg, Mn contents of 1.06–1.40 mg/kg, and Zn contents of 2.50–4.21 mg/kg, which were significantly higher ($p < 0.05$) in experimental treatments than in the control (10% > 5% > control). Such increases in heavy metals can ruin the whole bioremediation process if marigold flowers are hazardously disposed of into the environment after being used in temples for religious purposes. Therefore, the recycling of marigold flowers grown on SS-treated soils in various biological and chemical processes is highly required.

Table 4. Impact of sewage sludge amendment on heavy metal accumulation in different plant parts of cultivated marigold.

Heavy Metal	Plant Part	Arable Soil	Sewage Sludge Treatments	
			5%	10%
Cd	Root	0.21 ± 0.02^{a}	0.48 ± 0.05^{b}	0.72 ± 0.07^{c}
	Shoot	0.12 ± 0.03^{a}	0.16 ± 0.02^{a}	0.19 ± 0.05^{b}
	Flowers	0.01 ± 0.00^{a}	0.02 ± 0.01^{a}	0.02 ± 0.01^{a}
Cr	Root	0.90 ± 0.07^{a}	1.37 ± 0.10^{b}	1.52 ± 0.19^{bc}
	Shoot	0.74 ± 0.05^{a}	0.83 ± 0.04^{a}	1.10 ± 0.08^{b}
	Flowers	0.16 ± 0.02^{a}	0.19 ± 0.03^{ab}	0.22 ± 0.02^{b}
Cu	Root	3.50 ± 0.09^{a}	6.94 ± 0.25^{b}	8.40 ± 0.71^{c}
	Shoot	2.17 ± 0.04^{a}	4.63 ± 0.16^{b}	5.73 ± 0.32^{c}
	Flowers	1.88 ± 0.11^{a}	2.10 ± 0.08^{b}	2.50 ± 0.29^{c}
Fe	Root	12.40 ± 1.82^{a}	26.62 ± 3.01^{b}	34.02 ± 4.35^{c}
	Shoot	17.37 ± 0.60^{a}	20.10 ± 1.25^{ab}	22.30 ± 0.94^{b}
	Flowers	3.53 ± 0.20^{a}	4.86 ± 0.52^{b}	5.10 ± 0.17^{c}
Mn	Root	5.71 ± 0.19^{a}	12.18 ± 2.40^{b}	14.09 ± 1.64^{bc}
	Shoot	4.02 ± 0.08^{a}	8.49 ± 0.65^{b}	10.67 ± 1.28^{bc}
	Flowers	1.06 ± 0.04^{a}	1.34 ± 0.07^{b}	1.40 ± 0.05^{c}
Zn	Root	7.62 ± 0.30^{a}	9.91 ± 0.55^{b}	11.83 ± 1.02^{bc}
	Shoot	5.27 ± 0.12^{a}	8.02 ± 0.19^{b}	9.35 ± 0.24^{c}
	Flowers	2.50 ± 0.09^{a}	3.98 ± 0.07^{b}	4.21 ± 0.11^{bc}

Values are mean followed by the standard deviation of ten replicates; a–c: the same letters indicate no significant difference between treatment groups at $p < 0.05$.

The bioaccumulation factor (BAF) evaluation is an important step to detect which heavy metal can be harmful or cause deleterious impacts on the grown crops [34]. In this study, Cd bioaccumulation was safe (BAF < 1.00) except for 5% and 10% treatments' roots (1.26 and 1.60), whereas Cr bioaccumulation was safe (BAF < 1.00) in all treatments and plant parts (Figure 1). Cu bioaccumulation was safe (BAF < 1.00) except for 5% treatment's roots (BAF: 1.05). Fe bioaccumulation showed to be highest in 5% of treatment shoots (BAF: 1.92). Mn bioaccumulation was generally moderate (BAF < 1.00), except in experimental treatments' roots (BAF: 1.15). However, the highest threat was attributed to Zn, which showed to be heavily bioaccumulated in all treatments' roots (1.06 < 1.23 < 2.01). In this context, Madanan et al. [66] reported an unsafe BAF (BAF > 1) for Cd of marigold roots grown on lateritic-polluted soils. These findings corroborate our obtained results from the experimental treatments. Further, the same authors also denoted high BAF (BAF > 1) for Zn of the same species' shoots. However, such an observation was not found in the experimental treatments of the present study.

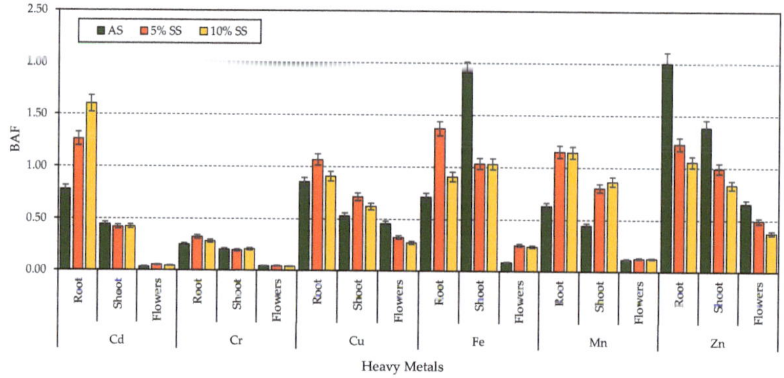

Figure 1. Bioaccumulation factor (BAF) of six heavy metals in different plant parts of marigold cultivated in different treatments (AS: arable soil as control, SS: sewage sludge).

3.5. Prediction Models for Heavy Metal Uptake by Marigold

Table 5 shows the results of prediction models for the uptake of heavy metal by different vegetative parts of marigold plants cultivated on SS-amended soils. In particular, the models were able to accurately predict the uptake amount of selected heavy metals, i.e., Cd, Cr, Cu, Fe, Mn, and Zn. As explained from the MLR model, it was observed that pH showed a negative association with the uptake of all heavy metals except for Cd (shoot and flowers), Cr (shoot), Cu (flower), and Mn (root). Additionally, OM showed a negative influence except for Cd (root and shoot), Cr (shoot), Fe (shoot), and Mn (root). However, the total heavy metal content showed a positive influence on their uptake by marigold plants except for Cr (shoot), and Mn (flowers). Cu accumulation in marigold flowers may be related to the presence of Cu transporters in their cells. In addition, the concentration of Cu in the plant may vary depending on the stage of growth and development. The amount of heavy metals that plants absorb depends significantly on the pH of the soil. Low soil pH enhances the availability of heavy metals and facilitates their uptake by plants. Similarly, the amount of OM in the soil can affect how well plants absorb heavy metals. Heavy metals can attach to OM, which makes them more accessible for plants to absorb [36]. The ANOVA test explained the suitability of models since p-values remain within the significance level of <0.05 except for the developed model for Cd prediction in the flower region. Notwithstanding, the models showed relatively high coefficient of determination (R^2) values ranging from 0.73 to 0.99, which were acceptable. With root mean square error (RMSE) values of 0.01–3.25, the models were able to correctly forecast the amount of heavy metal uptake by the marigold plant. This was further supported by the range of predicted response values ($y_{\min}$ and $y_{\max}$).

MLR is a widely accepted approach for monitoring heavy metal contamination in both edible and non-edible crops. A recent study by Zhou et al. [67] constructed prediction models for the uptake of Cd by wheat-rice rotation crop system. The authors reported that the adopted approach was used for precise prediction of Cd contents in root, stem, and leaf, while pH and OM showed significant association with the uptake process. Similarly, Yu et al. [68] utilized the MLR method for heavy metal uptake by wheat crops fertigated with SS. The authors found that the developed models were efficient to predict the contents of six heavy metals, including Cd, Cu, Pb, Zn, Cr, and Ni while taking pH and OM as independent variables. Recently, Eid et al. [35] utilized MLR equations for uptake prediction of nine heavy metals (Cd, Co, Cr, Cu, Fe, Mn, Ni, Pb, and Zn) by forage sorghum (*Sorghum bicolor* L.) crop. Their results showed the metal tissues in plant parts exhibited a negative correlation with soil pH, whereas OM, EC, and plant metal contents were all found to be positively correlated. The findings reported in the above studies are in line with those obtained in the present study; thus, the developed equations can be used for the prediction of heavy metal content absorbed by different parts of the marigold plant (root, shoot, and flower).

Table 5. Prediction models for uptake of heavy metal by different vegetative parts of marigold cultivated on sewage sludge amended soils.

Heavy Metals	Plant Parts	Model Equation	y_{min}	y_{max}	R^2	ANOVA		ME	RMSE
						F-Value	*p*-Value		
Cd	Root	$y = 0.40 - 0.08\ pH_{soil} + 0.13\ OM_{soil} + 0.73\ Cd_{soil}$	0.17	0.76	0.99	297.64	<0.01	0.98	0.04
	Shoot	$y = -0.76 + 0.10\ pH_{soil} + 0.01\ OM_{soil} + 0.46\ Cd_{soil}$	0.09	0.21	0.88	12.55	<0.01	0.75	0.12
	Flowers	$y = -0.09 + 0.01\ pH_{soil} - 0.01\ OM_{soil} + 0.12\ Cd_{soil}$	0.01	0.03	0.73	4.66	0.06	0.49	0.01
Cr	Root	$y = 13.62 - 2.18\ pH_{soil} - 0.80\ OM_{soil} + 0.87\ Cr_{soil}$	0.89	1.58	0.84	9.04	0.04	0.98	0.06
	Shoot	$y = -9.04 + 1.47\ pH_{soil} + 0.66\ OM_{soil} - 0.30\ Cr_{soil}$	0.71	1.18	0.99	236.72	<0.01	0.88	0.08
	Flowers	$y = 1.59 - 0.27\ pH_{soil} - 0.14\ OM_{soil} + 0.16\ Cr_{soil}$	0.14	0.23	0.95	36.95	<0.01	0.72	0.01
Cu	Root	$y = 39.40 - 6.00\ pH_{soil} - 5.65\ OM_{soil} + 2.07\ Cu_{soil}$	3.27	9.20	0.96	46.87	<0.01	0.94	0.80
	Shoot	$y = 29.48 - 4.45\ pH_{soil} - 3.06\ OM_{soil} + 1.73\ Cu_{soil}$	2.10	6.16	0.06	53.32	<0.01	0.97	0.43
	Flowers	$y = -5.22 + 0.86\ pH_{soil} - 0.69\ OM_{soil} + 0.53\ Cu_{soil}$	1.71	2.74	0.98	112.21	<0.01	0.55	0.24
Fe	Root	$y = 177.89 - 29.78\ pH_{soil} - 0.87\ OM_{soil} + 2.12\ Fe_{soil}$	8.93	37.27	0.96	53.08	<0.01	0.90	3.25
	Shoot	$y = 16.93 - 1.17\ pH_{soil} + 1.08\ OM_{soil} + 0.39\ Fe_{soil}$	16.50	23.31	0.97	71.26	<0.01	0.85	1.01
	Flowers	$y = 27.93 - 4.14\ pH_{soil} - 0.64\ OM_{soil} + 0.26\ Fe_{soil}$	3.16	5.40	0.88	13.06	<0.01	0.81	0.30
Mn	Root	$y = 37.16 - 7.13\ pH_{soil} - 0.72\ OM_{soil} + 2.02\ Mn_{soil}$	5.61	16.60	0.85	0.68	<0.01	0.82	2.51
	Shoot	$y = 48.54 - 8.50\ pH_{soil} - 1.16\ OM_{soil} + 1.61\ Mn_{soil}$	3.76	12.16	0.96	43.56	<0.01	0.89	1.49
	Flowers	$y = 3.96 - 0.54\ pH_{soil} - 0.10\ OM_{soil} - 0.10\ Mn_{soil}$	1.05	1.49	0.86	11.08	<0.01	0.85	0.09
Zn	Root	$y = -34.96 + 5.77\ pH_{soil} + 0.27\ OM_{soil} + 0.86\ Zn_{soil}$	7.04	12.60	0.98	147.54	<0.01	0.89	0.77
	Shoot	$y = 5.57 - 0.32\ pH_{soil} - 0.99\ OM_{soil} + 0.86\ Zn_{soil}$	5.00	9.63	0.99	301.78	<0.01	0.98	0.28
	Flowers	$y = 3.21 - 0.19\ pH_{soil} - 1.42\ OM_{soil} + 0.69\ Zn_{soil}$	2.52	4.35	0.98	120.04	<0.01	0.98	0.14

y: predicted heavy metal content (mg/kg); R^2: coefficient of determination; ME: model efficiency; RMSE: root mean square error.

4. Conclusions

The results of this study concluded that SS can be effectively utilized for marigold cultivation. The maximum values for the growth, yield, and biochemical constituents of marigold were reported using 10% SS treatment. Moreover, the heavy metal analysis showed that the marigold accumulated the maximum contents of selected heavy metals in root parts followed by shoot and flowers. The bioaccumulation factor values >1 indicated that the marigold plant can be a useful candidate for the reclamation of SS-treated soils. Moreover, the multiple linear regression-based prediction models showed good fitness and high accuracy, as indicated by the validation results. Therefore, the most feasible improvement in the productivity of the marigold crop was reported using a 10% SS application, which suggests its sustainable suitability for floriculture. Additional research is required to better understand the mechanisms involved in the heavy metal uptake process by marigold plants as well as to test the suitability of SS for the cultivation of other flower species.

Author Contributions: Conceptualization, P.K.; Data curation, S.A.F.; Formal analysis, P.K. and M.G.; Funding acquisition, A.A.A.-H. and M.A.T.; Investigation, P.K. and M.G.; Methodology, P.K.; Project administration, A.A.A.-H., M.A.T. and E.M.E.; Resources, P.K.; Supervision, P.K. and E.M.E.; Validation, A.A.A.-H., S.A.F., B.A., I.Š., M.G., K.S.C., M.A.T., A.S.E.-K. and E.M.E.; Writing—original draft, P.K. and S.A.F.; Writing—review and editing, A.A.A.-H., B.A., I.Š., M.G., K.S.C., M.A.T., A.S.E.-K. and E.M.E. All authors have read and agreed to the published version of the manuscript.

Funding: This research was funded by Princess Nourah bint Abdulrahman University Researchers Supporting Project number (PNURSP2023R93), Princess Nourah bint Abdulrahman University, Riyadh, Saudi Arabia. This research was funded by the Deanship of Scientific Research at King Khalid University, Abha, Saudi Arabia (grant number RGP.2/220/44).

Data Availability Statement: Not applicable.

Acknowledgments: The authors are grateful to their host institutes for providing the necessary facilities to conduct this study. This is joint work from the members of the Sustainable Agro-Environment International Research Group (SAEIRG). The authors express their gratitude to Princess Nourah bint Abdulrahman University Researchers Supporting Project number (PNURSP2023R93), Princess Nourah bint Abdulrahman University, Riyadh, Saudi Arabia. The authors extend their appreciation to the Deanship of Scientific Research at King Khalid University, Abha, Saudi Arabia, for funding this work through the Research Group Project under grant number RGP.2/220/44. All individuals included in this section have consented to the acknowledgement.

Conflicts of Interest: The authors declare no conflict of interest.

References

1. Mateo-Sagasta, J.; Raschid-Sally, L.; Thebo, A. Global Wastewater and Sludge Production, Treatment and Use. In *Wastewater: Economic Asset in an Urbanizing World*; Springer: Dordrecht, The Netherlands, 2015; pp. 15–38. [CrossRef]
2. Shanmugam, K.; Gadhamshetty, V.; Tysklind, M.; Bhattacharyya, D.; Upadhyayula, V.K.K. A Sustainable Performance Assessment Framework for Circular Management of Municipal Wastewater Treatment Plants. *J. Clean. Prod.* **2022**, *339*, 130657. [CrossRef]
3. Page, S.; Renner, R.; Science, E.; Technology, V.I.; Mcgrath, S. Sewage Sludge, Pros & Cons. *Environ. Sci. Technol.* **2000**, *34*, 430A–435A.
4. Manzoor, J.; Sharma, M. Impact of Biomedical Waste on Environment and Human Health. *Environ. Claims J.* **2019**, *31*, 311–334. [CrossRef]
5. Chu, L.; He, W.; Xu, F.; Tong, Y.; Xu, F. Ecological Risk Assessment of Toxic Metal(Loid)s for Land Application of Sewage Sludge in China. *Sci. Total Environ.* **2022**, *836*, 155549. [CrossRef]
6. Zhu, X.; Xu, Y.; Zhen, G.; Lu, X.; Xu, S.; Zhang, J.; Gu, L.; Wen, H.; Liu, H.; Zhang, X.; et al. Effective Multipurpose Sewage Sludge and Food Waste Reduction Strategies: A Focus on Recent Advances and Future Perspectives. *Chemosphere* **2023**, *311*, 136670. [CrossRef]
7. Bin Khawer, M.U.; Naqvi, S.R.; Ali, I.; Arshad, M.; Juchelková, D.; Anjum, M.W.; Naqvi, M. Anaerobic Digestion of Sewage Sludge for Biogas & Biohydrogen Production: State-of-the-Art Trends and Prospects. *Fuel* **2022**, *329*, 125416. [CrossRef]
8. Ragoobur, D.; Huerta-Lwanga, E.; Somaroo, G.D. Reduction of Microplastics in Sewage Sludge by Vermicomposting. *Chem. Eng. J.* **2022**, *450*, 138231. [CrossRef]
9. Dregulo, A.M.; Bobylev, N.G. Integrated Assessment of Groundwater Pollution from the Landfill of Sewage Sludge. *J. Ecol. Eng.* **2020**, *22*, 68–75. [CrossRef]
10. Jamil Khan, M.; Qasim, M.; Umar, M. Utilization of Sewage Sludge as Organic Fertiliser in Sustainable Agriculture. *J. Appl. Sci.* **2006**, *6*, 531–535. [CrossRef]
11. Pascual, I.; Azcona, I.; Aguirreolea, J.; Morales, F.; Corpas, F.J.; Palma, J.M.; Rellán-Álvarez, R.; Sánchez-Díaz, M. Growth, Yield, and Fruit Quality of Pepper Plants Amended with Two Sanitized Sewage Sludges. *J. Agric. Food Chem.* **2010**, *58*, 6951–6959. [CrossRef] [PubMed]
12. Eid, E.M.; Hussain, A.A.; Taher, M.A.; Galal, T.M.; Shaltout, K.H.; Sewelam, N. Sewage Sludge Application Enhances the Growth of *Corchorus olitorius* Plants and Provides a Sustainable Practice for Nutrient Recirculation in Agricultural Soils. *J. Soil Sci. Plant Nutr.* **2020**, *20*, 149–159. [CrossRef]
13. Kirchmann, H.; Börjesson, G.; Kätterer, T.; Cohen, Y. From Agricultural Use of Sewage Sludge to Nutrient Extraction: A Soil Science Outlook. *Ambio* **2017**, *46*, 143–154. [CrossRef] [PubMed]
14. The Council of the European Communities. *Council Directive of 12 June 1986 on the Protection of the Environment, and in Particular of the Soil, When Sewage Sludge Is Used in Agriculture (86/278/EEC)*; The Council of the European Communities: Brussels, Belgium, 1986; Volume L.
15. Solanki, P.; Kalavagadda, B.; Akula, B.; Sharma, S.H.K.; Reddy, D.J. Effect of Sewage Sludge on Marigold (*Tagetes erecta*). *Int. J. Curr. Microbiol. Appl. Sci.* **2017**, *6*, 825–831. [CrossRef]
16. Ahmadpour, P.; Nawi, A.M.; Abdu, A.; Abdul-Hamid, H.; Singh, D.K.; Hassan, A.; Majid, N.M.; Jusop, S. Uptake of Heavy Metals by Jatropha Curcas L. Planted in Soils Containing Sewage Sludge. *Am. J. Appl. Sci.* **2010**, *7*, 1291–1299. [CrossRef]
17. Ghademian Sorboni, A.; Bahmanyar, M.; GHajar, M. Effect of Different Periods of Application of Sewage Sludge and Municipal Solid Wastes Compost on the Amount of Cadmium and Nickel Content of Soil and Borage (*Borago officinadis*). *J. Water Wastewater* **2013**, *24*, 80–89.
18. Fijalkowski, K.; Rosikon, K.; Grobelak, A.; Hutchison, D.; Kacprzak, M.J. Modification of Properties of Energy Crops under Polish Condition as an Effect of Sewage Sludge Application onto Degraded Soil. *J. Environ. Manag.* **2018**, *217*, 509–519. [CrossRef]

19. Kumari, S.; Kothari, R.; Kumar, V.; Kumar, P.; Tyagi, V.V. Kinetic Assessment of Aerobic Composting of Flower Waste Generated from Temple in Jammu, India: A Lab Scale Experimental Study. *Environ. Sustain.* **2021**, *4*, 393–400. [CrossRef]
20. Kaur, H.; Singh, J.; Singh, B. Importance and Prospects of Marigold. *Just Agric.* **2021**, *2*, 1–5.
21. Greinert, A.; Drab, M.; Węclewski, S. Phytotoxic Effects of Sewage Sludges on Decorative Plants. *Pol. J. Environ. Stud.* **2009**, *6*, 41–48.
22. Greinert, A. Die Schlammverwertung Als Dünger Für Die Rabatte-Zierpflanzen (The Sewage Sludge Utilization as a Fertilizer for the Ornamental Bedding Plants). In Proceedings of the Studies and Materials: Ecology of the Borderland: VI Polish-German Scientific Conference, IBEN, Gorzów Wielkopolski, Poland, 23 November 2001; pp. 18–19, 211–216.
23. Walkley, A.; Black, I.A. An Examination of the Degtjareff Method for Determining Soil Organic Matter, and a Proposed Modification of the Chromic Acid Titration Method. *Soil Sci.* **1934**, *37*, 29–38. [CrossRef]
24. Chromý, V.; Vinklárková, B.; Šprongl, L.; Bittová, M. The Kjeldahl Method as a Primary Reference Procedure for Total Protein in Certified Reference Materials Used in Clinical Chemistry. I. A Review of Kjeldahl Methods Adopted by Laboratory Medicine. *Crit. Rev. Anal. Chem.* **2015**, *45*, 106–111. [CrossRef]
25. Kumar, P.; Kumar, V.; Adelodun, B.; Bedeković, D.; Kos, I.; Širić, I.; Alamri, S.A.M.; Alrumman, S.A.; Eid, E.M.; Abou Fayssal, S.; et al. Sustainable Use of Sewage Sludge as a Casing Material for Button Mushroom (*Agaricus bisporus*) Cultivation: Experimental and Prediction Modeling Studies for Uptake of Metal Elements. *J. Fungi* **2022**, *8*, 112. [CrossRef]
26. Širić, I.; Eid, E.M.; Taher, M.A.; El-Morsy, M.H.E.; Osman, H.E.M.; Kumar, P.; Adelodun, B.; Abou Fayssal, S.; Mioč, B.; Andabaka, Ž.; et al. Combined Use of Spent Mushroom Substrate Biochar and PGPR Improves Growth, Yield, and Biochemical Response of Cauliflower (*Brassica oleracea* var. botrytis): A Preliminary Study on Greenhouse Cultivation. *Horticulturae* **2022**, *8*, 830. [CrossRef]
27. Latimer, G.W. *Official Methods of Analysis of AOAC INTERNATIONAL*, 21st ed.; AOAC International: Rockville, MD, USA, 2019.
28. Shah, S.H.; Houborg, R.; McCabe, M.F. Response of Chlorophyll, Carotenoid and SPAD-502 Measurement to Salinity and Nutrient Stress in Wheat (*Triticum aestivum* L.). *Agronomy* **2017**, *7*, 61. [CrossRef]
29. Yang, T.; Poovaiah, B.W. Hydrogen Peroxide Homeostasis: Activation of Plant Catalase by Calcium/Calmodulin. *Proc. Natl. Acad. Sci. USA* **2002**, *99*, 4097–4102. [CrossRef] [PubMed]
30. Mafakheri, A.; Siosemardeh, A.; Bahramnejad, B.; Struik, P.C.; Sohrabi, Y. Effect of Drought Stress and Subsequent Recovery on Protein, Carbohydrate Contents, Catalase and Peroxidase Activities in Three Chickpea (*Cicer arietinum*) Cultivars. *Aust. J. Crop Sci.* **2011**, *5*, 1255–1260.
31. de Carvalho, L.M.J.; Gomes, P.B.; de Oliveira Godoy, R.L.; Pacheco, S.; do Monte, P.H.F.; de Carvalho, J.L.V.; Nutti, M.R.; Neves, A.C.L.; Vieira, A.C.R.A.; Ramos, S.R.R. Total Carotenoid Content, α-Carotene and β-Carotene, of Landrace Pumpkins (*Cucurbita moschata* Duch): A Preliminary Study. *Food Res. Int.* **2012**, *47*, 337–340. [CrossRef]
32. Kumar, V.; Kumar, P.; Singh, J.; Kumar, P. Kinetics of Nutrients Remediation from Sugar Industry Effluent-Treated Substrate Using *Agaricus bisporus*: Mushroom Yield and Biochemical Potentials. *3 Biotech* **2021**, *11*, 164. [CrossRef] [PubMed]
33. Šivel, M.; Kráčmar, S.; Fišera, M.; Klejdus, B.; Kubáň, V. Lutein Content in Marigold Flower (*Tagetes erecta* L.) Concentrates Used for Production of Food Supplements. *Czech J. Food Sci.* **2014**, *32*, 521–525. [CrossRef]
34. Al-Huqail, A.A.; Kumar, P.; Eid, E.M.; Singh, J.; Arya, A.K.; Goala, M.; Adelodun, B.; Fayssal, S.A.; Kumar, V.; Širić, I. Risk Assessment of Heavy Metals Contamination in Soil and Two Rice (*Oryza sativa* L.) Varieties Irrigated with Paper Mill Effluent. *Agriculture* **2022**, *12*, 1864. [CrossRef]
35. Eid, E.M.; Hussain, A.A.; Alamri, S.A.M.; Alrumman, S.A.; Shaltout, K.H.; Sewelam, N.; Shaltout, S.K.; El-Bebany, A.F.; Ahmed, M.T.; Al-Bakre, D.A.; et al. Prediction Models Based on Soil Characteristics for Evaluation of the Accumulation Capacity of Nine Metals by Forage Sorghum Grown in Agricultural Soils Treated with Varying Amounts of Poultry Manure. *Bull. Environ. Contam. Toxicol.* **2023**, *110*, 40. [CrossRef]
36. Zeng, F.; Ali, S.; Zhang, H.; Ouyang, Y.; Qiu, B.; Wu, F.; Zhang, G. The Influence of PH and Organic Matter Content in Paddy Soil on Heavy Metal Availability and Their Uptake by Rice Plants. *Environ. Pollut.* **2011**, *159*, 84–91. [CrossRef] [PubMed]
37. Yavitt, J.B.; Pipes, G.T.; Olmos, E.C.; Zhang, J.; Shapleigh, J.P. Soil Organic Matter, Soil Structure, and Bacterial Community Structure in a Post-Agricultural Landscape. *Front. Earth Sci.* **2021**, *9*, 590103. [CrossRef]
38. Zhang, T.; Chen, H.Y.H.; Ruan, H. Global Negative Effects of Nitrogen Deposition on Soil Microbes. *ISME J.* **2018**, *12*, 1817–1825. [CrossRef]
39. Chaudhary, P.; Xu, M.; Ahamad, L.; Chaudhary, A.; Kumar, G.; Adeleke, B.S.; Verma, K.K.; Hu, D.-M.; Širić, I.; Kumar, P.; et al. Application of Synthetic Consortia for Improvement of Soil Fertility, Pollution Remediation, and Agricultural Productivity: A Review. *Agronomy* **2023**, *13*, 643. [CrossRef]
40. Okereafor, U.; Makhatha, M.; Mekuto, L.; Uche-Okereafor, N.; Sebola, T.; Mavumengwana, V. Toxic Metal Implications on Agricultural Soils, Plants, Animals, Aquatic Life and Human Health. *Int. J. Environ. Res. Public Health* **2020**, *17*, 2204. [CrossRef] [PubMed]
41. Antolín, M.C.; Muro, I.; Sánchez-Díaz, M. Application of Sewage Sludge Improves Growth, Photosynthesis and Antioxidant Activities of Nodulated *Alfalfa* Plants under Drought Conditions. *Environ. Exp. Bot.* **2010**, *68*, 75–82. [CrossRef]
42. Iticescu, C.; Georgescu, P.-L.; Arseni, M.; Rosu, A.; Timofti, M.; Carp, G.; Cioca, L.-I. Optimal Solutions for the Use of Sewage Sludge on Agricultural Lands. *Water* **2021**, *13*, 585. [CrossRef]
43. Bai, J.; Sun, X.; Xu, C.; Ma, X.; Huang, Y.; Fan, Z.; Cao, X. Effects of Sewage Sludge Application on Plant Growth and Soil Characteristics at a *Pinus sylvestris* var. Mongolica Plantation in Horqin Sandy Land. *Forests* **2022**, *13*, 984. [CrossRef]

44. Bi, G.; Evans, W.B.; Spiers, J.M.; Witcher, A.L. Effects of Organic and Inorganic Fertilizers on Marigold Growth and Flowering. *HortScience* **2010**, *45*, 1373–1377. [CrossRef]
45. Erwin, J. Factors Affecting Flowering in Ornamental Plants. In *Flower Breeding and Genetics: Issues, Challenges and Opportunities for the 21st Century*; Springer: Dordrecht, The Netherlands, 2006; pp. 7–48. [CrossRef]
46. Grigatti, M.; Giorgioni, M.; Ciavatta, C. Compost-Based Growing Media: Influence on Growth and Nutrient Use of Bedding Plants. *Bioresour. Technol.* **2007**, *98*, 3526–3534. [CrossRef] [PubMed]
47. James, E.C.; Van Iersel, M.W. Fertilizer Concentration Affects Growth and Flowering of Subirrigated Petunias and Begonias. *HortScience* **2001**, *36*, 40–44. [CrossRef]
48. Alzamel, N.M.; Taha, E.M.M.; Bakr, A.A.A.; Loutfy, N. Effect of Organic and Inorganic Fertilizers on Soil Properties, Growth Yield, and Physiochemical Properties of Sunflower Seeds and Oils. *Sustainability* **2022**, *14*, 12928. [CrossRef]
49. Ming, X.; Tao, Y.-B.; Fu, Q.; Tang, M.; He, H.; Chen, M.-S.; Pan, B.-Z.; Xu, Z.-F. Flower-Specific Overproduction of Cytokinins Altered Flower Development and Sex Expression in the Perennial Woody Plant *Jatropha curcas* L. *Int. J. Mol. Sci.* **2020**, *21*, 640. [CrossRef]
50. War, A.R.; Paulraj, M.G.; War, M.Y.; Ignacimuthu, S. Role of Salicylic Acid in Induction of Plant Defense System in Chickpea (*Cicer arietinum* L.). *Plant Signal. Behav.* **2011**, *6*, 1787–1792. [CrossRef]
51. Zahid, A.; Yike, G.; Kubik, S.; Fozia; Ramzan, M.; Sardar, H.; Akram, M.T.; Khatana, M.A.; Shabbir, S.; Alharbi, S.A.; et al. Plant Growth Regulators Modulate the Growth, Physiology, and Flower Quality in Rose (*Rosa hybirda*). *J. King Saud. Univ. Sci.* **2021**, *33*, 101526. [CrossRef]
52. Peralta-Sánchez, M.G.; Gómez-Merino, F.C.; Tejeda-Sartorius, O.; Trejo-Téllez, L.I. Nitrogen Nutrition Differentially Affects Concentrations of Photosynthetic Pigments and Antioxidant Compounds in Mexican Marigold (*Tagetes erecta* L.). *Agriculture* **2023**, *13*, 517. [CrossRef]
53. Tian, Z.; Wang, F.; Zhang, W.; Liu, C.; Zhao, X. Antioxidant Mechanism and Lipid Peroxidation Patterns in Leaves and Petals of Marigold in Response to Drought Stress. *Hortic. Environ. Biotechnol.* **2012**, *53*, 183–192. [CrossRef]
54. Ohmiya, A.; Hirashima, M.; Yagi, M.; Tanase, K.; Yamamizo, C. Identification of Genes Associated with Chlorophyll Accumulation in Flower Petals. *PLoS ONE* **2014**, *9*, e113738. [CrossRef]
55. Ghadyeh Zarrinabadi, I.; Razmjoo, J.; Abdali Mashhadi, A.; Karim Mojeni, H.; Boroomand, A. Physiological Response and Productivity of Pot Marigold (*Calendula officinalis*) Genotypes under Water Deficit. *Ind. Crops Prod.* **2019**, *139*, 111488. [CrossRef]
56. Sun, T.; Rao, S.; Zhou, X.; Li, L. Plant Carotenoids: Recent Advances and Future Perspectives. *Mol. Hortic.* **2022**, *2*, 3. [CrossRef]
57. Johari, M.A.; Khong, H.Y. Total Phenolic Content and Antioxidant and Antibacterial Activities of Pereskia Bleo. *Adv. Pharmacol. Sci.* **2019**, *2019*, 7428593. [CrossRef] [PubMed]
58. Chkhikvishvili, I.; Sanikidze, T.; Gogia, N.; Enukidze, M.; Machavariani, M.; Kipiani, N.; Vinokur, Y.; Rodov, V. Constituents of French Marigold (*Tagetes patula* L.) Flowers Protect Jurkat T-Cells against Oxidative Stress. *Oxid. Med. Cell. Longev.* **2016**, *2016*, 4216285. [CrossRef]
59. Smirnoff, N.; Wheeler, G.L. Ascorbic Acid in Plants: Biosynthesis and Function. *CRC Crit. Rev. Plant Sci.* **2000**, *19*, 267–290. [CrossRef]
60. Krzymińska, A.; Frąszczak, B.; Gąsecka, M.; Magdziak, Z.; Kleiber, T. The Content of Phenolic Compounds and Organic Acids in Two *Tagetes patula* Cultivars Flowers and Its Dependence on Light Colour and Substrate. *Molecules* **2022**, *27*, 527. [CrossRef]
61. Alotaibi, H.N.; Anderson, A.K.; Sidhu, J.S. Influence of Lutein Content of Marigold Flowers on Functional Properties of Baked Pan Bread. *Ann. Agric. Sci.* **2021**, *66*, 162–168. [CrossRef]
62. Gansukh, E.; Mya, K.K.; Jung, M.; Keum, Y.S.; Kim, D.H.; Saini, R.K. Lutein Derived from Marigold (*Tagetes erecta*) Petals Triggers ROS Generation and Activates Bax and Caspase-3 Mediated Apoptosis of Human *Cervical carcinoma* (HeLa) Cells. *Food Chem. Toxicol.* **2019**, *127*, 11–18. [CrossRef]
63. Ricachenevsky, F.K.; de Araújo Junior, A.T.; Fett, J.P.; Sperotto, R.A. You Shall Not Pass: Root Vacuoles as a Symplastic Checkpoint for Metal Translocation to Shoots and Possible Application to Grain Nutritional Quality. *Front. Plant Sci.* **2018**, *9*, 00412. [CrossRef] [PubMed]
64. Yan, A.; Wang, Y.; Tan, S.N.; Mohd Yusof, M.L.; Ghosh, S.; Chen, Z. Phytoremediation: A Promising Approach for Revegetation of Heavy Metal-Polluted Land. *Front. Plant Sci.* **2020**, *11*, 00359. [CrossRef]
65. Biswal, B.; Singh, S.K.; Patra, A.; Mohapatra, K.K. Evaluation of Phytoremediation Capability of French Marigold (*Tagetes patula*) and African Marigold (*Tagetes erecta*) under Heavy Metals Contaminated Soils. *Int. J. Phytoremediat.* **2022**, *24*, 945–954. [CrossRef]
66. Madanan, M.T.; Shah, I.K.; Varghese, G.K.; Kaushal, R.K. Application of Aztec Marigold (*Tagetes erecta* L.) for Phytoremediation of Heavy Metal Polluted Lateritic Soil. *Environ. Chem. Ecotoxicol.* **2021**, *3*, 17–22. [CrossRef]
67. Zhou, Y.; Jia, Z.; Wang, J.; Chen, L.; Zou, M.; Li, Y.; Zhou, S. Heavy Metal Distribution, Relationship and Prediction in a Wheat-Rice Rotation System. *Geoderma* **2019**, *354*, 113886. [CrossRef]
68. Yu, X.; Wang, Z.; Lynn, A.; Cai, J.; Huangfu, Y.; Geng, Y.; Tang, J.; Zeng, X. Heavy Metals in Wheat Grown in Sewage Irrigation: A Distribution and Prediction Model. *Pol. J. Environ. Stud.* **2016**, *25*, 413–418. [CrossRef]

horticulturae

Article

Black Soldier Fly (*Hermetia illucens*) Frass on Sweet-Potato (*Ipomea batatas*) Slip Production with Aquaponics

Nicholas Romano [1,*], Carl Webster [2], Surjya Narayan Datta [3], Gde Sasmita Julyantoro Pande [4], Hayden Fischer [5], Amit Kumar Sinha [5], George Huskey [2], Steven D. Rawles [2] and Shaun Francis [6]

1 Virginia Cooperative Extension, College of Agriculture, Virginia State University, Petersburg, VA 23806, USA
2 USDA-ARS Harry K. Dupree Stuttgart National Aquaculture Research Center, Stuttgart, AR 72160, USA
3 Department of Fisheries Resource Management, College of Fisheries, Guru Angad Dev Veterinary and Animal Sciences University, Ludhiana 141004, India
4 Department of Aquatic Resource Management, Faculty of Marine Science and Fisheries, Udayana University, Kuta Selatan 80361, Bali, Indonesia
5 Department of Aquaculture & Fisheries, University of Arkansas at Pine Bluff, 1200 N University Dr., Pine Bluff, AR 71601, USA
6 1890 Cooperative Extension Program, University of Arkansas at Pine Bluff, 1200 N University Dr., Pine Bluff, AR 71601, USA
* Correspondence: nromano@vsu.edu; Tel.: +1-804-524-5620

Abstract: Nutrient supplementations are often added to aquaponic systems to optimize plant production, and black soldier fly larvae frass is a promising organic fertilizer. However, the mineral composition of the frass is substantially influenced by the initial substrate. In an 8-week study, sweetpotato slips were cultured at commercial stocking densities in an aquaponic system which received weekly additions of either BSFL frass made from high-nitrogen expired fish diets or low-nitrogen fruits/vegetables. The sweetpotato slips (≥8 nodes) were harvested weekly. Despite differences in the mineral composition between the frass types, the water quality as well as slip production/sugar content were unaffected by frass type. The results indicate that a wide array of substrates may be suitable for producing black soldier fly larvae frass as a fertilizer in aquaponic systems. Lastly, aquaponics is a viable system to commercially produce sweetpotato slips.

Keywords: sweetpotato slips; insect frass; black soldier fly larvae; organic fertilizer

Citation: Romano, N.; Webster, C.; Datta, S.N.; Pande, G.S.J.; Fischer, H.; Sinha, A.K.; Huskey, G.; Rawles, S.D.; Francis, S. Black Soldier Fly (*Hermetia illucens*) Frass on Sweet-Potato (*Ipomea batatas*) Slip Production with Aquaponics. *Horticulturae* **2023**, *9*, 1088. https://doi.org/10.3390/horticulturae9101088

Academic Editors: Francesco De Mastro, Gennaro Brunetti, Karam Farrag and Huadong Zang

Received: 25 August 2023
Revised: 27 September 2023
Accepted: 28 September 2023
Published: 29 September 2023

1. Introduction

Sweetpotatoes (*Ipomea batatas*) are the sixth-most produced crop in the world and are increasingly being recognized as a 'super food' due to their high content of health-promoting carotenoids, vitamins, and minerals [1,2]. The storage roots are grown in soil and while the roots can be used again as planting material, there is a point when this practice is no longer viable due to the accumulation of viruses and mutations that limit their production [3]. In response, virus-indexed planting material, known as 'slips', are produced via apical meristem culture and are then vegetatively multiplied [4]. These first-generation slips are often cultured in greenhouses in order to accelerate their growth and thus allow a longer growing season for their later storage root production. Sweetpotato farmers rely on obtaining a sufficient amount of slips and thus, sweetpotato slip production is itself an industry. Nevertheless, shortages are still common, and it has been suggested that the nutritional requirements of sweetpotato slips are not fully known and further research on the most appropriate fertilizer is an area that could be improved to optimize production [5,6].

One promising method to grow slips under controlled conditions is with aquaponics, which is the symbiotic fusion of aquaculture with hydroponics [7]. In aquaponics, the waste excreted from the fish acts as nutrients for plants to enhance sustainability and profitability due to the production of two marketable food items. Recently, it was shown that sweetpotato slips grown in an aquaponic system led to a five-fold and thousand-fold increase in the amount of slips and total weight, respectively, compared to those grown in soil [8]. The constant supply of water and nutrients, particularly dissolved nitrogen, were suggested as the contributors to this result. However, it is common practice to supplement additional minerals, most notably iron (Fe), calcium (Ca), and potassium (K), to the system to enhance plant production. Aquaponic supplements are often added in synthetic forms, such as potash, rock phosphate, Epsom salt (magnesium sulfate), and CalMag (calcium magnesium), but farmers interested in organic farming could be interested in an organic fertilizer that contains a variety of essential nutrients.

A by-product of insect farming is 'frass', which is the mineral-rich excrement of insects [9]. In particular, BSFL frass is relatively high in essential minerals compared to the frass of other edible insect species [10], while the accumulated chitin in the frass may also act as a plant prebiotic [11]. The use of BSFL frass on terrestrial plants has, in some cases, exceeded production compared to synthetic fertilizers [10,12,13]. In an aquaponic context, adding BSFL frass tea enhanced the sugar content in sweet banana peppers and the manganese content in sweetpotato slips grown aquaponically but had no influence on the overall production [14]. However, the growth of collard greens significantly increased when higher amounts of BSFL frass were added in an aquaponic system [15]. While the amount of BSFL frass additions likely plays a role in plant growth, the type of the initial substrate used to produce the frass may also be a factor. This is because the initial substrate greatly influences the mineral composition of the BSFL frass [16]. For example, the use of vegetable waste led to significantly higher phosphorus and potassium content in BSFL frass compared to either fruit waste or starches [17].

Nile tilapia (*Oreochromis niloticus*) is a tropical cichlid native to the Middle East (Jordan, Egypt, and Israel) and parts of Africa and is the third-most cultured fish in the world [18]. The fish has many highly desirable culture traits which include rapid growth rate, excellent flesh taste and quality, resistance to numerous diseases, ability to reproduce easily in captivity, possession of dietary requirements on the lower end of the food chain (herbivorous/omnivorous), and the ability to tolerate varied environmental and production conditions [19]. As Nile tilapia are a popular choice for use in aquaponic systems, growth and survival of the fish can serve as a benchmark when conducting aquaponic research.

The aim of this study was to compare the growth and mineral composition of sweetpotato slips under commercial stocking conditions in an aquaponic system receiving supplementations of BSFL frass produced with expired fish diets (EFD) or from fruits/vegetables (FV). It was hypothesized that the different frass types would have a different elemental composition and thus additions of these to an aquaponic system would influence the production and/or composition of the sweetpotato slips.

2. Materials and Methods

2.1. Source of Plants, Fish, and BSFL Frass

Virus-indexed sweetpotato slips were produced and provided by the Agriculture Department at the University of Arkansas at Pine Bluff (UAPB), which had at least 7 nodes. The all-male tilapia used in this study were purchased from AZGardens and upon arriving at UAPB, these were kept in a 1000 L acclimation tank. The fish were fed once daily with a floating commercial pellet (32% protein) designed for tilapia. The BSFL frass were produced in the lab according to Fischer et al. [20]. Briefly, the eggs of BSFL were placed on top of spoiled fish feeds designed for catfish (Rangen; 32% protein) or a combination of fruits (orange peels, banana peels, apple cores, and strawberries) and vegetables (sweetpotato and peas). The approximate composition of the SF and FV was measured using the standard Association of Official Analytical Chemists [21] methods and results are presented in Table 1. Hereafter, the frass made from spoiled feeds or fruits/vegetables will be referred to as SF frass and FV frass, respectively.

Table 1. Approximate composition (% dry matter) of expired fish diets (EFD) and fruit/vegetables (FV) that were provided to the black soldier fly larvae.

	EFD	FV
Moisture	5.23	85.95
Crude protein	32.13	9.32
Crude lipid	8.47	3.28
Crude ash	7.64	14.52
Crude fiber	4.31	9.27

The BSFL converted these into frass, which took approximately three weeks. The frass was then dried in a forced air oven (Despatch; LBB Series 2–12-3, Illinois Tool Works, Inc., Minneapolis, MN, USA) at 100 °C for two days and then ground into a fine powder with a hammer mill. Frass nitrogen was measured using a Leco N analyzer while the mineral content was measured at the Fayetteville Agricultural Diagnostic Laboratory at the University of Arkansas with inductively coupled plasma (ICP) analysis. The results are shown in Table 2.

Table 2. Nutrient composition of black soldier fly larvae frass produced with expired fish diets (EFD frass) or fruits/vegetable (FV frass).

	%						mg/kg					
	N	P	K	Ca	Mg	S	Na	Fe	Mn	Zn	Cu	B
EFD frass	4.64	2.54	2.95	5.28	0.44	0.75	13,561	463	87	200	30.7	23
FV frass	3.37	1.16	4.12	6.38	0.38	0.50	11,815	295	63	104	22.2	33

2.2. Aquaponic Systems and Experimental Design

There were a total of six identical aquaponic systems (5110 L capacity) that are described in detail in Romano et al. [14]. A total of 40 tilapia (initial weight of 45.7 g) were added into each aquaponic system and fed twice daily to apparent satiation with commercial floating feeds (Rangen; 32% protein). Each tank received gentle aeration with an air stone and the tanks were covered with netting to provide shade (to minimize algae growth) and prevent any escapees. The amount of food provided to each system was recorded.

After one week of feeding the fish, a total of 200 sweetpotato slips were planted in each of the aquaponic media beds and each slip was spaced 2.5 cm (or 1 inch) apart from each other in media bed (145 cm × 75 cm) filled with expanded lava rock. This stocking density is the same used in commercial settings [5]. After adding the slips, a total of 10 g of SF frass or FV frass were sprinkled on top of the media bed containing the slips. This was performed to potentially encourage frass mineralization but also because, based on past experience, adding frass to the sump encouraged filamentous algae growth. Every week, waste that settled in the sump was siphoned out followed by adding 5 mL of iron chelate (Iron-gluconate; SEACHEM Flourish, Root 98 Warehouse, Lakeland, Florida). No buffers were added to adjust pH.

2.3. Water Quality Analysis

The ammonia-N, nitrite-N, and nitrate-N levels were measured with an API master test kit once a week. The water temperature, dissolved oxygen and pH were measured with a digital multimeter probe (YSI Professional Plus). On week 2, 4, 6, and 8, a water sample was collected from the sump for later mineral analysis and stored at −20 °C. The minerals were measured with a flame atomic absorption spectrophotometer (AAS, iCE 3000 series, Thermo Scientific, Santa Clara, CA, USA) with deuterium lamp background correction. Calibrations were made using single element standard solutions (CPI International, Santa Rosa, CA, USA). However, for phosphorus (P), the persulfate digestion method (HACH method 8190) was used because the concentrations were too low for the AAS.

2.4. Aquaponic Sampling

By week 2 of adding the cuttings, the majority grew to be considered a slip (>6 nodes) and were harvested to allow at least 2 nodes remaining in the media bed. The total biomass of all the slips were weighed among the treatments while 40 were used to measure the total length, number of nodes, and stem diameter. These were then placed in a zip lock plastic bag and stored at −20 °C for later mineral and sugar analysis. After 8 weeks, the slips were harvested 7 times in total and all the remaining slips were counted to determine the overall survival. The tilapia were also harvested after 8 weeks, counted, and were individually measured for weight and length using a digital scale and metric ruler, respectively.

2.5. Mineral and Sugar Analysis

For mineral analysis, the sweetpotato leaves from each replicate were oven-dried at 60 °C for 24 h then digested in a heat block (Environmental Express, Charleston, SC, USA) at 115 °C for 30 min in 4.0 mL trace metal-grade HNO_3 (69%; Sigma-Aldrich). After digestion, 0.1 mL of H_2O_2 (30%) was added and then 40 mL of Milli-Q water was added. Samples were measured for iron, calcium, zinc, magnesium, and manganese using a flame atomic absorption spectrophotometer (AAS) (iCE 3000 series, Thermo Scientific, Santa Clara, CA, USA), while phosphorus was measured using the persulfate digestion method (HACH method 8190) as described above. Each replicate sample was measured in triplicate.

For sugar analysis, total sugar was estimated from 100 mg of sweetpotato leaves from each replicate that were ground in liquid nitrogen using a mortar and pestle. Ground samples were transferred to 1 mL 100% acetone and kept overnight at 4 °C. Samples were centrifuged and the residue was repeatedly washed with hot 80% ethanol to remove all traces of soluble sugars. This filtrate was used for the determination of soluble sugars while the residue/pellet was used for the determination of the insoluble sugar content. To the residue, 2 mL of 0.2 N H_2SO_4 was added, followed by heating at 100 °C in a water bath for 30 min, which was then centrifuged and the supernatant was collected. Anthrone reagent (150 μL) was added to each microplate well containing 50 mL of glucose standard solutions, blanks, and samples (soluble and insoluble sugars). Plates were then placed for 10 min at 4 °C and then incubated for 20 min at 100 °C. A cooling step for 20 min at room temperature was completed before reading absorbance at 620 nm triplicate in a microplate reader (PowerWave XS, BioTek Instruments, Winooski, VT, USA). A standard curve was obtained with different concentrations of glucose. Each replicate sample was measured in triplicate.

2.6. Statistical Analysis

Sweetpotato slip composition and performance in response to frass type (SF vs. FV) and harvest period (first versus final composition, or slip performance among harvests 1–7) were analyzed by factorial mixed model (MIXED) analysis of variance (SAS version 9.4, SAS Institute, Inc., Cary, NC, USA). Aquaponic system water quality and tilapia growth performance in response to frass type were analyzed by one-way mixed model analysis of variance. All response values were natural log-transformed prior to analyses. Differences among response means were considered significant at $p \leq 0.05$.

3. Results

3.1. Frass Characteristics

The approximate compositions of the two initial substrates and subsequent frass are presented in Table 1. The frass made from the expired fish diet (EFD) contained 30.79% protein, 7.69% lipid, and 24.83% ash. This is in contrast to the frass made from fruits/vegetables (FV) which contained 22.73% protein, 4.03% lipid, and 31.16% ash.

The analyzed mineral compositions of each BSFL frass is shown in Table 2. There were differences in the mineral compositions for each frass with FV frass having higher levels of N, P, Mg, S, Na, Fe, Mn, Zn, and Cu compared to EFD frass, while EFD frass had higher levels of K, Ca, and B than FV frass.

3.2. Water Quality and Chemistry

The addition of the two types of BSFL frass to media beds led to only one difference in measured water quality parameters and all were within acceptable limits for fish. The aquaponic system water quality was optimum for tilapia culture and did not differ markedly in response to frass type (Table 3). Averages during the study were as follows: temperature, 27.4 °C; dissolved oxygen, 5.31 mg/L; hardness, 47.4 mg/L; total ammonia nitrogen (TAN), 0.33 mg/L; nitrite, 0.05 mg/L; nitrate, 42.08 mg/L. Only the pH was statistically higher in beds receiving the FV frass (7.54) compared to the SF frass (7.49), which was not biologically significant.

Table 3. Mean (±SE) water quality parameters in an aquaponic system receiving frass by black soldier fly (*Hermetia illucens*) larvae fed expired fish feeds (EFD) or fruits/vegetables (FV) over 8 weeks.

Parameter	EFD	FV
Temperature (°C)	27.3 ± 0.05	27.5 ± 0.09
Dissolved oxygen (mg/L)	5.14 ± 0.33	5.47 ± 0.01
pH	7.49 ± 0.01 ^{b}	7.54 ± 0.01 ^{a}
Hardness (mg/L)	45.5 ± 0.14	49.3 ± 0.27
TAN [1] (mg/L)	0.34 ± 0.00	0.31 ± 0.02
Nitrite (mg/L)	0.08 ± 0.04	0.03 ± 0.00
Nitrate (mg/L)	42.08 ± 0.41	42.08 ± 0.12

[1] Total ammonia nitrogen (TAN). Different superscripted letters indicate significant difference ($p < 0.05$).

3.3. Fish Growth

The tilapia growth performance was typical for this species and initial size grown in optimum conditions, and there were no significant differences between frass types (Table 4). The averages for length (cm), final weight (g), weight gain (%), specific growth rate (SGR; %/day), feed intake (g diet/fish), and feed conversion ratio (FCR) were: 21.09 cm, 193.1 g, 147.35 g, 322.2%, 2.95%/day, 162.15 g diet/fish, and 1.11, respectively.

Table 4. Growth performance of Nile tilapia (n = 3) in an aquaponics system receiving frass by black soldier fly (*Hermetia illucens*) larvae fed expired fish feeds (EFD) or fruits/vegetables (FV) after 8 weeks.

Response	EFD	FV
Length (cm)	21.23 ± 0.19	20.95 ± 0.29
Final weight (g)	197.3 ± 4.6	188.9 ± 8.0
Weight gain (%) [1]	332.6 ± 13.8	311.8 ± 13.4
SGR [2]	3.00 ± 0.56	2.89 ± 0.42
Feed intake (g/fish)	164.6 ± 5.8	159.7 ± 2.3
FCR [3]	1.09 ± 0.06	1.12 ± 0.04

[1] Percent gain from initial weight. [2] Specific growth rate (SGR) = (ln final weight—ln initial weight)/time. [3] Feed conversion ratio (FCR) = g dry feed fed/g gained.

3.4. Sweetpotato Slip Production and Mineral Composition

The sweetpotato slip production and quality were unaffected by the frass type but differed among partial harvests (Table 5). Generally, the slip diameter decreased, whereas the node length and number of nodes per slips increased between the initial (#1) and final harvests (#7). The slip length and weight differed by harvest with no discernible pattern.

The sweetpotato slip mineral and sugar composition was unaffected by the frass type but differed between the initial and final harvests (Table 6). The concentrations of Fe, Mn, and Zn increased at the final harvest from the initial content, whereas the Ca and Na concentrations decreased at the final harvest. The concentrations of Mg, P, and K in the slips were unchanged between the initial and final harvests and were unaffected by frass type. The soluble and insoluble sugars decreased significantly by three-fold and six-fold, respectively, from the initial concentrations.

Table 5. Mean (±SE) production parameters of sweetpotato slips (n = 3) grown in an aquaponics system with Nile tilapia for 8 weeks when black soldier fly (*Hermetia illucens*) larvae frass was produced from expired fish feeds (EFD) or fruits/vegetables (F/V). Main effects of least squares means with different letters are significantly different ($p < 0.05$).

Treatments		Response Variables				
Frass	**Harvest**	**Length (cm)**	**Weight (g)**	**Diameter (mm)**	**Nodes/Slip**	**Nodes/Length**
EFD	1	31.15	912.7	4.37	7.69	0.263
	2	27.75	259.7	3.21	8.88	0.343
	3	50.18	1029.7	3.10	9.35	0.213
	4	29.52	381.7	3.06	8.69	0.333
	5	39.89	644.6	2.97	10.33	0.300
	6	22.23	205.1	2.90	8.78	0.420
	7	31.03	633.3	2.88	11.14	0.397
FV	1	30.28	758.7	4.24	7.57	0.260
	2	31.81	486.6	3.30	8.88	0.310
	3	48.03	1102.9	3.11	9.49	0.213
	4	30.30	413.7	3.24	8.74	0.310
	5	42.85	649.9	2.98	10.08	0.273
	6	29.68	299.3	2.89	9.50	0.367
	7	31.41	828.5	2.95	11.59	0.393
Pooled SE		4.44	138.7	0.12	0.45	0.026
Main effects of means						
EFD		33.11	581.0	3.21	9.27	0.324
FV		34.91	648.5	3.24	9.40	0.304
	1	30.72 bc	835.7 a	4.31 a	7.63 d	0.262 bc
	2	29.78 c	373.2 c	3.25 b	8.88 bc	0.327 ab
	3	49.11 a	1066.3 a	3.10 bc	9.40 bc	0.213 c
	4	29.91 c	397.7 bc	3.15 bc	8.72 cd	0.322 b
	5	41.37 ab	647.3 ab	2.97 bc	10.20 a	0.287 b
	6	25.95 c	252.2 c	2.89 c	9.14 bc	0.393 a
	7	31.22 bc	730.9 ab	2.91 c	11.36 a	0.395 a
ANOVA Source, Pr > F						
Frass		0.638	0.694	0.792	0.777	0.463
Harvest		<0.001	<0.001	<0.001	<0.001	<0.001
F × H		0.818	0.922	0.881	0.907	0.838

Table 6. Mean (±SE) mineral (mg/g dry weight) and sugar composition (mg/g dry weight) of sweetpotato slips (n = 3) grown in an aquaponics system with Nile tilapia for 8 weeks when black soldier fly (*Hermetia illucens*) larvae frass was produced from expired fish feeds (EFD) or fruits/vegetables (F/V). Main effects of least squares means with different letters are significantly different ($p < 0.05$).

Treatments		Macronutrients (mg/g)				Micronutrients (mg/g)				Sugar (mg/g)	
Harvest	**Frass**	**P**	**K**	**Ca**	**Mg**	**Na**	**Fe**	**Mn**	**Zn**	**Soluble**	**Insoluble**
First	EFD	9.47	109.24	7.49	10.94	0.071	0.121	0.137	0.071	30.08	26.20
	FV	9.50	109.80	7.47	11.17	0.090	0.118	0.134	0.090	37.63	35.29
Last	EFD	9.79	92.73	6.71	10.78	0.175	0.148	1.500	0.175	8.83	5.55
	FV	9.17	90.95	6.89	10.46	0.187	0.142	1.499	0.187	13.55	6.01
Pooled SE		0.385	8.36	0.165	0.21	0.047	0.006	0.014	0.015	4.25	4.83
First		9.48	109.52	7.48 a	11.05	0.080 b	0.120 b	0.135 b	0.080 b	33.86 a	30.75 a
Last		9.48	91.84	6.80 b	10.62	0.181 a	0.145 a	1.499 a	0.181 a	11.19 b	5.78 b
	EFD	9.63	100.99	7.10	10.86	0.123	0.134	0.819	0.123	49.46	15.87
	FV	9.33	100.38	7.18	10.82	0.138	0.130	0.816	0.138	25.59	20.65

Table 6. *Cont.*

Treatments		Macronutrients (mg/g)				Micronutrients (mg/g)				Sugar (mg/g)	
Harvest	**Frass**	**P**	**K**	**Ca**	**Mg**	**Na**	**Fe**	**Mn**	**Zn**	**Soluble**	**Insoluble**
ANOVA Source, Pr > F											
Time		0.934	0.062	0.003	0.071	0.035	0.004	<0.001	<0.001	<0.001	<0.001
Frass		0.456	0.456	0.646	0.837	0.165	0.528	0.860	0.324	0.096	0.422
T X F		0.427	0.752	0.558	0.893	0.537	0.826	0.893	0.774	0.757	0.506

4. Discussion

One of the criticisms of aquaponics is that the production of plants is often limited in scale compared to terrestrial farming, where hundreds or thousands of acres of crops can be grown outside and subsequently harvested with tractors and other heavy machinery. Thus, aquaponics is often viewed as being more suitable for growing niche crops and/or farming in an urban environment where space is limited [7]. However, in the case of virus-indexed sweetpotato slips, these are often cultivated anyway under controlled environmental conditions, like greenhouses, to optimize growth before being transplanted outdoors in the soil [5]. Consequently, obtaining a sufficient amount of slips is a bottleneck in the sweetpotato industry and any method to optimize slip growth would help extend the season for storage root production in soil [4].

Aquaponics appears to be a viable method for sweetpotato slip production in which it has been previously shown that sweetpotato slips grew substantially faster in aquaponic conditions compared to those grown in soil [8]. It is known that an abundance and consistent supply of water and nitrogen in aquaponic systems can promote leafy growth in a variety of plants [7], and inhibit storage root production in sweetpotatoes [6,22]. Indeed, under aquaponic conditions, storage root growth was not observed and thus it was suggested that more energy could be diverted to leafy growth [14]. However, the stocking density used in Romano et al. [14] was low (three cuttings in a 145 cm $\times$ 75 cm plant culture bed) and not representative of commercial conditions. Thus, a higher stocking density of cuttings was adopted in this study in which a total of 200 cuttings were planted in each of the media beds (145 cm $\times$ 75 cm).

After the cuttings were planted in this study, it took about a week before these began to discernibly start growing, which indicates a period of acclimation to the new culture conditions. By the second week, the cuttings were sufficiently long enough to be considered slips because they had at least six nodes; thus, the slips were subsequently harvested by week two of this study. Each week thereafter the slips were harvested another six times (for a total of seven harvests) until the study concluded when the fall season was approaching. Typically, sweetpotato slips grown in soil are harvested between 10 to 14 days to provide a total of nine harvests in a year, first starting in April [6]. In this study, harvesting was up to two-fold faster where it could be possible to harvest up to 18 or more times before the season ends. Additionally, the number of nodes on each slip gradually increased with each harvest, which is considered desirable because more nodes means more planting material for storage root production in the soil.

While nitrogen is generally abundant in aquaponics systems, the most common limiting nutrients include K, Ca, and Fe and these are often added in the forms of potassium bicarbonate, CalMag, and iron chelate, respectively [14,20,23]. However, there are other essential macro- and micro-nutrients that may be at insufficient levels for the optimal growth and well-being of plants in aquaponic systems. In this common scenario, adding a mineral-rich fertilizer consisting of various essential minerals may be effective. It was previously shown that directly adding BSFL frass tea to the water of an aquaponic system had no effect on sweetpotato slip production [14], but when added in a quantity more than two-fold higher this significantly increased collard green growth [15]. More recently, dietary inclusions of BSFL frass at 10% increased the growth of catfish (*Ictalurus punctatus*) as well as stevia (*Stevia rebaudiana*), and lavender (*Lavaridula angustifolia*) in an aquaponic system [24].

While there appears to be strong indications that adding BSFL frass can provide benefits to aquaponic plants, it is known that the composition of BSFL frass greatly depends on the initial substrate provided. However, to date, the efficacy of BSFL frass made from different substrates has not been compared in aquaponics. The two different types of BSFL frass that were compared in this study were produced with high-nitrogen expired fish diets (EFD) while the other with low-nitrogen fruits/vegetables (FV). The EFD frass did indeed have a higher nitrogen content, but the difference was not as remarkable as the difference between the initial substrates. In terms of the limiting nutrients in aquaponic systems, the EFD frass had less K and Ca, but more Fe than the FV frass. Despite the different mineral composition of the BSFL frass, additions of these different frass types led to no difference in sweetpotato slip production. It could be argued that the amounts added were insufficient to make a difference. While this could be a factor, it is perhaps worthy to note that 10 g of BSFL was added weekly in this study compared to 2.5 g each week which was sufficient to enhance collard green growth in the same system with similar stocking densities of fish [15]. Even though higher amounts were added in this study, the mean ammonia levels did not exceed 0.5 mg/L, while the other water quality parameters were similar.

Among the tested minerals in the water, P was significantly higher in the EFD treatment at week 8. It is tempting to attribute the higher P to the EFD frass having an over two-fold higher P content. However, this would not explain the K water content being significantly lower in both the water and slips from the FV frass treatment at week 6, because the K content was higher in the FV frass. Moreover, the Fe content of the slips was significantly higher in FV frass treatment, despite the FV frass having almost two-fold less Fe. Nevertheless, Fe as well as Mn were consistently at undetectable levels in the water. This was despite the weekly additions of iron chelate (along with the BSFL frass), indicating that Fe was being absorbed by the slips at a faster rate than the inputs of this nutrient. Indeed, this seems to be supported by the increased Fe and Mn (as well as Zn) content of the sweetpotato slips compared to their initial values. It is conceivable that sweetpotato slip production could be further enhanced by ensuring Fe is not limiting and perhaps should be monitored more closely. It is important to point out, however, that chlorosis, which is yellowing of the leaves and a symptom of Fe deficiency, was not observed in this study.

It has been demonstrated that fish grown in an aquaponic system have normal growth and survival compared to traditional production methods. Some of the species successfully grown aquaponically include largemouth bass (*Micropterus salmoides*) [20], channel catfish (*Ictalurus punctatus*) [24], Nile tilapia (*Oreochromis niloticus*) [25], rainbow trout (*Oncorhynchus mykiss*) [26,27], goldfish (*Carassius auratus*) [28], and white shrimp (*Litopenaeus vannamei*) [29]. In this study, the fish growth was acceptable and similar to the production parameters in other reports [30–33] and the tilapia growth was not adversely affected by the frass type.

5. Conclusions

Inclusions of BSFL frass, either directly to the water or in aquafeeds, has been previously shown to benefit plant production in an aquaponic system. The results of this study appear to indicate that the initial substrate used to make BSFL frass is not a major factor in sweetpotato slip production. Considering that the farming of BSFL is expected to increase in the coming years, this could also increase the availability of BSF frass as an option to aquaponic farmers, particularly those interested in using an organic fertilizer rather than traditionally relying on synthetic ones. Finally, the production of sweetpotato slips at commercially stocking densities appears to be a viable farming method that may improve slip availability and help extend the duration for storage root production. Further studies on optimizing the BSFL frass dose and potential ways to enhance the Fe and Mn in the BSFL frass may further improve plant production.

Author Contributions: Conceptualization, methodology, supervision, and writing original draft preparation, N.R.; formal analysis, S.N.D., G.S.J.P., G.H. and A.K.S.; data curation and final editing, N.R., S.D.R. and C.W.; assistance with conceptualization, H.F.; provide virus-index slips and technical assistance, S.F. All authors have read and agreed to the published version of the manuscript.

Funding: This study was funded by 1890's Institutional Global Food and Nutritional Security Grant and a non-assistance Cooperative Agreement (58-6028-2-005). This research was supported, in part, by funds provided by the USDA/ARS CRIS Project 6028-31630-009-00D and is a research component of the USDA-ARS Grand Challenge project entitled "Debugging a new mini livestock commodity: Developing a model of insect production to demonstrate their value as a safe solution for food waste and sustainable fish and livestock production".

Data Availability Statement: Data can be made available with reasonable request.

Acknowledgments: We would like to thank John Brewer for the technical assistance.

Conflicts of Interest: The authors declare no conflict of interest.

References

1. Alam, M.K. A compreshensive review of sweet potato (*Ipomoea batatas* (L.) Lam): Revisiting the associated health benefits. *Trends Food Sci. Technol.* **2021**, *115*, 512–529. [CrossRef]
2. Laveriano-Santos, E.P.; López-Yerena, A.; Jaime-Rodríguez, C.; González-Coria, J.; Lamuela-Raventós, R.M.; Vallverdú-Queralt, A.; Romanyá, J.; Pérez, M. Sweet potato is not simply an abundant food crop: A comprehensive review of its phytochemical constituents, biological activities, and the effects of processing. *Antioxidants* **2022**, *11*, 1648. [CrossRef]
3. Clark, C.A.; Valverde, R.A.; Fuentes, S.; Salazar, L.F.; Moyer, J.W. Research for improved management of sweetpotato pests and diseases: Cultivar decline. *Acta Hortic.* **2002**, *583*, 103–112. [CrossRef]
4. Loebenstein, G. Chapter two—Control of sweet potato virus diseases. *Adv. Virus Res.* **2015**, *91*, 33–45. [PubMed]
5. Tedesco, D.; Moreira, B.R.D.A.; Barbosa Júnior, M.R.; Maeda, M.; Silva, R.P.D. Sustainable management of sweet potatoes: A review on practices, strategies, and opportunities in nutrition-sensitive agriculture, energy security, and quality of life. *Agric. Syst.* **2023**, *210*, 103693. [CrossRef]
6. Jiang, C.; Pesic-VanEsbroeck, Z.; Osborne, J.A.; Schultheis, J.R. Factors affecting greenhouse sweetpotato slip production. *Int. J. Veg. Sci.* **2017**, *23*, 185–194. [CrossRef]
7. Hager, J.; Bright, L.A.; Dusci, J.; Tidwell, J. *Aquaponics Production Manual. A Practical Handbook for Growers*; Kentucky State University: Frankfort, KY, USA, 2021.
8. Romano, N.; Francis, S.; Islam SPowell, A.; Fischer, H. Aquaponics substantially improved sweetpotato (*Ipomoea batatas*) slip production compared to soil but decreased phenol and antioxidant capacity. *Aquac. Int.* **2022**, *30*, 2603–2610. [CrossRef]
9. Fischer, H.; Romano, N.; Sinha, A.K. Conversion of spent coffee and donuts by black soldier fly (*Hermetia illucens*) larvae into potential resources for animal and plant farming. *Insects* **2021**, *12*, 332. [CrossRef]
10. Beesigamukama, D.; Subramanian, S.; Tanga, C.M. Nutrient quality and maturity status of frass fertilizer from nine edible insects. *Sci. Rep.* **2002**, *12*, 7182. [CrossRef]
11. Quilliam, R.S.; Nuku-Adeku, C.; Pierre-Olivier, M.; Little, D.C. Integrating insect frass biofertilisers into sustainable peri-urban agro-food systems. *J. Insects Food Feed* **2020**, *6*, 315–322. [CrossRef]
12. Agustiyani, D.; Agandi, R.; Nugroho, A.A.; Antonius, S. The effect of application of compost and frass from black soldier fly larvae (*Hermetia illucens* L.) on growth of pakchoi (*Brassica rapa* L.). *IOP Conf. Ser. Earth Environ. Sci.* **2020**, *762*, 012036. [CrossRef]
13. Romano, N.; Fischer, H.; Powell, A.; Sinha, A.K.; Islam, S.; Deb, U.; Francis, S. Applications of black soldier fly (*Hermetia illucens*) larvae frass on sweetpotato slip production, mineral content and benefit-cost analysis. *Agronomy* **2022**, *12*, 928. [CrossRef]
14. Romano, N.; Powell, A.; Islam, S.; Fischer, H.; Renukdas, N.; Sinha, A.K.; Francis, S. Supplementing aquaponics with black soldier fly (*Hermetia illucens*) larvae frass tea: Effects on the production and composition of sweetpotato slips and sweet banana peppers. *Aquaculture* **2022**, *555*, 738160. [CrossRef]
15. Romano, N.; Islam, S. Productivity and elemental/chlorophyll composition of collard greens in an aquaponic system at different combinations of media and black soldier fly (*Hermetia illucens*) larvae frass supplementations. *Aquac. Res.* **2023**, *2023*, 3308537. [CrossRef]
16. Chia, S.Y.; Tanga, C.M.; Osuga, I.M.; Cheseto, X.; Ekesi, S.; Dicke, M.; van Loon, J.J.A. Nutritional composition of black soldier fly larvae feeding on agro-industrial by-products. *Entomol. Exper. Appl.* **2020**, *168*, 472–481. [CrossRef]
17. Fischer, H.; Romano, N. Fruit, vegetable, and starch mixtures on the nutritional quality of black soldier fly (*Hermetia illucens*) larvae and resulting frass. *J. Insects Food Feed* **2021**, *7*, 319–327. [CrossRef]
18. Geletu, T.T.; Zhao, J. Genetic resources of Nile tilapia (*Oreochromis niloticus* Linnaeus, 1758) in its native range and aquaculture. *Hydrobiologia* **2023**, *850*, 2425–2445. [CrossRef]
19. Lim, C.E.; Webster, C.D. Nutrient requirements. In *Tilapia, Biology, Culture, and Nutrition*; Lim, C., Webster, C.D., Eds.; Haworth Press: Binghamton, NY, USA, 2006; pp. 469–501.

20. Fischer, H.; Romano, N.; Jones, J.; Howe, J.; Renukdas, N.; Sinha, A.K. Comparing water quality/bacterial composition and productivity of largemouth bass *Micropterus salmoides* juveniles in a recirculating aquaculture system versus aquaponics as well as plant growth/mineral composition with or without media. *Aquaculture* **2021**, *538*, 736554. [CrossRef]
21. AOAC (Association of Official Analytical Chemists). *Official Methods of Analysis of AOAC International, 18th*, 211st ed.; Cunniff, P.A., Ed.; AOAC International: Arlington, VA, USA, 2006.
22. Tucker, M.R. *Essential Plant Nutrients: Their Presence in North Carolina Soils and Role in Plant Nutrition*; North Carolina Agronomic Services Division, Department of Agriculture and Consumer Services: Raleigh, NC, USA, 1999.
23. Yang, T.; Kim, H.-J. Characterizing nutrient composition and concentration in tomato-, basil-, and lettuce-based aquaponic and hydroponic systems. *Water* **2020**, *12*, 1259. [CrossRef]
24. Romano, N.; Datta, S.N.; Pande, G.S.J.; Sinha, A.K.; Yamamoto, F.Y.; Beck, B.H.; Webster, C.D. Dietary inclusions of black soldier fly (*Hermetia illucens*) larvae frass enhanced production of channel catfish (*Ictalurus punctatus*) juveniles, stevia (*Stevia rebaudiana*), and lavender (*Lavaridula angustifolia*) in an aquaponic system. *Aquaculture* **2023**, *575*, 739742. [CrossRef]
25. Al-Zahrani, M.S.; Hassanien, H.A.; Alsaade, F.W.; Wahsheh, H.A.M. Effect of stocking density on sustainable growth performance and water quality of Nile tilapia-spinach in NFT aquaponic system. *Sustainability* **2023**, *15*, 6935. [CrossRef]
26. Birolo, M.; Bordignon, F.; Trocino, A.; Fasulato, L.; Pascual, A.; Godoy, S.; Nicoletto, C.; Maucier, C.; Xicato, G. Effects of stocking density on the growth and flesh quality of rainbow trout (*Onchorynchus mykiss*) reared in a low-tech aquaponic system. *Aquaculture* **2020**, *529*, 735653. [CrossRef]
27. Bordignon, F.; Gasco, L.; Birolo, M.; Trocino, A.; Caimi, C.; Ballarin, C.; Bortoletti, M.; Nicoletto, C.; Maucieri, C. Performance and fillet traits of rainbow trout (*Onchorynchus mykiss*) fed different levels of Hermetia illucens meal in a low-tech aquaponic system. *Aquaculture* **2022**, *546*, 737279. [CrossRef]
28. Nuwansi, K.K.T.; Verma, A.K.; Prakash, C.; Tiwari, V.K.; Chandrakant, M.H.; Shete, A.P.; Prabhata, G.P.W.A. Effect of water flow on polyculture of koi carp (*Cyprinus carpio* var. koi) and goldfish (*Carassius auratus*) with water spinach (*Iponopa aquatica*) in recirculating aquaponic system. *Aquac. Int.* **2016**, *24*, 385–393. [CrossRef]
29. Fierro-Sañudo, J.F.; Rodriguez-Montes de Oca, G.A.; León-Cañedo, J.A.; Alacrón-Silvas, S.G.; Mariscal-Lagarda, M.M.; Díaz-Valdés Páez-Osuna, F. Production and management of shrimp (*Penaeus vannamei*) in co-culture with basil (*Ocimum basilicum*) using two sources of low-salinity water. *Lat. Am. J. Aquat. Res.* **2017**, *46*, 63–71. [CrossRef]
30. Azaza, M.S.; Khiari, N.; Dhraief, M.N.; Aloui, N.; Kraiem, M.M.; Elfeki, A. Growth performance, oxidative stress indices and hepatic carbohydrate metabolic enzymes activities of juvenile Nile tilapia, *Oreochromis niloticus* L., in response to dietary starch to protein ratios. *Aquac. Res.* **2015**, *46*, 14–27. [CrossRef]
31. El-Naggar, K.; Mohamed, R.; El-Katcha, M.I.; Abdo, S.E.; Soltan, M.A. Plant ingredient diet supplemented with lecithin as fish meal and fish oil alternative affects growth performance, serum biochemical, lipid metabolism and growth-related gene expression in Nile tilapia. *Aquac. Res.* **2021**, *52*, 6308–6321. [CrossRef]
32. Qin, Y.; Naumovski, N.; Ranadheera, C.S.; D'Cunha, N.M. Nutrition-related health outcomes of sweet potato (*Ipomoea batatas*) consumption: A systematic review. *Food Biosci.* **2022**, *50*, 102208. [CrossRef]
33. El-Sayed, A.-F.M.; Figueiredo-Silva, C.; Zeid, S.M.S.; Makled, S.O. Metal-amino acid complexes (Zn, Se, Cu, Fe, and Mn) as a replacement of inorganic trace minerals in commercial diets for Nile tilapia (*Oreochromis niloticus*) reared under field conditions: Effects on growth, feed efficiency, gut microbiota, intestinal histology, and economic return. *Aquaculture* **2023**, *567*, 739223.

horticulturae

MDPI

Article

Nutrient Solution from Aqueous Extracts as an Alternative to Fertigation in Hydroponic

María Carmen Salas-Sanjuán [1,*], José Luis Ruíz-Zubiate [2], Juan Luis Valenzuela [3] and Antonio Xavier Campos [4]

1 Department of Agronomy, Campus de Excelencia Internacional Agroalimentario, ceiA3, Almeria University, La Cañada, 04120 Almería, Spain
2 Doctoral Program in Protected Agriculture, Campus de Excelencia Internacional Agroalimentario, ceiA3, Almeria University, La Cañada, 04120 Almería, Spain; jl.ruizz@hotmail.com
3 Department of Biology & Geology, Higher Engineering School, Research Centres CIAIMBITAL, ceiA3, University of Almería, 04120 Almería, Spain; jvalenzu@ual.es
4 Faculdade de Agronomia e Medicina Veterinária, Universidade de Brasília, Brasília 70910-000, DF, Brazil; antonioxavier.campos@gmail.com
* Correspondence: csalas@ual.es

Abstract: The reintegration of agro-waste into the same agriculture site fulfils the objective of the European Bio-Economy Strategy: to reduce transport costs, waste volume, and the need for mineral fertilizers. One of the fundamental principles in sustainable agriculture is the recycling of crop residues through composting or vermicomposting. From this process, it is possible to obtain organic matter for the production of aqueous extracts (tea) that can be used as a source of nutrients in fertigation as an alternative to mineral fertilizers. The objective of this research was to evaluate the use of an aerated or non-aerated aqueous extract as a recirculating nutrient solution in a hydroponic culture (NFT) of lettuce. For this, the test method was compared to hydroponic cultivation with a conventional nutrient solution. The conventional nutrient solution contained minerals or synthetic fertilizers and the aqueous extracts of vermicompost from vegetal residues of horticultural crops. The evolution of the chemical composition of the nutrient solutions during cultivation was analyzed, obtaining adequate concentrations of NO_3^-, K^+, and Ca^{2+} and taking possible imbalances in nutrients such as P-$H_2PO_4^-$ into consideration. Plants fertigated with an organic and aerated nutrient solution obtained good yields and improvements in quality by having six times less N-NO_3^- in edible leaves compared to plants exposed to the mineral treatment. The preparation of aqueous extracts as a source of nutrients opens the door to circular agriculture to make processes in intensive production systems more efficient.

Keywords: organic agriculture; vermicompost; *Lactuca sativa* L.; circular economy; organic fertilizer; ammonification; nitrification; N-mineralization; nutrient film technique

Citation: Salas-Sanjuán, M.C.; Ruíz-Zubiate, J.L.; Valenzuela, J.L.; Campos, A.X. Nutrient Solution from Aqueous Extracts as an Alternative to Fertigation in Hydroponic. *Horticulturae* **2023**, *9*, 1281. https://doi.org/10.3390/horticulturae9121281

Academic Editors: Francesco De Mastro, Gennaro Brunetti, Karam Farrag and Huadong Zang

Received: 9 October 2023
Revised: 19 November 2023
Accepted: 23 November 2023
Published: 28 November 2023

1. Introduction

Hydroponics and/or soil-less cultivation is a technique that grows roots out of the soil in a container with a substrate or in a nutrient solution. It is a very widespread technique that is used for the production of vegetables under protected conditions that has made it possible to increase the yields and quality of fruits [1]. Hydroponic systems for the production of horticultural species require the use of soluble, mineral, and/or synthetic fertilizers, which, when dissolved in water, ensure the availability of nutrients [2,3]. In hydroponic systems without a substrate, the chemical characteristics of the nutrient solution take on special importance and must contain all of the nutrients at appropriate and balanced concentrations. These systems are sensitive to imbalances in the nutrient solution due to their low or almost no buffering capacity [4].

New regulations limit the use of mineral or synthetic fertilizers to reduce the environmental impact of agricultural systems [5]. In this sense, as an alternative to synthetic fertilizers, it is necessary to resort to the use of organic matter from the composting and/or vermicomposting of plant and/or animal waste [6,7]. However, supplementing nutrients by the application of nutrient solutions via fertigation is relatively simple when using mineral or synthetic fertilizers, but it is more challenging in systems where only organic nutrient materials are used [8,9].

We have compost or vermicompost teas (aqueous extracts) that have been shown to promote plant health and increase the yield and quality of fruits, aromatic plants, and flowers because they contain beneficial microorganisms that promote the absorption of essential nutrients in the form of ions [9,10]. Additionally, they are used as organic fertilizers because they contain high microbial loads (plant growth promotion microorganisms, PGPM) that increase the solubility of nutritional elements and promote plant growth [8,11,12].

However, although aqueous extracts have a high microbial load and a high concentration of nutritional elements [9], in general, they are not balanced as nutrient solutions for fertigation [3], and a large number of nutritional elements are in their organic forms and are not available to plants [13–15]. Recent work carried out by our research team [9,12,16] shows that it is possible to find a more or less balanced nutrient solution for horticultural crops based on modifications in the preparation process of the aqueous extract depending on the type of organic matter used for preparation. According to the results, it is possible to solubilize and/or mineralize the nutrients present in organic forms in the aqueous extracts during the production process [9] and use them as a nutrient solution for soilless crops in containers [8,12,16].

In hydroponic crops that do not use a substrate for root growth, continuous recirculation of the nutrient solution and strict control of the ion and O_2 concentrations in it is necessary. In hydroponic systems without a substrate, the use of organic extracts becomes more complex since it is difficult to guarantee that the nutrient solution will provide nutrients in an ionic and balanced form to allow the correct development of the crop. It is also difficult to ensure that there are high levels of O_2 in the solution, since aerobic bacteria and fungi consume most of the oxygen during their metabolic processes. Therefore, it is necessary to ensure that the supply of oxygen is greater to maintain aerobic conditions and avoid the appearance of substances such as valeric, butyric acid, and phenolic compounds, which can be harmful to crops and beneficial to microorganisms [10]. The aeration process of the aqueous extracts generates favorable conditions for the proliferation and activity of microorganisms that are naturally present in the organic matter and the mineralization of the organic matter [8].

To determine the effectiveness of the use of aqueous extracts as organic nutrient solutions in hydroponics and to analyze the response of a crop commonly used in soil-less cultivation, lettuce, a test was carried out in a hydroponic system (NFT—nutrient film technique). The NFT system was developed in the 1960s and was primarily intended for the production of high-quality vegetables in greenhouses [17]. The NFT cultivation system is a hydroponic technique that uses a thin film of nutrient solution to deliver nutrients to plants. A very shallow stream of water containing the dissolved nutrients is recirculated past the bare roots [18]. Figure 1 shows a scheme of the system. The nutrient solution is stored in a tank (C), from where it is recirculated to the crop, and the characteristics of the nutrient solution are controlled (EC, pH, ion concentrations, and O_2). The amount consumed by the crop is completed from another tank (A), where the concentrated nutrient solution is stored. In the tank (C), where the nutrient solution is used to irrigate the crop or in the stock tank (A), it is possible to incorporate a pump that allows the nutrient solution to be aerated, helping to maintain higher O_2 concentrations.

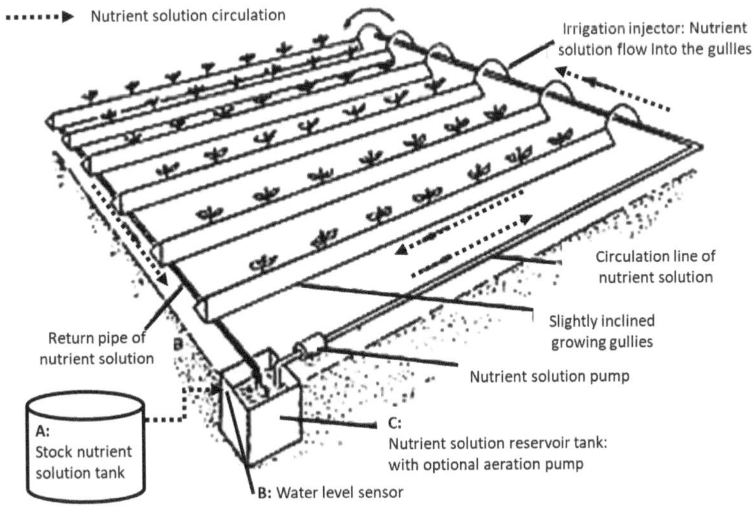

Figure 1. Nutrient film technique. Hydroponic system: The nutrient solution is pumped up to the top of the gullies. The solution passes down the gullies in a thin film and is recirculated to the reservoir tank (C) with an optional submersible pump to increase aeration. A: Stock nutrient solution tank used to refill the nutrient solution tank (C). C: Recirculation tank where it is possible to control the EC, pH, and ion and O_2 concentrations in the nutrient solution during crop growth (adapted from Cooper [17]).

Starting from the hypothesis that it is possible to achieve a sufficiently balanced nutrient solution from aqueous extracts according to the results of [9] and that improving aeration in the organic nutrient solution during cultivation will improve the concentration of nutrients in assimilable forms, it follows that production will not be affected.

The objective of this work was to evaluate the use of aerated or non-aerated aqueous extracts as recirculating nutrient solutions in a hydroponic culture (NFT) of lettuce. For this, it was compared with hydroponic cultivation with a conventional nutrient solution. This work evaluated the variation of nutrient content in the hydroponic cultivation system and the yield and quality responses of the lettuce crop.

2. Materials and Methods

2.1. Culture Establishment and Experiment Description

The experiment was established in a multi-tunnel polycarbonate greenhouse located at the University of Almería (Southeast Spain). The lettuce used was of the Trocadero type (cv. Rex-Rijk Zwaan) at a planting density of 30 plants m^{-2}. The plants were transplanted to the hydroponic system (NFT—nutrient film technique).

Three treatments were established depending on the nutrient solution used (Table 1).

Table 1. Chemical compositions of the nutrient solutions (initial nutrient solution and that used to refill that consumed by the crop) used in growing lettuce in a hydroponic–NFT system. Ion concentration (mmol L^{-1}), pH, EC (dS m^{-1}), and O_2 concentration (mg L^{-1}).

Treatment [1]	NO_3^-	$H_2PO_4^-$	Cl^-	NH_4^+	K^+	Ca^{2+}	Mg^{2+}	Na^+	pH	EC	O_2
T0	12	2.0	6.1	2.1	5.0	4.8	2.0	5.0	6.50	2.5	6.95
T1/T2	7.9	0.5	6.7	0.2	4.2	3.3	1.2	6.3	8.25	2.1	7.67

[1] T0, conventional nutrient solution; T1 aqueous extract with supplementary aeration in a recirculation tank; T2, aqueous extract without supplementary aeration in a recirculation tank.

Photographs of the experiment are included in Figure 2. Photograph (a) shows the materials used to make the aqueous extract, the vermicompost of vegetable waste from the horticultural crops, and (b) the 500 L tank used in the greenhouse for the preparation of the aerated aqueous extracts for 12–16 days to be used as an initial nutrient solution and filler for treatments T1 and T2. Additionally, there is a photograph of the NFT cultivation table (one per repetition and three tables per treatment) with the recirculation tank with a submersible pump inside to push the irrigation solution to the top of the cultivation channels (c). In (d), there is a photograph of the recirculation tank (100 L) with an extra submersible pump for supplementary aeration of the irrigation solution of treatment T1. Additionally, there are two photographs of the NFT cultivation table with three channels planted (seven plants per channel and three rows per table: 21 plants per table and repetition. In (e) are the seedlings, and in (f) are the plants used for harvest. The experiment used nine NFT growing tables, one growing table (21 plants per table) per repetition and three per treatment. The recirculation tank was topped up every one or two days and refreshed with nutrient solution to maintain the level at 100 L, and a sample was collected to monitor the EC, pH, and ion concentrations. The water consumption that occurs throughout the crop was replaced, but since it is a short cycle, it was not necessary to make a complete change.

Figure 2. (**a**) Vermicompost of vegetable waste of the horticultural crops used to prepare the aqueous extracts. (**b**) Greenhouse preparation of aerated aqueous extract for 12–16 days to later be used as an initial nutrient solution and to fill the T1 and T2 recirculation tanks. (**c**) NFT cultivation tables with the details of the recirculation tank with a submersible pump inside to irrigate the solution to the top of the channels. (**d**) Details of the recirculation tank with an extra submersible pump for supplementary aeration of the irrigation solution (T1). (**e**) Growing table with three growing channels with seven plants each (3 × 7: 21 plants per table) and detail of the seedlings. (**f**) Plants in the canals for harvest.

The harvest of lettuce was carried out when the plants reached commercial size: this occurred at 35 days from transplantation (DAT) in T0, while in T1 and T2, harvest occurred at 50 DAT.

2.2. Characterization of the Nutrient Solution

For the treatment used as a control (T0), a mineral nutrient solution [19] was prepared with the following fertilizers: $CaNO_3$, NPK with a 15–5–30 balance, and chelated microelements (microstep®® complex). The pH was adjusted in the tank to 6.5 with HNO_3. This same solution was used as a solution to fill the tank during the crop growth.

For the preparation of the aqueous extract used as a nutrient solution, we proceeded according to the method published by [16]. According to [16], vermicompost from horticultural waste was dissolved in water until reaching an electrical conductivity (EC) of 2.0 dS m^{-1}. Then, it was aerated for 12–16 days before being used as a nutrient solution in T1 and T2 (b; Figure 2). It was aerated by recirculation with a submersible pump with a flow rate of 6500 L h^{-1} in a 500 L tank. For T1 and T2, no acid was applied to the nutrient solution, so the pH remained at alkaline values. This same solution was used as a solution to fill the tank during cultivation.

Table 1 includes the initial chemical concentrations (ion composition), pH, EC, and O_2 concentrations of the nutrient solutions used for the hydroponic cultivation of lettuce. These nutrient solutions were the nutritional source throughout the production cycle. It is also the composition of the nutrient solution used to fill the tank and supply the plants' consumption. This operation was carried out every 1 or 2 days, not allowing it to go below 50 L. The water consumption that occurs throughout the crop is replaced, but since it is a short cycle, it is not necessary to make a complete change. The water consumption of each treatment is different, and therefore, the changes in concentration in the irrigation solution according to treatments are included.

To characterize the chemical compositions of the aqueous extracts (T1 and T2) and the conventional solution (T0), the pH and ionic concentration (NO_3^-, NH_4^+, K^+, Ca^{2+}, Mg^{2+}, Cl^-, and Na^+) were analyzed using the selective electrode potentiometry technique (Imacimus-NT Sensor, Spain). The concentration of phosphates (P-$H_2PO_4^-$) was determined by the spectrophotometric analysis method based on the colorimetry of molybdenum blue obtained for P extraction in an aqueous medium [20] modified by [21]. The oxygen concentration in the nutrient solutions was determined using a portable Crison model Oxi45+ oximeter. The EC was determined with a Tec-Hu TE52 conductivity meter.

We used vermicompost of compost made from vegetable waste from horticultural crops from greenhouses in SE Spain whose average characteristics, according to the manufacturer (Tecomsa S.L. (Almería-Spain)), are as follows: organic matter: 32%; N: 2%; humic acids: 7.40%; fulvic acids: 11.10%; humic extract: 17.5%; Phosphoric anhydride: 1.30%; humidity: 39%; pH: 8; organic carbon: 18.6%; granulometry: 90% > 25 mm.

2.3. Nutritional Characteristics of the Crop: Analysis of the Cell Petiole Extract of or Sap

With a press, the juice from the petioles of the leaves (cell petiole extract—ECP) was obtained. By analyzing the ionic composition of the ECP, we can determine the nutritional status of the crop [22]. To conduct the ionic analysis of the ECP, a random and representative sample of fully developed young leaves was obtained. The petiole was separated and crushed with a press to obtain the greatest amount of extract and was analyzed by the potentiometry method with an ion-selective electrode (Imacimus, NTsensor, Spain), obtaining the concentrations of the ions NO_3-, NH_4^+, K^+, Ca^{2+}, Mg^{2+}, Cl^-, and Na^+. The phosphate content (P-$H_2PO_4^-$) was analyzed by the same colorimetry method as that carried out for the nutrient solution [20], modified by [21].

2.4. Yield and Leaf Nitrate Content

As a determining parameter of the quality of leafy vegetables, the nitrate content was determined once harvested. The nitrate content in leaves was analyzed with the method described by [23]. The complex formed by the nitration of salicylic acid under highly acidic conditions absorbs maximally at 410 nm in basic (pH > 12) solutions. The absorbance of the chromophore is directly proportional to the amount of nitrate, N, present.

Another important factor is the percentage of dry matter in the leaves and roots, which is why the aerial and root parts of a representative sample of each treatment group were weighed (three plants per repetition). Subsequently, the samples were placed in a forced-air oven at a temperature of 600 °C for 72 h to eliminate the water content present in the plant tissues and determine the percentage of dry matter.

The average fruit weight was also quantified using a sample of 12 lettuces (three per repetition) when the plants reached commercial size. They were weighed with an analytical scale with a precision of ±0.01 g. With the average weight of the commercial fruit (W) and according to the density of the plants (d), the commercial production was calculated (kg m^{-2}) (W (kg $fruit^{-1}$) × d (plant m^{-2})), based on the average weight of the biomass of the samples.

2.5. Statistical Analysis of the Results

The procedure chosen for the analysis of the evaluated parameters was the analysis of variance (ANOVA) at a confidence level of 95.0%. This was followed by multiple range tests to determine the degree of significance using Fisher's least significant difference (LSD) at $p \leq 0.05$.

3. Results and Discussion

3.1. Nutrient Solution for Lettuce Cultivation in an NFT System

Periodic analyses of the nutrient solution were carried out to evaluate the variation in the concentrations of nutrient ions during crop development. Table 2 shows the results obtained in each of the treatments structured over three fortnights with the average data for each period.

The highest concentrations of N-NO_3^- in the nutrient solution were achieved in the treatment with mineral fertilization (T0) throughout the crop (Table 2) with values close to 10 mmol L^{-1}, followed by the organic treatment with greater aeration (T1) with statistically significant differences. Among the organic treatments, T1 stands out, as it had a statistically significant mean concentration of N-NO_3^- that was greater than 27% compared to T2 (Table 2). The determining factor for the concentration of nitrates between organic treatments (T1 and T2) was the aeration system of the nutrient solution present in T1. Aeration generates a higher concentration of oxygen dissolved in the solution, helping to accelerate microbial activity for the mineralization of organic nitrogen to forms that are assimilable by the plant (nitrate) [9,10]. It also responds to the increase in aerobic fermentation that favors increases in the microbial populations present in vermicompost teas, allowing greater mineralization of organic matter [24]. Furthermore, the agitation that accompanies aeration leads to the increased solubility of minerals in water [25].

The higher concentration of N-NO_3^- in the organic solution is also justified by the fixation of atmospheric N_2 that passes into the ammoniacal form and is subsequently converted to NO_2^- and finally to NO_3^- by the actions of nitrosomonas and nitrobacter bacteria [26]. Previously conducted studies state that the aqueous extract of vermicompost tea contains high populations of N-fixing bacteria that reach populations of 12 CFU mL^{-1} (log10) after 12 days of aeration [9].

Goto et al. [27] indicated that at an O_2 concentration of 2 mg L^{-1}, the growth rate of nitrifying bacteria significantly decreases, therefore indicating that the minimum need for dissolved O_2 to promote nitrification processes is 4.5 mg L^{-1}. In our experiment, all three treatments maintained oxygen concentrations above 6 mg L^{-1}. T1 stands out by showing significant statistical differences (LSD 95%) with concentrations greater than 7 mg L^{-1}, and it is evident that aeration favors the mineralization of extracts made with organic materials.

The concentration of N-NH_4^+ was very similar during fortnights 1 and 2 (Table 2), without significant differences between the treatment groups. However, in the third fortnight, the concentrations in T1 and T2 decreased, both having significantly different (LSD 95%) ammonium concentrations to T0. The evolution of the data shows a decrease in the concentration of N-NH_4^+ over time, which is possibly an indicator of bacterial activity in the transformation of N-NH_4^+ to N-NO_3^- [24]. As indicated previously, the higher concentration of O_2 in T1 favors increases in the microbial populations present in the vermicompost tea, allowing greater availability of oxygen for microbial reactions for the conversion of ammonium to nitrate [24], decreasing the concentration of ammonium.

Table 2. Mean values and SDs (standard deviations) for the chemical composition (mmol L^{-1}), pH, EC (dS m^{-1}), and O_2 concentration (mg L^{-1}) of the nutrient solution analyzed by fortnight from transplant as functions of the treatments evaluated in a lettuce hydroponic crop. T0, conventional nutrient solution; T1, aqueous extract with supplementary aeration in a recirculation tank; T2, aqueous extract without supplementary aeration in a recirculation tank.

Fortnight		T0	T1	T2
1	NO_3^-	11.25 ± 1.06 a	7.10 ± 0.21 b	6.05 ± 0.49 c
	$H_2PO_4^-$	2.60 ± 0.02 a	0.41 ± 0.91 b	0.21 ± 0.93 c
	Cl^-	10.85 ± 1.63 a	8.35 ± 2.19 a	7.15 ± 2.19 a
	NH_4^+	0.25 ± 0.21 a	0.20 ± 0.14 a	0.20 ± 0.14 a
	K^+	3.95 ± 1.48 a	3.15 ± 0.07 a	2.70 ± 0.28 a
	Ca^{2+}	4.25 ± 0.07 a	3.20 ± 0.14 a	3.35 ± 0.78 a
	Mg^{2+}	2.00 ± 0.57 a	1.20 ± 0.14 b	1.20 ± 0.14 b
	Na^+	9.55 ± 0.64 a	8.95 ± 0.64 a	7.10 ± 0.14 b
	pH	7.00 ± 0.47 b	8.54 ± 0.05 a	8.23 ± 0.16 a
	EC	2.63 ± 0.28 a	2.11 ± 0.14 b	1.91 ± 0.65 b
	O_2	6.89 ± 0.14 b	7.62 ± 0.22 a	6.91 ± 0.13 b
2	NO_3^-	11.03 ± 2.10 a	7.25 ± 0.37 b	5.38 ± 0.41 c
	$H_2PO_4^-$	2.20 ± 0.04 a	0.38 ± 0.05 b	0.36 ± 0.06 b
	Cl^-	9.83 ± 1.90 a	9.85 ± 1.98 a	10.28 ± 1.66 a
	NH_4^+	0.43 ± 0.10 a	0.28 ± 0.15 a	0.30 ± 0.13 a
	K^+	4.50 ± 0.90 a	4.13 ± 0.99 a	4.00 ± 0.72 a
	Ca^{2+}	4.45 ± 2.40 a	3.82 ± 0.70 a	4.05 ± 0.68 a
	Mg^{2+}	1.60 ± 0.30 a	1.10 ± 0.30 a	1.20 ± 0.70 a
	Na^+	11.30 ± 2.34 a	9.35 ± 1.47 a	8.75 ± 4.37 a
	pH	6.87 ± 0.10 b	8.41 ± 0.30 a	8.26 ± 0.20 a
	EC	2.83 ± 0.15 a	2.36 ± 0.05 b	2.35 ± 0.16 b
	O_2	6.72 ± 0.14 b	7.36 ± 0.22 a	6.73 ± 0.13 b
3	NO_3^-	9.80 ± 1.27 a	6.92 ± 0.27 b	5.54 ± 0.56 c
	$H_2PO_4^-$	2.10 ± 0.06 a	0.38 ± 0.12 b	0.32 ± 0.03 b
	Cl^-	11.50 ± 3.54 a	11.88 ± 4.22 a	10.92 ± 4.05 a
	NH_4^+	0.50 ± 0.14 a	0.20 ± 0.09 b	0.20 ± 0.07 b
	K^+	5.05 ± 0.92 a	3.80 ± 0.93 ab	3.08 ± 1.02 b
	Ca^{2+}	3.85 ± 0.49 a	4.42 ± 1.15 a	5.04 ± 1.20 a
	Mg^{2+}	1.80 ± 0.14 a	1.98 ± 0.72 a	1.72 ± 0.73 a
	Na^+	9.00 ± 0.14 a	12.40 ± 3.81 a	10.90 ± 3.47 a
	pH	7.12 ± 0.06 b	8.30 ± 0.20 a	8.18 ± 0.20 a
	EC	2.59 ± 0.27 b	2.93 ± 0.05 a	2.85 ± 0.05 b
	O_2	6.60 ± 0.14 b	7.18 ± 0.22 a	6.67 ± 0.13 b

Different letters express significant differences (LSD ≤ 95%) between treatments.

Phosphorus is the second key element after nitrogen as a nutrient in qualitative terms for plants. Regarding the concentration of P-$H_2PO_4^-$ in the nutrient solution, there were highly significant differences between the mineral treatment group (T0) and the organic treatment groups (T1 and T2). As expected, T0 maintained an adequate concentration for lettuce cultivation. In the organic treatment groups (T1 and T2), the concentration of P-$H_2PO_4^-$ was deficient with respect to optimal nutritional requirements [19]. According to

Ruiz and Salas [9], the P present in the extract is in organic form, while some of the mineral forms may be precipitated with other cations, making them inaccessible and leaving a very deficient level for a nutrient solution. The high pH (>8) of the organic solution does not facilitate the solubilization of P-$H_2PO_4^-$. It is advisable, during specific moments, to lower the pH of the aqueous extract with acids authorized for use in organic agriculture, such as acetic acid. According to the results, it is necessary to increase the concentration of this element in the nutrient solution by adding P from other sources, such as mineral rock and/or the inoculation of phosphorus-solubilizing microorganisms [16,28].

In the initial stage of cultivation, there were no significant differences in the concentration of K^+ in the nutrient solutions between treatment groups. In all treatment groups, the nutrient solution had a concentration of 65% of that recommended by Valverde [29] for lettuce in a hydroponic system. During fortnights 1 and 2, the concentration of K^+ in the nutrient solution increased in all treatment groups. However, in the third fortnight, it decreased, possibly due to an increase in the crop's K^+ needs in this last stage. K^+ is a fundamental element in osmotic adjustment for cellular turgor [30]. During the final stage of cultivation, according to the results presented in this work, in the organic nutrient solution, it is recommended to control and, if necessary, increase the concentration of K^+ to avoid possible deficiencies.

In the third fortnight, in the aerated organic solution, the concentrations of nitrate and potassium increased significantly compared to the organic solution without aeration. The results of this experiment agree with those of Pant et al. [14], who found higher concentrations of nitrate and potassium when aeration was maintained during the extraction process compared to non-aerated compost extracts.

In terms of the concentration of Ca^{2+}, there were no significant differences between treatment groups in the three samplings carried out during the crop cycle. An increase in the Ca^{2+} concentration was observed over time in all treatment groups. This increase was generated because Ca is an element that is slowly assimilated by the plant [31], and consequently, its accumulation occurs in the tank of the recirculated nutrient solution.

Throughout the entire crop cycle, the concentration of Mg^{2+} in T0 is within acceptable values, while in the organic treatment groups (T1 and T2), it does not reach the recommended concentration for lettuce until the third fortnight. The increase is achieved by accumulation in the nutrient solution tank due to the slow absorption of Mg^{2+} due to competition with other cations [32], as occurs with Ca.

The concentrations of Na^+ and Cl^- did not register statistically significant differences between treatment groups during the crop cycle, except for the concentration of Na^+ in the first fortnight, which showed significantly lower values in T2 compared to T0 and T1 (Table 2). The greatest increase in the concentration of Na^+ in T1 was due to the effect of aeration, which increases the solubilization of elements such as Na^+ [24]. Control of the sodium concentration is important, since sodium is a non-essential element for plants that frequently causes stress or toxicity situations in crops when it is present in high concentrations and limits the absorption of K^+ [33].

The lower pH in the mineral treatment group (T0) was due to its control using HNO_3, as indicated in the Materials and Methods Section, a process that was not carried out in the organic treatment groups (T1 and T2). Organic treatments produce alkaline pH values. High pH values are largely associated with the mineralization processes of organic matter, which are favored in turn by the availability of O_2 [26].

The effect of using a high EC in the fertigation solution was an osmotic imbalance (dehydration), which hindered the flow of water and nutrients in the plant [34]. Consequently, for the preparation of aqueous extracts, conditioning the amount of vermicompost per volume of water to the EC limit values ensures that elements such as Na+ do not reach excessive concentrations. It is also assumed that it limits the concentrations of other elements that are as important as P. The concentrations of ions during the preparation of liquid solutions depend directly on the EC, which is determined by the types and amounts of solid materials dissolved in the water.

The content of dissolved oxygen in the nutrient solution was higher in T1 with more statistically significant differences (LSD 95%) than the rest of the treatments as a consequence of the aeration system included in this treatment. Our results agree with the data obtained by St Martín and Brathwaite [24], who mentioned that the solubility of nutrients in the composts, as well as the different rates of consumption and release of nutrients by microorganisms present in the compost teas, appear to be the main factors that affect the concentrations of nutrients across the brewing time.

3.2. Nutritional Status of the Crop: Petiole Cellular Extract Analysis (ECP)

During cultivation, the nutritional status of the crop was assessed through the rapid analysis of petiole cellular extract or plant sap analysis, as recommended by Cadahía [22], with the aim of assessing the effect of an unbalanced nutrient solution. To do this, we compared the concentrations of the main ions in the sap of lettuce grown with organic solutions (T1 and T2) and in the T0 treatment, which was considered to be a control.

The maximum concentrations of N-NO_3^- in the petiole cell extract (Table 3) were obtained in the T0 (mineral) and T1 (organic + aeration) plants without statistical differences between them, which shows the adequate availability and good assimilation of this element by cultivation. This result agrees with the analysis of the nutrient solution where the highest concentrations of N-NO_3^- were found in T0 and T1.

Table 3. Plant sap analysis: Ion concentration (mmol L^{-1}) according to the treatments evaluated in a lettuce hydroponic crop. T0, conventional nutrient solution; T1, aqueous extract with supplementary aeration in a recirculation tank; T2, aqueous extract without supplementary aeration in a recirculation tank.

	NO_3^-	$H_2PO_4^-$	Cl^-	K^+	Ca^{2+}	Mg^{2+}	Na^+
T0	59.0 ± 3.5 a	12.1 ± 3.0 a	49.3 ± 3.5 b	76.0 ± 5.0 a	1.4 ± 0.1 a	1.8 ± 0.4 a	29.0 ± 5.0 a
T1	53.6 ± 0.5 b	7.9 ± 1.5 b	69.0 ± 3.0 a	71.3 ± 2.0 a	0.5 ± 0.0 c	0.5 ± 0.0 b	26.6 ± 3.7 a
T2	50.0 ± 2.6 b	8.5 ± 1.1 b	64.0 ± 5.0 a	71.0 ± 3.6 a	0.7 ± 0.0 b	0.4 ± 0.0 b	18.6 ± 1.5 b

Values with the same letters within each column are statistically similar (LSD ≤ 95%).

The highest concentrations of P-$H_2PO_4^-$ were obtained in the mineral treatment group (T0) due to the higher concentration of this nutrient in the nutrient solution (T0). In the organic treatment groups (T1 and T2), the concentrations were significantly lower than in T0, without significant differences (LSD 95%) between them, coinciding with the already-mentioned low concentrations of P-$H_2PO_4^-$ in the organic nutrient solutions.

In terms of the concentration of K^+ in ECP, no statistically significant differences were observed between the treatment groups, which indicates that the different nutrient solutions and/or aeration treatments had no influence on this element. The behavior of K^+ concentrations in sap coincides with that of the concentration of this element in the nutrient solution.

The concentration of Ca^{2+} in sap presented significant statistical differences (LSD 95%) between the organic treatment groups (T1 and T2) and the mineral group (T0). It is important to mention that the Ca^{2+} content in the nutrient solutions did not show significant differences between the organic and mineral treatment groups, which is why the result obtained in the ECP differs from what was expected. The treatment with mineral fertilization (T0) led to the highest concentration of Ca^{2+} in sap, while the organic treatment groups presented considerably lower concentrations of Ca^{2+}. The low absorption of Ca^{2+} in the organic treatment groups may be a consequence of the low concentration of P-$H_2PO_4^-$ in the nutrient solution. According to Jakobsen [35], the Ca^{2+} assimilation capacity of plants is limited by the phosphate content, and the Ca^{2+} absorption capacity increases depending on the phosphate concentration. The concentration of P in the T1 and T2 solutions was limited, with concentrations lower than the needs of the crop, a limiting factor in the absorption of Ca^{2+}. This would also lead to limited Ca^{2+} availability due to the alkaline pH in the nutrient solution.

The highest concentration of Mg^{2+} in the petiole was obtained in the mineral treatment group (T0), which showed statistically significant differences compared with the organic treatment groups (T1 and T2). The lower concentration of Mg^{2+} in the organic treatment groups could be caused by the low concentration of Mg^{2+} in the first fortnight and/or the high concentration of Na^+ in the last fortnight in the nutrient solution, which may have had a direct effect on the absorption of Mg^{2+} [32].

The Cl^- content in sap was higher in the two organic treatment groups, with an average concentration of 66.5 mmol L^{-1}, while the culture with inorganic nutrition accumulated the lowest amount of Cl^- in the ECP, with an average value of 49.3 mmol L^{-1}. The competition, described by Cadahía [36], between the absorption of Cl^- and NO_3^- would also justify the lower petiole concentration of NO_3^- in the organic treatment groups.

The highest concentrations of Na^+ in the plant sap analysis were found in the crops irrigated with mineral solution (T0) and aerated organic solution (T1), significantly higher than those in the plants in treatment group T2.

3.3. Generation of the Aerial Part and Root Part Biomasses

The greatest fresh weight of the aerial part of the plant, without statistical differences, was obtained in the organic treatment group with aeration (T1) with an average of 170 g $plant^{-1}$ (Table 4). The increase in the total biomass production (aerial and root parts) in T1 is justified by the higher concentration of O_2 in the nutrient solution (Table 2) with respect to T2 and the more balanced nutrient solution. The high concentrations of O_2 in the nutrient solution prepared from aqueous extract (T1) allow for the maintenance of higher NO_3 concentrations than in T2. Regarding the root biomass (fresh weight), significantly greater fresh root weights found in the organic treatment groups (T1 and T2) also show that with lower nutrient concentrations in the irrigation solution, equal levels of weight could be obtained in the aerial part.

Table 4. Fresh weights and dry weights of the aerial and root parts of lettuce plants according to the treatments evaluated in a lettuce hydroponic crop. T0, conventional nutrient solution; T1, aqueous extract with supplementary aeration in a recirculation tank; T2, aqueous extract without supplementary aeration in a recirculation tank.

	Fresh Weight (g Plant^{-1})		**Dry Matter (%)**	
	Aerial Part	**Root**	**Aerial Part**	**Root**
T0	159.10 ± 28.44 a	26.67 ± 2.25 b	11.83 ± 1.54 a	25.97 ± 2.00 a
T1	170.61 ± 16.72 a	53.81 ± 9.20 a	6.84 ± 0.63 b	14.10 ± 1.17 b
T2	154.98 ± 12.85 a	42.97 ± 6.88 a	7.11 ± 0.74 b	15.95 ± 1.86 b

Values with the same letters within each column are statistically similar (LSD $\leq$ 95%).

The values obtained show that treatment with inorganic nutrition led to the highest percentages of dry matter in the aerial and root parts (Table 4), with statistically significant differences. The results do not coincide with those obtained by Masarirambi et al. [37], who mentioned that treatment with mineral nutrition leads to a lower percentage of dry matter compared to organic fertilization in a lettuce crop. The difference between both manuscripts may be caused by the type of organic matter used as amendment and its richness in nutrients. In our experiment, the aqueous extracts (T1 and T2) present low concentrations of P (Table 2), which justifies that the dry matter of the organic treatments is lower, even with aeration, as it does not reach the recommended values for lettuce (T0). The dry weight of the root and aerial part of the T0 treatment are significantly greater than those quantified in the organic treatments (T1 and T2). This increase, more than 4% and 10% in the dry matter of the aerial part and root, respectively, may be caused by the increase in the assimilation, translocation, and utilization of phosphorus as a consequence of the higher concentration of H_2PO_4 (Table 2) in the T0 (P concentration in T0, is six times higher than in T1 and T2), which is corroborated by the highest concentration of P in sap (Table 3). Phosphorus is essential for the processes of photosynthesis and respiration; in this regard, [38] Russo and Pappelis (1995) report that phosphorus promotes greater elongation and production of root hairs that lead to an increase in the dry weight of the plant.

3.4. Yield and Leaf Nitrate Content

No statistically significant differences were observed during production (kg m^{-2}) between the mineral treatment group (T0) and the organic treatment groups (T1 and T2), coinciding with the average weights of the plants (Figure 3). The weight of the harvested pieces (aerial part) was not affected by lower concentrations of some nutrients in the nutrient solution or in the plant sap analysis. The organic treatment group T1 was the most productive, reaching an average yield of 5 kg m^{-2} according to the plantation density. The results agree with the data reported by Masarirambi et al. [37], who mentioned that a higher yield is obtained with the application of organic fertilizers compared to conventional management in lettuce cultivation. They also coincide with the results of González et al. [39] and Xu et al. [40], who showed that differences between vegetables depend on whether they were grown with fertilizers from organic sources or through inorganic fertilization and the greatest quantity and quality are obtained in organic crops. Martínez et al. [41] showed that percentages of dissolved oxygen greater than 90% in nutrient solution generate the greatest increases in the production of lettuce crops. This coincides with the result showing that T1 had an average value of 7.39 mg L^{-1}, which is equivalent to 92.47% dissolved oxygen. T1, throughout the culture, has a significantly higher concentration than T0 and T2.

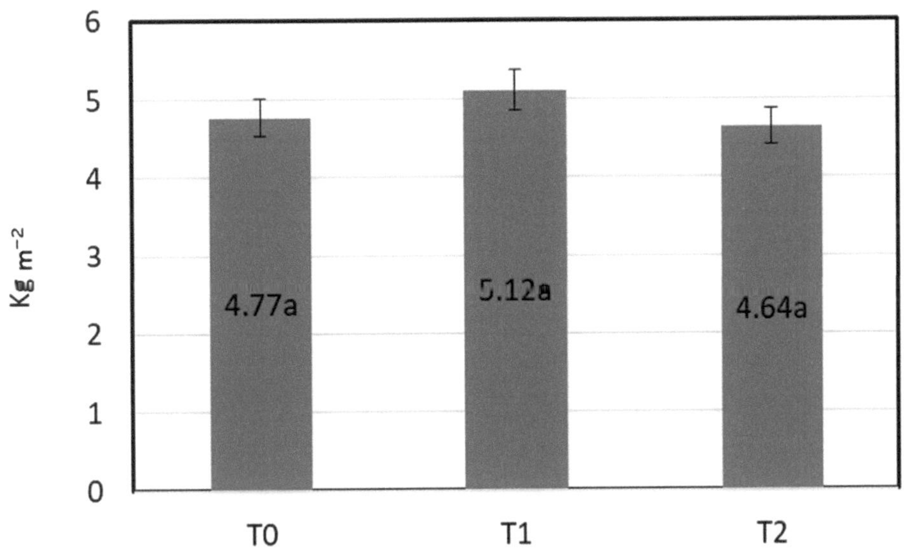

Figure 3. Yields of lettuce crops according to the treatment groups: T0, conventional nutrient solution; T1, aqueous extract with supplementary aeration in a recirculation tank; T2, aqueous extract without supplementary aeration in a recirculation tank. Numerical values followed by different letters indicate statistically significant differences (LSD $\leq$ 95%).

The main difference between the mineral and organic treatments lies in the time necessary to reach the harvest size. The crop cycle increased by 15 days in the T1 and T2 treatment groups compared to T0. The growing cycle increased the number of days to harvest by 40%.

3.5. Nitrate Content in Leaves

The nitrate content in the leaves of edible vegetables is one of the main factors that drive the production of organic vegetables. Minimizing the N-NO_3^- content in edible leaves contributes to an improvement in the quality of food in relation to consumer health In turn, the correct management of organic fertilizers contributes to a reduction in the environmental impact of inorganic fertilizers. In 2011, the European Commission [42] established the maximum permitted content of nitrates in vegetables for their marketing and consumption with the aim of protecting consumers from potential toxicological risks due to the consumption of foods with high nitrate contents.

The greatest accumulation of nitrates in leaves was found in lettuce plants grown with inorganic fertilization (T0), with values close to 3000 mg NO_3^- kg^{-1} m.f. (Table 5). This is below the limit established by the European Commission [42] (relative to the maximum nitrate content in lettuce produced in a greenhouse between April 1 and September 30).

Table 5. Nitrate content in lettuce leaves.

Treatment	mg NO_3 kg^{-1} f.m. [1]
T0	2946 ± 47 a
T1	557 ± 43 b
T2	509 ± 33 b

Values ± standard deviations. T0, conventional nutrient solution; T1, aqueous extract with supplementary aeration in a recirculation tank; T2, aqueous extract without supplementary aeration in a recirculation tank. Different letters express significant differences (LSD $\leq$ 95%). [1] f.m. Fresh matter.

The accumulation of nitrates in leaves treated with organic nutrient solution (T1 and T2) was notably lower than in the control group (T0), with values of around 500 mg NO_3^-·Kg^{-1} m.f. observed, six times lower than those produced with the mineral treatment. There were no significant statistical differences in the leaf nitrate content between plants treated with organic fertilizer; however, as with the compositions of nutrient solutions, the treatment group with greater aeration (T1) had the highest concentration of N-NO_3^- in the leaves. This was directly related to the ionic composition of the vermicompost tea used as a nutrient solution.

The concentration coincides with the values obtained by Pavlou et al. [7] in lettuce cultivated with organic fertilization. Magkos et al. [43] mentioned a significant difference between the nitrate contents in the organic and traditional agriculture systems, mainly regarding the cultivation of leafy vegetables, finding the accumulation of nitrates to be three times lower in organic systems compared to that in traditional production systems.

A lower nitrate concentration adds value to the quality of the product. Mineral fertilizer sources with high solubility allow easy availability of nutrients in assimilable form, sometimes leading to plants for luxury consumption.

The slow mineralization of organic N in organic treatments induces the progressive absorption of N by plants. This slow absorption provides the time necessary for the metabolic assimilation processes to be carried out in the plant tissues to form structures, so the accumulation of nitrates is lower [44]. However, the time necessary for the plants to reach the size considered to be appropriate for commercial harvest also increases, going from 35 days from transplant in the mineral treatment group to 50 days in the organic treatment group.

The lower leaf nitrate contents in the organic treatment groups are also justified by the relationship of nitrate with the dry weights of the harvested plants since the plants from treatment groups T1 and T2 had aerial part dry weights that were almost two times lower (Table 4) than those of the mineral treatment plants (T0).

According to the results of this work, it is possible to work towards the replacement of mineral nutrient solutions through hydroponics or soil-less cultivation techniques, either totally or partially, with solutions generated from aqueous extracts of local plant remains, a process known as biofertigation [9]. The reintegration of agro-waste into the same agriculture site fulfills the objective of the European Bio-Economy Strategy 2018–2030 [5] and reduces transport costs, waste volume, and the need for mineral fertilizers [45,46]. Many alternatives to the disposal of agro-waste have been evaluated, with the use of organic matter as a source of crop nutrients being the most competitive solution in an area in which intensive agriculture activity is prevalent [6,45,47].

4. Conclusions

The aqueous extract vermicompost tea from horticultural plants is used as a nutrient solution in a hydroponic system, providing concentrations of NO_3^-, K^+, and Ca^{2+} that are sufficient for the development of the lettuce crop. However, the $H_2PO_4^-$ and Mg^{2+} concentrations in the aqueous extract are lower than the needs of the culture, as corroborated by the analyses in sap.

For the preparation of the aqueous extract, conditioning the amount of vermicompost per volume of water to the EC limit values ensures that elements such as Na^+ do not reach excessive concentrations. It is assumed that this preparation methodology limits the concentrations of other elements that are as important as P.

The high concentration of dissolved O_2 (>7 mg L^{-1}) in the organic nutrient solution achieved through aeration throughout the crop cycle allows a commercial production level similar to that obtained with treatment with mineral fertilizers with greater generation of root biomass. It allows adequate nutritional levels in sap for the crop to be obtained.

There was an improvement in the quality of the harvest of plants fertigated with organic nutrient solution, with a leaf N-NO_3^- content that was six times lower than that in the leaves produced with the traditional system involving irrigation with synthetic fertilizer. However, the duration of the growing cycle until harvest increased when an organic nutrient solution was used.

The vermicompost tea prepared from horticultural vegetable waste could be an alternative to mineral nutrient solutions for the production of short-cycle hydroponic crops such as lettuce.

Author Contributions: Conceptualization, M.C.S.-S. and J.L.R.-Z.; methodology, M.C.S.-S. and J.L.R.-Z.; validation, M.C.S.-S.; formal analysis, M.C.S.-S., J.L.R.-Z., J.L.V. and A.X.C.; investigation, M.C.S.-S., J.L.R.-Z., J.L.V. and A.X.C.; resources, M.C.S.-S., J.L.R.-Z., J.L.V. and A.X.C.; data curation, J.L.R.-Z.; writing—original draft preparation, M.C.S.-S. and J.L.R.-Z.; writing—review and editing, M.C.S.-S., J.L.R.-Z., J.L.V. and A.X.C.; supervision, M.C.S.-S. and A.X.C.; project administration, M.C.S.-S., J.L.V. and A.X.C.; funding acquisition, M.C.S.-S. All authors have read and agreed to the published version of the manuscript.

Funding: This research received no external funding.

Data Availability Statement: The data presented in this study are available on request from the corresponding author. The data are not publicly available due to the inclusion of information that could compromise the privacy of research participants.

Conflicts of Interest: The authors declare no conflict of interest.

References

1. Preciado, P.; Fortis, H.M.; García-Hernández, J.L.; Rueda, E.O.; Esparza, J.R.; Lara, A.; Segura, M.A.; Orozco, J.A. Evaluación de soluciones nutritivas orgánicas en la producción de tomate en invernadero. *Interciencia* **2011**, *36*, 689–693.
2. Shinohara, M.; Aoyama, C.; Fujiwara, K.; Watanabe, A.; Ohmori, H.; Uehara, Y.; Takano, M. Microbial mineralization of organic nitrogen into nitrate to allow the use of organic fertilizer in hydroponics. *Soil Sci. Plant Nutr.* **2011**, *57*, 190–203. [CrossRef]
3. Pardossi, A.; Incrocci, L.; Salas, M.C.; Gianquinto, G. Managing Mineral Nutrition in Soilless Culture. In *Rooftop Urban Agriculture*; Orsini, F., Dubbeling, M., de Zeeuw, H., Gianquinto, G., Eds.; Springer: Cham, Switzerland, 2017.
4. Jensen, M.H. Principales sistemas hidropónicos: Principios, ventajas y desventajas. En: Hidroponía comercial. In Proceedings of the Conferencia Internacional, 6–8 August 1997; Universidad Agraria La Molina y Hidroponic Society of America: Lima, Perú, 1997; pp. 35–48.
5. Reglamento (UE) 2019/1009 del Parlamento Europeo y del Consejo de 5 de junio de 2019, por el que se establecen disposiciones relativas a la puesta a disposición en el mercado de los productos fertilizantes UE. Available online: https://www.boe.es/doue/2019/170/L00001-00114.pdf (accessed on 22 November 2023).

6. Carricondo Martínez, I.; Berti, F.; Salas-Sanjuán, M.d.C. Different Organic Fertilisation Systems Modify Tomato Quality: An Opportunity for Circular Fertilisation in Intensive Horticulture. *Agronomy* **2022**, *12*, 174. [CrossRef]
7. Pavlou, C.G.; Constantinos, D.; Ehaliotis, V.; Kavvadias, A. Effect of organic and inorganic fertilizers applied during successive crop seasons on growth and nitrate accumulation in lettuce. *Sci. Hortic.* **2007**, *111*, 319–325. [CrossRef]
8. Mejia, P.A.; Salas, M.C.; López, M.J. Evaluation of physicochemical properties and enzymatic activity of organic substrates during four crop cycles in soilless containers. *Food Sci. Nutr.* **2018**, *6*, 2066–2078. [CrossRef]
9. Ruiz, J.; Salas, M.C. The use of plant growth promoting bacteria for biofertigation; effects on concentrations of nutrients in the inoculated aqueous vermicompost extract and on the yield and quality of tomatoes. *Biol. Agric. Hortic.* **2022**, *38*, 145–161. [CrossRef]
10. Ingham, E.R. The Compost Tea Brewing Manual: Latest Methods and Research. *Soil Food Web* **2005**, 13–86.
11. Atiyeh, R.M.; Subler, S.; Edwards, C.A.; Bachman, G.; Metzger, J.D.; Shuster, W. Effects of vermicompost and composts on plant growth in horticultural container media and soil. *Pedo Biol.* **2000**, *44*, 579–590. [CrossRef]
12. Mejía, P.A.; Ruíz-Zubiate, J.L.; Correa-Bustos, A.; López-López, M.J.; Salas-Sanjuán, M.d.C. Effects of Vermicompost Substrates and Coconut Fibers Used against the Background of Various Biofertilizers on the Yields of *Cucumis melo* L. and *Solanum lycopersicum* L. *Horticulturae* **2022**, *8*, 445. [CrossRef]
13. Hargreaves, J.C.; Adl, M.S.; Warman, P.R. Are compost teas an effective nutrient amendment in the cultivation of strawberries? Soil and plant tissue effects. *J. Sci. Food Agric.* **2009**, *89*, 390–397. [CrossRef]
14. Pant, A.P.; Radovich, T.J.K.; Hue, N.V.; Paull, R.E. Biochemical properties of compost tea associated with compost quality and effects on pak choi growth. *Sci. Hortic.* **2012**, *148*, 138–146. [CrossRef]
15. González, S.K.; Rodríguez, M.M.; Trejo, T.L.; Sánchez, E.J.; García, C.J. Propiedades químicas del té de vermicompost. *Cienc. Agrícolas* **2013**, *5*, 901–911.
16. Ruiz, J.; Salas, M.C. Evaluation of organic substrates and microorganisms as bio-fertilisation tool in container crop production. *Agronomy* **2019**, *9*, 705. [CrossRef]
17. Cooper, A. *The ABC of NFT*; Casper Publications Pty Ltd.: Narrabeen, Australia, 1996; pp. 145–196.
18. Williams, K.A.; Nelson, J.S. Challenges of using organic fertilizers in hydroponic production systems. *Acta Hortic.* **2016**, *1112*, 365–370. [CrossRef]
19. Steiner, A.A. A universal method for preparing nutrient solutions of certain desired composition. *Plant Soil* **1961**, *15*, 134–154. [CrossRef]
20. Murphy, J.; Riley, J.P. A modified single solution method for the determination of phosphate in natural waters. *Anal. Chim. Acta* **1962**, *27*, 31–36. [CrossRef]
21. Guzmán, M. Equilibrios Nutricionales en Condiciones de Invernadero: Corrección, y Mejora de la Cosecha. Ph.D. Thesis, Universidad de Granada, Granada, Spain, 1987.
22. Cadahía, C. *La Savia Como Índice de Fertilización. Cultivos Agroenergéticos, Hortícolas, Frutales y Ornamentales*; Mundi-Prensa: Madrid, Spain, 2008; pp. 35–68.
23. Cataldo, M.; Schrader, L.E.; Youngs, V.L. Rapid Colorimetric Determination of Nitrate in Plant Tissue by Nitration of Salicylic Acid. *Soil Sci. Plant Anal.* **1975**, *6*, 71–80. [CrossRef]
24. St. Martín, C.C.G.; Brathwaite, R.A.I. Compost and compost tea: Principles and prospects as substrates and soil-borne disease management strategies in soil-less vegetable production. *Biol. Agric. Hortic.* **2012**, *28*, 1–33. [CrossRef]
25. Kannangara, T.; Forget, B.; Dang, B. Effects of aeration, molasses, kelp, compost type, and carrot juice on the growth of Escherichia coli in compost teas. *Compost Sci. Util.* **2006**, *14*, 40–47. [CrossRef]
26. Sánchez, M.M.; Roig, C.; Bernal, M.P. Nitrogen transformation during organic waste composting by the Rutgers system and and it effects on pH, EC, and maturity of the composting mixtures. *Bioresour. Technol.* **2001**, *78*, 301–308. [CrossRef]
27. Goto, E.; Both, A.J.; Albright, L.D.; Langhans, R.W.; Leed, A.R. Effect of dissolved oxygen concentration on lettuce growth in floating hydroponics. *Acta Hortic.* **1996**, *440*, 205–210. [CrossRef]
28. Kumar, V.; Singh, K.P. Enriching vermicompost by nitrogen fixing and phosphate solubilizing bacteria. *Bioresour. Technol.* **2001**, *76*, 173–175. [CrossRef] [PubMed]
29. Valverde, K.; Chang, M.; Rodríguez, D.A. Effect of the Light Quality on the Nitrate Reductase Activity in Lettuce Plants Grown in NFT. *Acta Hortic.* **2009**, *843*, 89–96. [CrossRef]
30. Afzal, I.; Hussain, B.; Ahzmed, S.M.; Ullah, S.H.; Shakeel, Q.; Kamran, M. Foliar application of potassium improves fruit quality and yield of tomato plants. *Acta Sci. Pol. Hortorum Cultus* **2015**, *14*, 3–13.
31. Monge, E.; Val, J.; Sanz, M.; Blanco, A.; Montañés, L. El calcio nutriente para las plantas. Bitter pit en manzano. *An. Estac. Exp. Aula Dei* **1994**, *21*, 189–201.
32. Mengel, K. Potassium. In *Handbook of Plant Nutrition*; Barker, A.V., Pilbeam, D.J., Eds.; CRC Press: Boca Raton, FL, USA, 2007; pp. 91–120.
33. Wu, S.C.; Cao, Z.H.; Li, Z.G.; Cheung, K.C.; Wong, M.H. Effects of biofertilizer containing N-fixer, P and K solubilizers and AM fungi on maize growth: A greenhouse trial. *Geoderma* **2005**, *125*, 155–166. [CrossRef]
34. Numan, M.; Bashir, S.; Khan, Y.; Mumtaz, R.; Shinwari, Z.K.; Khan, A.L. Plant growth promoting bacteria as an alternative strategy for salt tolerance in plants: A review. *Microbiol. Res.* **2018**, *209*, 21–32. [CrossRef] [PubMed]

35. Jakobsen, T.S. Interaction between Plant Nutrients: IV. Interaction between Calcium and Phosphate. *Acta Agric. Scand. Sect. B Soil Plant Sci.* **1993**, *43*, 6–10. [CrossRef]
36. Cadahia, C. (Ed.) *Fertirrigación. Cultivos Hortícolas, Frutales y Ornamentales*, 3rd ed.; Ediciones Mundi-Prensa: Madrid, Spain, 2008; pp. 75–76.
37. Masarirambi, T.M.; Hlawe, M.M.; Oseni, T.O.; Thokozile, E. Effects of organic fertilizers on growth, yield, quality and sensory evaluation of red lettuce (*Lactuca sativa* L.) 'Venezaoxa'. *Agric. Biol. J. N. Am.* **2010**, *6*, 1319–1324. [CrossRef]
38. Russo, V.; Pappelis, A. Senescence in sweet corn as influenced by phosphorous nutrition. *Plant Nutr.* **1995**, *18*, 707–717. [CrossRef]
39. González, S.K.; Rodríguez, M.M.; Trejo, T.L.; Sánchez, E.J.; García, C.J. Efluente y té de vermicompost en la producción de hortalizas de hoja en sistema NFT. *Interciencia* **2013**, *28*, 863–869.
40. Xu, H.L.; Wang, R.; Xu, R.Y.; Mridha, M.A.U.; Goyal, S. Yield and quality of leafy vegetables grown with organic fertilizations. *Acta Hortic.* **2005**, *627*, 25–33. [CrossRef]
41. Martínez, G.G.; Ortiz, H.Y.; López, P.R. Oxigenación de la solución nutritiva recirculante y su efecto en tomate y lechuga. *Rev. Fitotec. Mex.* **2012**, *35*, 49–52. [CrossRef]
42. Reglamento (UE) No 1258/2011 de la Comisión. de 2 de diciembre de 2011. Modifica el Reglamento (CE) no 1881/2006 por lo que respecta al contenido máximo de nitratos en los productos alimenticios. *D. Of. De La Unión Eur.* **2011**, *L 320*, 15–17.
43. Magkos, F.; Arvaniti, F.; Zampelas, A. Putting the safety of organic food into perspective. *Nutr. Res. Rev.* **2003**, *16*, 211–221. [CrossRef] [PubMed]
44. Hernández, T.; Chocano, C.; Moreno, J.L.; García, C. Use of compost as an alternative to conventional inorganic fertilizers in intensive lettuce (*Lactuca sativa* L.) crops—Effects on soil and plant. *Tillage Res.* **2016**, *160*, 14–22. [CrossRef]
45. López, M.J.; Masaguer, A.; Paredes, C.; Perez, L.; Muñoz, M.; Salas, M.C.; Hernandez, R. De resíduos a recursos. El camino hacia la Sostenibilidad. *Red Española Compost.* **2015**, 91–121.
46. Janssen, B.H.; Oenema, O. Global economics of nutrient cycling. *Turk. J. Agric. For.* **2008**, *32*, 165–176.
47. Aznar-Sánchez, J.A.; Velasco-Muñoz, J.F.; García-Arca, D.; López-Felices, B. Identification of opportunities for applying the circular economy to intensive agriculture in Almería (South-East Spain). *Agronomy* **2020**, *10*, 1499. [CrossRef]

horticulturae

Essay

Enhancing Soil Fertility and Elevating Pecan Fruit Quality through Combined Chemical and Organic Fertilization Practices

Yinhao Tong [1,†], Zhaocheng Wang [1,2,†], Duxin Gong [1], Cheng Huang [1], Xiaomin Ma [1], Xiaoxiang Ma [1], Feiyang Yuan [1], Songling Fu [1,*] and Chun Feng [1,*]

1 School of Forestry & Landscape Architecture, Anhui Agricultural University, Hefei 230036, China; tongyinhao@163.com (Y.T.); wzc@ahau.edu.cn (Z.W.); linxu1127@126.com (D.G.); huangcheng0103@163.com (C.H.); maxiaomin0808@163.com (X.M.); maxiaoxiang027@163.com (X.M.); feiyang0012022@163.com (F.Y.)
2 Huzhou Vocational & Technical College, Huzhou 313000, China
* Correspondence: fusongling@ahau.edu.cn (S.F.); fengchun_2000@ahau.edu.cn (C.F.)
† These authors contributed equally to this work.

Citation: Tong, Y.; Wang, Z.; Gong, D.; Huang, C.; Ma, X.; Ma, X.; Yuan, F.; Fu, S.; Feng, C. Enhancing Soil Fertility and Elevating Pecan Fruit Quality through Combined Chemical and Organic Fertilization Practices. *Horticulturae* **2024**, *10*, 25. https://doi.org/10.3390/horticulturae10010025

Academic Editors: Francesco De Mastro, Gennaro Brunetti, Karam Farrag and Huadong Zang

Received: 9 November 2023
Revised: 19 December 2023
Accepted: 22 December 2023
Published: 26 December 2023

Correction Statement: This article has been republished with a minor change. The change does not affect the scientific content of the article and further details are available within the backmatter of the website version of this article.

Abstract: This study focused on 6-year-old 'Pawnee' pecan trees to elucidate the differential responses of physicochemical properties of orchard soil and pecan fruit quality when combining chemical and organic fertilizers. The aim was to unveil the mechanisms that underlie the effects of different fertilization treatments on soil fertility, soil enzyme activities, and pecan fruit quality. Four treatments were established: sole chemical fertilizer (CF; $N{:}P_2O_5{:}K_2O$ is 15:15:15), chemical fertilizer combined with cake fertilizer (CF+CC), chemical fertilizer combined with manure fertilizer (CF+M), and chemical fertilizer combined with cake and manure fertilizer (CF+CC+M). Measurements were taken to assess the soil nutrient content, soil enzyme activities, and fruit growth quality in some orchards under different fertilization treatments. The results revealed that the combined application could increase yield and enhance pecan quality. Among these, the CF+M+CC treatment demonstrated the most favorable outcomes, with the pecan kernel oil and unsaturated fatty acid contents reaching 72.33% and 97.54%, respectively. The combined fertilization treatments had no significant impacts on soil trace elements such as Mg, Cu, and Mn; however, it significantly increased the Available Phosphorus (AP), Total Nitrogen (TN), Soil Organic Matter (SOM) and S-ACP (soil acid phosphatase) activities. In summary, the combined application of chemical and organic fertilizers can significantly increase the soil nutrient content and enzyme activities in pecan orchards, to promote the enhancement of fruit quality and economic aspects.

Keywords: pecan; chemical fertilizer; organic fertilizers; soil fertility; quality of pecan fruits

1. Introduction

Fertilizers can quickly provide plants with the nutrients they require to support high crop yields and contribute to feeding an expanding global population. However, certain chemical fertilizers can bind to the soil and cause acidification [1], or are lost via leaching, runoff, and erosion, which reach waterbodies and contribute to eutrophication [2]. Phosphorus (P)-induced modification of microbial communities alters both microbial activities and carbon (C) cycling rates [3,4], which leads to a reduction in the SOM content and soil fertility [5]. This inhibits plant growth, reduces fruit yields, and reduces their quality [6,7]. Thus, it is necessary to develop prudent and sustainable management practices that can mitigate any deleterious environmental costs.

Organic fertilizers have several advantages over chemical fertilizers, as they are rich in active organic matter that can improve the soil structure, increase the sequestration of soil C, and promote the growth and reproduction of soil microorganisms. The metabolic processes of these microorganisms can facilitate the conversion and decomposition of mineral nutrients in the soil. Moreover, organic fertilizers themselves contain a significant quantity of organic matter and nutrients, which can help to restore or maintain soil structure and fertility levels, leading to sustainable development [8,9]. Further, they can mitigate issues such as low nutrient utilization rates that may result from the application of only chemical fertilizers [10].

Compared with the sole use of chemical fertilizers, the synergistic effects of organic fertilizers in conjunction with chemical fertilizers result in greater accumulation of TN and organic matter in the soil. Research involving the proportional substitution of chemical fertilizers with different types of organic fertilizers indicated that the mixed application of chemical and organic fertilizers plays a pivotal role in crop growth, soil fertility, and sustainability, as opposed to the application of chemical fertilizer by itself. Additionally, this approach safeguards the local environment, making it an environmentally compatible management strategy for achieving sustainable development [11,12].

In recent years, research on the combined application of chemical and organic fertilizers has shown that in contrast to using chemical or organic fertilizers alone, their mutual application enhances the efficacy of nutrient absorption and utilization. This combined application of fertilizers has a significant positive impact on the soil pH, enzyme activities, and nutrient levels [13–15], which enhance sustainable soil productivity and inevitably lead to increased crop yields [16]. Furthermore, in other studies, this application has shown to improve the protein content, unsaturated fatty acids, and oil content of walnuts, while enhancing fruit quality. Simultaneously, it has a significant impact on economic traits such as yields, fruit diameters, and other measurable parameters [17,18].

Organic fertilizers have the capacity to increase the soil microbial biomass and enhance soil biodiversity [19]. They can provide soil microbes with more readily available C and N sources, improve the ecological environment for soil microorganisms, and promote microbial activities [20,21]. As vital components of organic matter, soil microorganisms respond more quickly to fertilization compared to overall SOM [22,23]. Previous research has verified that replacing a higher proportion of chemical fertilizers with organic counterparts can significantly reduce soil acidification, facilitate nutrient decomposition, increase microbial C and N content in the soil, enhance soil fertility, and boost soil enzyme activities [9,24]. Soil enzymes play critical roles in the decomposition of organic matter and nutrient cycling [25,26]. When crops require higher levels of nutrients than the soil can provide, increased soil enzyme activities can assist in breaking down microbes to release nutrients that were previously fixed in the soil, making them available for crop uptake and utilization [27].

Pecan (*Carya illinoinensis*) is a tree of the genus *Carya* in the Juglandaceae, which is native to Northern Mexico and the Southern United States [28]. Pecans have a thin shell and are prized for their high oil content, averaging 72.0% in unroasted kernels [29–31]. They are rich in unsaturated fatty acids and antioxidants, which makes them highly valuable for the market [32]. Thus, various fertilization management strategies are employed to increase and optimize pecan yield. For instance, fertilization plays a crucial role in enhancing the supply of soil nutrients, ensuring walnut growth, and improving fruit quality [33]. Currently, the research on the application of pecan fertilizer is limited. Therefore, this experiment focused on the 'Pawnee' pecan and primarily investigated the effects of the mixed application of chemical and organic fertilizers on fruit growth, quality, as well as soil fertility and enzyme activities in pecan orchard soils. The main objectives of this research were twofold: 1. To identify suitable fertilizer types that effectively enhance pecan quality and soil fertility. 2. To analyze the interrelations between various ecological factors that affect pecan growth, with the aim of providing a basis for rational fertilization practices to optimize its cultivation.

2. Materials and Methods

2.1. Study Area

The experimental site is in Jiulong Town (115°33′44″ E, 32°57′25″ N), Yingzhou District, in Fuyang City, of Anhui Province, China. This area receives an average annual rainfall of

837.5 mm and temperature of 15.3 °C, with 2400 h of full sunlight per year, and a 210 day frost-free period. The main types of soil are sandy. The long-term fertilization experiment began in the spring of 2017 and has been applied for 5 years. chemical fertilizer with ratios of N:P_2O_5:K_2O is 15:15:15 (STANLEY Compound fertilizer); Cake fertilizer is rapeseed cake, the organic matter content in cake meal fertilizer was 60%, N: 114 g·kg^{-1}, P: 95 g·kg^{-1}, K: 80 g·kg^{-1}; Manure fertilizer is cow and goat manure compost, the organic matter and amino acid content in the manure were 40% and 10%, N: 71 g·kg^{-1}, P: 66 g·kg^{-1}, K: 53 g·kg^{-1}.The fundamental physicochemical properties of the soil as 2017: pH = 7.63, SOM 7.52 g·kg^{-1}, TN 50.24 mg·kg^{-1}, AP 1.12 mg·kg^{-1}, and K 6.78 mg·kg^{-1}.

2.2. Experimental

Design. 6-year-old 'Pawnee' Pecan trees were selected for the experiment. The trees were spaced at intervals of 5 m by 6 m, and their growth was uniform. There were four treatments with four replicates each: (1) Chemical fertilizer (CF) (100% chemical fertilizer); (2) Chemical fertilizer + cake fertilizer (CF+CC) (50% chemical fertilizer and 50% cake fertilizer); (3) Chemical fertilizer + manure fertilizer (CF+M) (50% chemical fertilizer and 50% manure fertilizer); (4) Chemical fertilizer + manure fertilizer + cake fertilizer (CF+M+CC) (50% chemical fertilizer, 25% *cake* fertilizer, and 25% manure fertilizer). For the CF+CC, CF+M, and CF+M+CC treatments the amounts of chemical fertilizer were halved, and the reduced fertilizer amount was replaced with organic fertilizer in terms of N, P, and K. At the local custom chemical fertilizer application amount. Each treatment was repeated three times and covered an area of 667 m^2 (Table 1). This fertilization experiment began in 2017 and was conducted annually annually after the walnut harvest. Prior to application, the different ratios of fertilizers were thoroughly mixed and placed in trench-like grooves, which were then uniformly mixed with the topsoil and covered. All other fertilizer, water, and management conditions remained consistent.

Table 1. Experimental design.

Treatment	Chemical Fertilizer (kg·hm^{-2})			Cake Fertilizer (kg·hm^{-2})	Manure Fertilizer (kg·hm^{-2})
	N	P_2O_5	K_2O		
CF	150	150	150	0	0
CF+CC	75	75	75	720	0
CF+M	75	75	75	0	1350
CF+M+CC	75	75	75	360	675

Note: 100% chemical fertilizer (CF); 50% chemical fertilizer and 50% cake fertilizer (CF+CC); 50% chemical fertilizer and 50% manure fertilizer (CF+M); 50% chemical fertilizer, 25% Manure fertilizer, and 25% cake fertilizer (CF+M+CC).

2.3. Laboratory Analysis

Individual trees with consistent growth, free of pests and diseases, were selected on 10 October 2022. From each tree, five fruits were randomly selected from three different levels within the canopy (upper, middle, and lower), as well as from the outer perimeter for a total of 30 sample fruits. Following the removal of the green skin when the fruits were ripe, they were naturally air-dried and their fruit and kernels were weighed. Additional measurements were made to quantify the individual fruit height, fruit diameter, and shell thickness. The oil was measured using the Soxhlet extraction technique [34], fatty acid content using the GC-MS external standard method [35], Sample 0.2 g of oil saponification and methyl ester treatment, GC-MS analysis was performed by Agilent 7890A and 5975C (Agilent Technology, Santa Clara, CA, USA). Chromatographic analysis conditions: DB-225MS column (30 m × 0.25 mm, 0.25 μm; Agilent Technology, USA). The carrier gas was helium, and the purge flow rate of the spacer was 3.0 mL·min^{-1}. Heating procedure: initial temperature 50 °C, 1 min; The temperature was first raised to 200 °C at the rate of 5 °C·min^{-1}, and then to 230 °C at the rate of 2 °C·min^{-1} for 10 min. Forward sample temperature 250 °C, transmission line temperature 250 °C, ion source temperature 230 °C, four-stage rod temperature 150 °C, ionization voltage −70 eV. Four replicate plots were established for each treatment, and soil samples were extracted from multiple sampling points within each plot, the soil sampling was carried out up to a depth of 40 cm, excluding the superficial vegetative layer. The mixed soil samples were subjected to physicochemical analysis, Soil pH was determined with a glass electrode (soil:water = 1:2.5) pH instrument(Mettler Toledo, Shanghai, China), Soil EC was determined with a glass electrode (soil:water = 1:5) EC meter (Mettler Toledo, Shanghai, China),with soil AP, TP, NH_4^+-N, NO_3^--N, TN, using an automated continuous chemical analyzer (De Chem-Tech, Clever Chem Anna, Germany) extractants by vitriol [36]. To measure available K, Mg, Fe, Zn, Mn, and copper (Cu), soil samples were extracted by Ammonium bicarbonate diethylenetriaminepentaacetic acid (DTPA) (AB-DTPA) [37]. The soil enzyme activity was assessed following the methods of Sun [38], Kandeler [39], and Taylor [40], with enzyme activities expressed in international units per gram of soil consumed ($U{\cdot}g^{-1}$).

2.4. Data Analysis

All data analysis was performed in R 4.1.3. to test whether they met the normal distribution, and then complete statistical analysis. The resulting data were presented as mean values and standard errors. Data were statistically analyzed via one-way ANOVA with repeated measures and the post hoc method (Duncan's method), for multiple comparisons at a 5% significance level. A Pearson's correlation matrix was generated between the fruit quality and soil parameters. Redundancy analysis (RDA) in Canoco5.0 was applied to explore the correlations between environmental factors, soil enzyme activities, and soil nutrient elements.

3. Results

3.1. Subsection Effects of Different Fertilization Treatments on Pecan Yield

With the use of fertilizers combined with manure, cake fertilizer and the other three kinds of fertilizer combinations, the yield of thin-shell pecan showed an increasing trend (Table 2). The combined application of the three fertilizers increased yields to 598.34 $kg{\cdot}hm^{-2}$, 583.73 $kg{\cdot}hm^{-2}$ and 618.87 $kg{\cdot}hm^{-2}$, respectively, in 2022, as compared with 2018. Compared with 2018 and 2020, the output of thin-shell pecan in 2022 increased significantly, and the output of each treatment in 2022 increased successively: CF+M+CC > CF+M > CF+CC > CF. By 2022, CF+M, CF+CC and CF+M+CC treatments had the highest yield increases of 13.02% for CF+M+CC, followed by 10.31% for CF+M and 8.03% for CF+CC.

Table 2. Effects of different fertilization treatments on pecan yield.

Treatment	2018 ($kg{\cdot}hm^{-2}$)	2020 ($kg{\cdot}hm^{-2}$)	2022 ($kg{\cdot}hm^{-2}$)	2022 Comparison with CF	
				Production Increase ($kg{\cdot}hm^{-2}$)	Yield Increase Rate (%)
CF	756.51 ± 4.82 a	839.96 ± 4.78 a	1083.36 ± 10.42 d	-	-
CF+M	597.14 ± 3.99 bc	768.67 ± 3.19 c	1195.48 ± 4.63 b	112	10.34
CF+CC	587.07 ± 2.58 c	782.14 ± 5.58 b	1170.8 ± 4.39 c	87	8.03
CF+M+CC	606.1 ± 6.03 b	792.53 ± 2.93 b	1224.97 ± 5.91 a	141	13.02

Different letters indicate significant differences between the treatments.

3.2. Effects of Different Fertilization Treatments on Fruit Traits of Pecan Nuts

Among the four different fertilization treatments applied to pecan, economic parameters such as nut mass, kernel weight, nut vertical and horizontal diameters exhibited significant differences between the sole application of chemical fertilizers and mixed application of inorganic and chemical fertilizers ($p < 0.05$). However, the kernel yield and shell thickness did not exhibit significant differences (Figure 1). In terms of the kernel weight (Figure 1B) and nut mass (Figure 1C), the CF+CC+M treatment yielded the highest values (7.18 g and 12.75 g, respectively), with increases of 1.65 g and 3.38 g, respectively, as compared with the CF treatment. Furthermore, the CF+CC+M treatment demonstrated the highest kernel yield, exceeding that of the CF treatment by 7.64% (Figure 1A). This difference may have been related to the thickness of the pecan shells. Regarding the nut vertical and horizontal diameters, the application of chemical fertilizers on their own resulted in values lower than those observed under treatments with additional organic fertilizers.

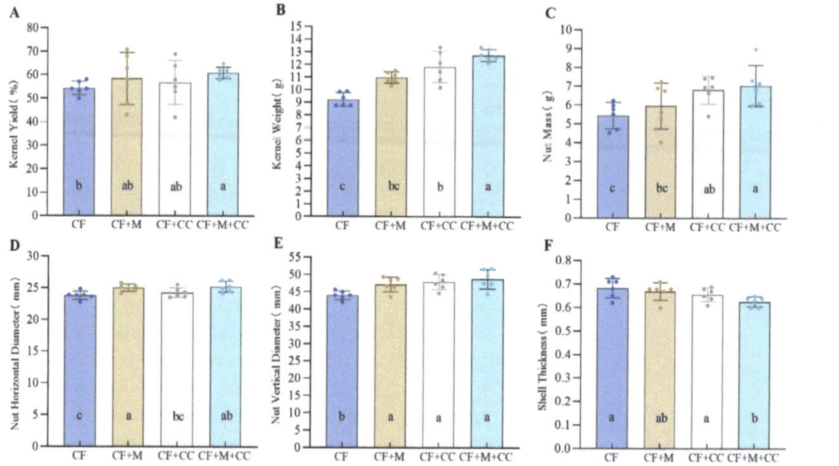

Figure 1. (**A–F**) Effects of different fertilization treatments on economic characteristics of pecan. Dots around the top of the bars represent the specific values of the five repetitions in the drawn bar chart. All data are the standard error means of five replicates±. The error bar shows the standard error. Different letters indicate significant differences between the treatments.

3.3. *Effects of Different Fertilization Treatments on Nucleolar Nutritional Quality*

Different fertilization treatments had varying degrees of influence on the oil content and unsaturated fatty acid content of pecans (Figure 2). Both were highest under the CF+CC+M treatment, which showed increases of 2.50% and 7.28%, respectively, in contrast to the CF treatment. When considering the contents of octadecenoic acid (C18:1), octadecadienoic acid (C18:2), octadecatrienoic acid (C18:3), and gondoic acid (C20:1), it was observed that the monounsaturated fatty octadecenoic acid decreased (Figure 2D), while the contents of polyunsaturated fatty octadecadienoic acid (Figure 2E) and octadecatrienoic acid (Figure 2F) increased. The fertilization treatments were beneficial for increasing the unsaturated fatty acid content.

3.4. *Effects of Different Fertilization Treatments on Soil pH, EC, and SOM*

Under different fertilization conditions, the combined application of inorganic and chemical fertilizers significantly impacted the pH, EC, and SOM ($p < 0.05$) (Table 3). In contrast to the sole application of chemical fertilizers (CFs), treatments that incorporated organic fertilizers (e.g., CF+CC, CF+M, and CF+M+CC) demonstrated notable increases in pH of 1.42, 1.3, and 1.41, decreases in EC of 15.56%, 18.52%, and 14.07%, and improvements in SOM of 64.3%, 66.5%, and 122%, respectively.

Table 3. Effects of different fertilization treatments on soil pH, EC, and SOM.

Treatment	pH	EC (ds·m^{-1})	SOM (g·kg^{-1})
CF	6.93 ± 0.08 b	1.35 ± 0.02 a	12.67 ± 1.42 b
CF+M	8.23 ± 0.06 a	1.10 ± 0.03 b	21.09 ± 3.68 a
CF+CC	8.35 ± 0.06 a	1.14 ± 0.03 b	21.24 ± 1.25 a
CF+M+CC	8.34 ± 0.10 a	1.16 ± 0.03 b	28.33 ± 2.52 a

Different letters indicate significant differences between the treatments.

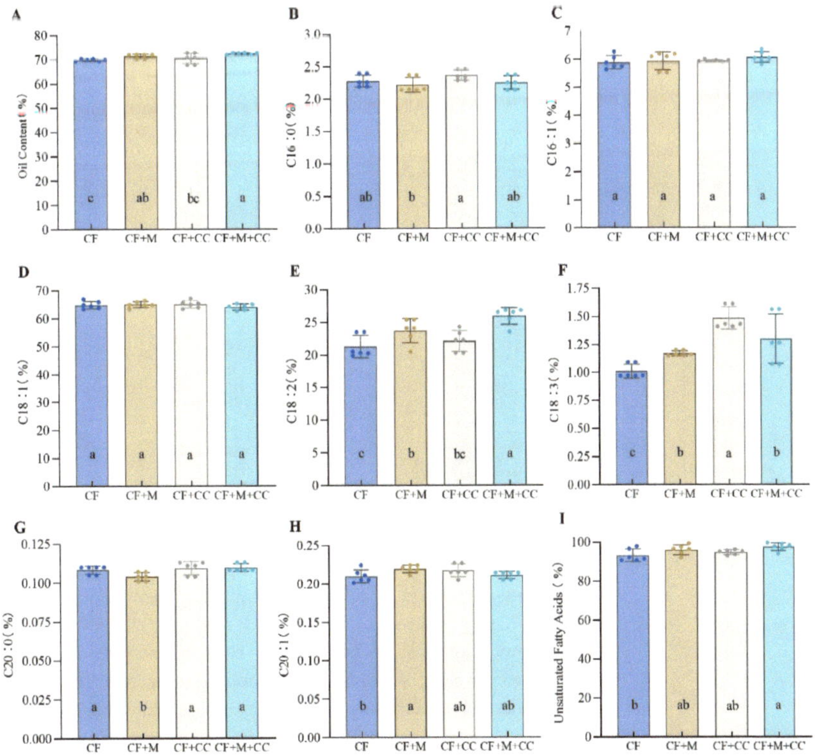

Figure 2. Effects of different fertilization treatments on nucleolar nutritional quality. (**A**) Oil content; (**B**) C16:0—hexadecanoic acid; (**C**) C16:1—palmitoleic acid; (**D**) C18:1—octadecenoic acid; (**E**) C18:2—octadecadienoic acid; (**F**) C18:3—octadecatrienoic acid; (**G**) C20:0—eicosanoic acid; (**H**) C20:1—gondoic acid; (**I**) unsaturated fatty acids. Dots around the top of the bars represent the specific values of the five repetitions in the drawn bar chart. All data are the standard error means of five replicates±. The error bar shows the standard error. Different letters indicate significant differences between the treatments.

3.5. Effects of Different Fertilization Treatments on Soil Nutrients

Under different fertilization conditions, the combined application of inorganic and chemical fertilizers significantly impacted the AP, NH_4^+-N, NO_3^--N, TN, and K levels ($p < 0.05$), while the TP content showed no significant differences (Figure 3). The CF treatment exhibited the lowest AP content (Figure 3C), at only 13.1% of CF+M+CC. Furthermore, NH_4^+-N (Figure 3A), NO_3^--N (Figure 3B), TN (Figure 3E), TP (Figure 3D), and K (Figure 3F) exhibited the highest levels under the CF+M+CC treatment, with increases of 75.5%, 465.1%, 15.5%, 7.8%, and 33.5%, respectively, as compared to the CF.

Treatments that involved the sole application of CF and combined application of chemical and organic fertilizers revealed significant differences in the soil Ca and Zn contents ($p < 0.05$), while the Mg and Cu contents showed no noticeable distinctions (Figure 4). The soil Zn content (Figure 4A) and Ca content (Figure 4D) reached their highest levels under the CF+M+CC treatment (4.21 mg·kg^{-1} and 1.54 mg·kg^{-1}, respectively). The soil Mg (Figure 4B) and Fe (Figure 4F) contents were highest under the CF+CC treatment, representing increases of 18% and 16% as compared to the CF. The soil Cu (Figure 4C) and Mn (Figure 4E) contents showed negligible differences between treatments; however, the CF+M showed the highest levels with increases of 7% and 26.8%, respectively, as compared to CF.

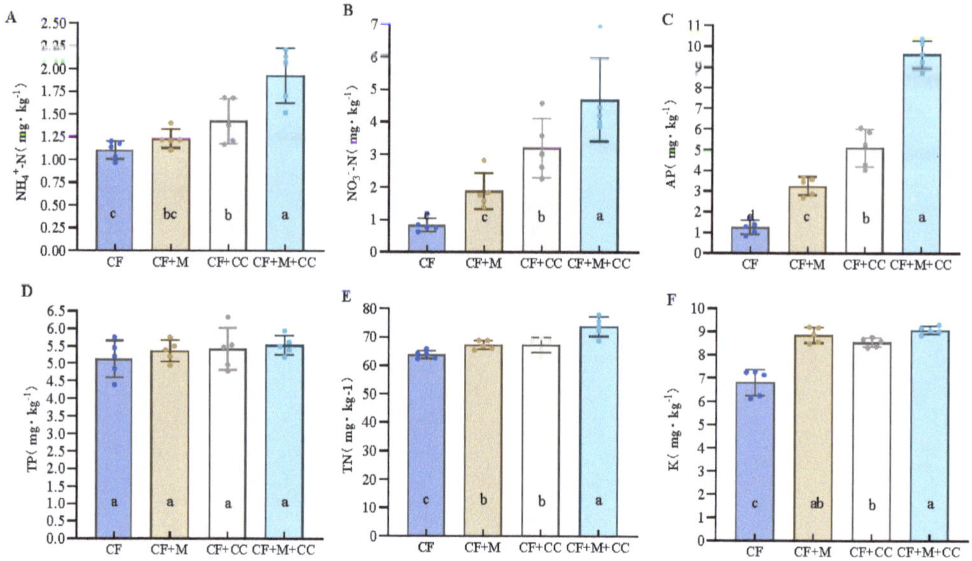

Figure 3. Effects of different fertilization treatments on soil nutrients. (**A**) NH_4^+-N content; (**B**) NO_3^--N content; (**C**) available P content; (**D**) total P content; (**E**) total N content; (**F**) total K content. Dots around the top of the bars represent the specific values of the five repetitions in the drawn bar chart. All data are the standard error means of five replicates±. The error bar shows the standard error. Different letters indicate significant differences between the treatments.

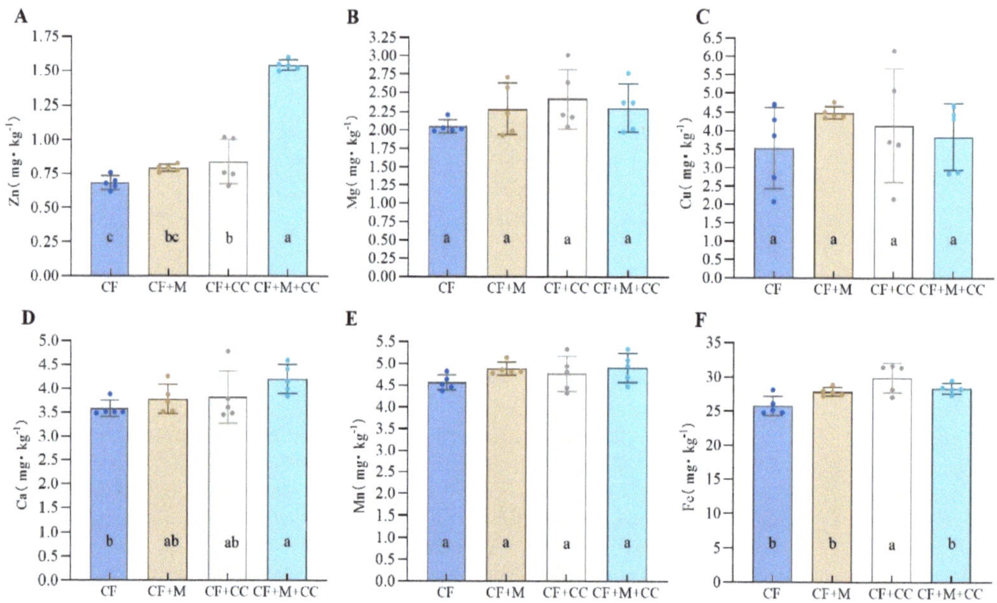

Figure 4. Effects of different fertilization treatments on soil nutrients. (**A**) Total Zn content; (**B**) total Mg content; (**C**) total Cu content; (**D**) total Ca content; (**E**) total Mn content; (**F**) total Fe content. Dots around the top of the bars represent the specific values of the five repetitions in the drawn bar chart. All data are the standard error means of five replicates±. The error bar shows the standard error. Different letters indicate significant differences between the treatments.

3.6. Effects of Different Fertilization Treatments on Soil Enzyme Activity

The various fertilization treatments exhibited varying degrees of influence on soil enzyme activities in the 0–20 cm soil layer (Figure 5). Significant differences ($p < 0.05$) were observed in soil urease (S-UE), soil peroxidase (S-POD), acid phosphatase (S-ACP), and saccharase (S-SC) between the treatments that involved inorganic fertilization by itself, and those that combined inorganic and organic fertilizers. Notably, the CF treatment exhibited the lowest activity levels in S-POD (Figure 5A), S-UE (Figure 5B), S-SC (Figure 5C), and S-ACP (Figure 5D), with differences that ranged from 28.9% to 153.0%, compared to the highest levels observed under the CF+M, CF+M+CC, CF+CC, and CF+M treatments.

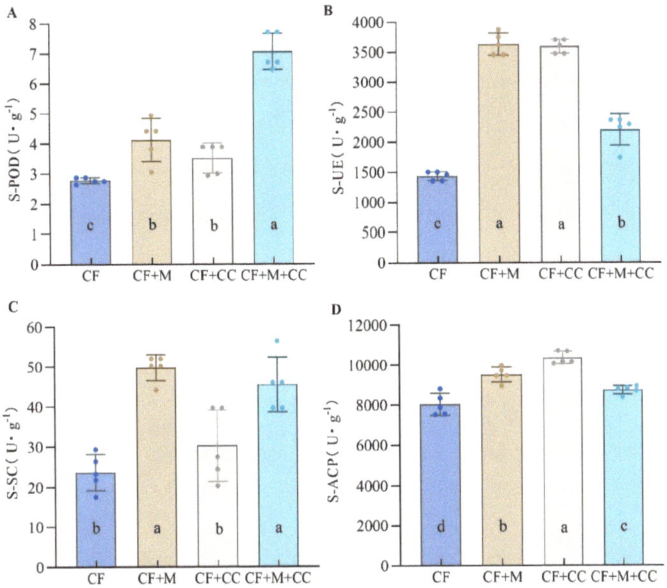

Figure 5. Effects of different fertilization treatments on soil enzyme activities. (**A**) S-POD: soil peroxidase; (**B**) S-UE: soil urease; (**C**) S-SC: saccharase; (**D**) S-ACP: acid phosphatase. Dots around the top of the bars represent the specific values of the five repetitions in the drawn bar chart. All data are the standard error means of five replicates±. The error bar shows the standard error. Different letters indicate significant differences between the treatments.

3.7. Correlation Analysis of Soil Nutrient and Enzyme Activities in Pecan

The yield exhibited significantly negative correlations ($p < 0.01$) with AP, NO_3^--N, Fe, S-UE, S-ACP, and showed significantly positive correlations ($p < 0.05$) withNH_4^+-N, K, Mg (Figure 6). The kernel weight exhibited significantly positive correlations with pH, SOM, AP, NH_4^+-N, NO_3^--N, TN, K, Ca, Fe, Zn, S-POD, S-UE and Mn, and negative correlations with EC. The kernel yields showed a significantly positive correlation with TN and no significant correlations with other nutrients and enzyme activities. The nut mass exhibited a significantly positive correlation with AP, as well as NH_4^+-N, NO_3^--N, Fe, and Zn. The nut vertical diameter exhibited significantly positive correlations with pH, AP, NO_3^--N, and S-POD, as well as with SOM, NH_4^+-N, TN, K, Zn, S-UE, and S-SC. The nut horizontal diameter exhibited significantly positive correlations with TN, K, Zn, S-POD, and S-SC; and negative correlations with EC. Furthermore, the shell thickness revealed significantly negative correlations with AP, NH_4^+-N, NO_3^--N, TN, TP, K, Ca, Mg, Fe, Mn, Zn, and S-POD, and negative correlations with pH. The oil content exhibited significantly positive correlations with AP, NO_3^--N, TN, K, Zn, and S-POD, as well as with SOM and NH_4^+-N, and showed negative correlations with EC. Unsaturated fatty acids showed a significantly positive correlation with S-POD, as well as with Zn and S-SC.

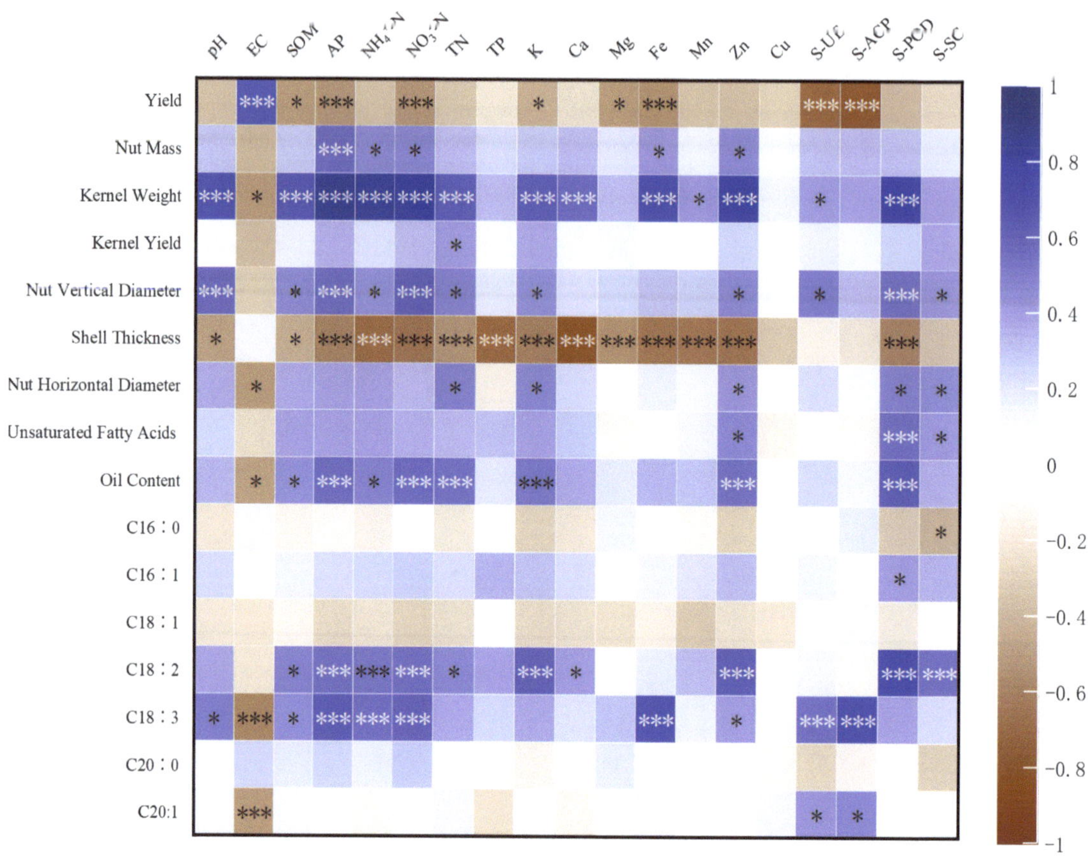

Figure 6. Analysis of fruit quality, soil nutrients, and enzyme activities of pecan under different fertilization treatments. ($p < 0.05$ *, $p < 0.01$ **, $p < 0.001$ ***).

C16:0 (hexadecanoic acid) and C16:1 (palmitoleic acid) revealed significantly positive correlations with S-SC and S-POD, respectively, and no significant correlations with other nutrients and enzyme activities. C18:2 exhibited significantly positive correlations with AP, NH_4^+-N, NO_3^--N, K, Zn, S-POD, and S-SC, as well as with SOC, TN, and Ca. C18:3 exhibited significantly positive correlations with AP, NH_4^+-N, NO_3^--N, Fe, S-UE, and S-ACP, as well as with pH, SOM and Zn, and showed significantly negative correlations with EC. C20:1 showed significantly positive correlations with S-UE and S-ACP, and significantly negative correlations with EC.

3.8. RDA Analysis of Soil and Quality of Pecan

Based on RDA analysis, we discerned the levels of influence of the fruit economic traits, soil nutrient factors, and soil enzyme activities on the quality of pecan fruits (Figure 7). For fruit quality and economic traits, the explanatory powers for the first and second axes were 34.77% and 3.49%, respectively, with a cumulative explanatory power of 38.26% (Figure 7A). For fruit quality and soil nutrients, the explanatory powers for the first and second axes were 54.01% and 7.15%, respectively, with a cumulative explanatory power of 61.16% (Figure 7B). For fruit quality and soil enzyme activities, the explanatory powers for the first and second axes were 19.34% and 9.02%, respectively, with a cumulative explanatory power of 28.36% (Figure 7C). Notably, the Zn, S-POD and nut vertical diameter, exhibited significant correlations with the pecan fruit quality ($p < 0.01$) (Table 4), while S-UE, NO_3^--N and the nut mass exhibited relatively weaker correlations with the pecan fruit quality.

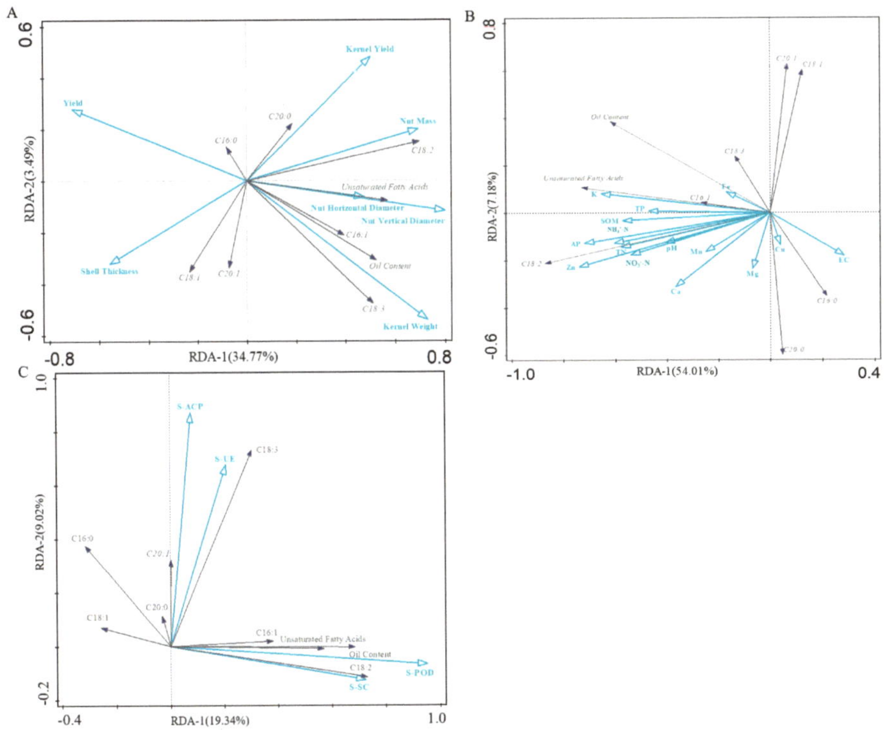

Figure 7. RDA plots showing the relationships between environmental factors and fruit traits (**A**), soil nutrients (**B**) and soil enzyme activities (**C**). Gray lines represent environmental variables, while blue lines represent fruit traits, soil nutrients, and soil enzyme activities.

Table 4. RDA data analysis table.

Name	Explains %	Contribution %	Pseudo-F	p
Zn	30.3	46.9	9.6	0.002
S-POD	18.1	52.5	4.9	0.002
Nut Vertical Diameter	22.3	55.8	6.3	0.006
Nut Mass	8.9	21.9	2.7	0.064
S-UE	7.8	22.5	2.2	0.07
NO_3^--N	7.3	11.2	2.5	0.088

4. Discussion

The application of organic fertilizers has direct impacts on both the accumulation and supply of soil nutrients, which is a critical step toward the reduction in our reliance on chemical fertilizers, and contributes to enhancing fruit quality and yield [16,41]. In this study, compared to CF, the CF+CC, CF+M, and CF+M+CC treatments significantly increased walnut yield, with the following yield ranking: CF+M+CC > CF+M > CF+CC. The introduction of organic fertilizers increases the activities of soil microorganisms, thereby accelerating the decomposition of organic compounds and enhancing the biological utilization of certain plant growth regulators, such as auxins, gibberellins, and essential nutrients [42,43]. Organic fertilizers are more effective than chemical fertilizers in increasing SOM content. As the SOM content increases, nitrogen fixation, sulfurization, and oxidation processes accelerate, which leads to the breakdown and release of certain less soluble nutrients [44–46]. The enriched organic matter improved soil structure and increased soil permeability. Under the joint action of the above aspects, the mineral nutrient content of the soil is increased, root growth and development is promoted, and the environment promotes the growth of pecan, thereby increasing yield [26]. The experimental results also showed that with the continuous use of organic fertilizer, the walnut yield of each fertilizer treatment would increase year by year. Studies have shown that soils with high nutrient content significantly improve leaf characteristics, photosynthetic rates, and stomatal conductance, thus leading to increased carbon accumulation, stabilized dry matter, and ultimately higher yields [47]. Correlation analysis in this study also indicates a highly significant relationship between soil N, P, and pecan yield. Mg and Fe play a crucial role in the quality of fruit tree yield. Balancing the supply and demand of nutrients in the soil for fruit trees throughout their entire growth period is essential to achieve high yields and quality.

The higher the content of unsaturated fatty acids and essential amino acids that cannot be synthesized by the human body, the better the nutritional quality of the nucleolar [48]. In this study, a comparison between mixed fertilization (CF+M+CC, CF+CC, and CF+M) treatments with exclusive chemical fertilizer (CF) treatments revealed significant improvements in the fruit oil and unsaturated fatty acid contents (Figure 2). While the content of the monounsaturated fatty acid C18:1 (octadecenoic acid) decreased in the kernel, there were increases in the contents of polyunsaturated fatty acids such as C18:2 (octadecadienoic acid) and C18:3 (octadecatrienoic acid). These results aligned with previous research findings [18,49]. In summary, the mixed application of organic and chemical fertilizers can provide balanced nutrition for pecan at different growth stages, thereby increasing pecan yield and improving kernel quality.

The most relevant chemical properties of soils are pH, electrical conductivity, fertility level, cation exchange capacity, and organic matter content. SOM is the most crucial foundational substance for soil fertility and a primary source of plant nutrition. Organic fertilizers contains abundant organic matter, which can promote the mineralization process of organic nutrients, improve soil pH, enhance soil fertility, and ameliorate soil quality [41,50]. In this study, it was observed that treatments involving the mixed application of chemical and organic fertilizers (CF+M+CC, CF+CC, and CF+M) significantly increased the SOM content and improved soil pH, in contrast to exclusive chemical fertilizer (CF) treatments. Furthermore, the mixed fertilizer treatments significantly elevated the levels of nutrients and micronutrients of the soil (Figures 3 and 4). Generally, the P content was significantly higher in neutral and alkaline than in highly acidic, acidic and slightly acidic soils. This is most likely because at higher pH, release of P from Al and Fe phosphates occurs. However, in highly alkaline soils, phosphate ions can easily form calcium phosphate precipitation with calcium ions, thus reducing the availability of phosphorus [51]. Plants can thrive within a broad range of soil pH values. However, the effectiveness of trace nutrients is closely associated with soil pH, and at higher pH levels, the availability of most trace nutrients to plant roots tends to be lower [52].

The combination of various organic fertilizers in the soil leads to nutritional advantages and potential. Studies have shown that mixed application of organic fertilizer increased the nutrient content of geranium (*Pelargonium* spp.). Soils where organic fertilizers were used tended to exhibit higher levels of N, P, K, Ca, and Mg [53,54]. The high SOM content of organic fertilizers may serve as chelating agents for certain elements, while enhancing the solubility and availability of nutrients in the soil [55,56]. Our findings were consistent in this respect, as they revealed higher levels of nutrient release through mixed organic fertilizer treatments for SOM and soil nutrients.

There is no doubt that the presence of microorganisms and their activity are vital for the normal and sustainable state of the soil. The introduction of organic fertilizers also improves the soil structure and quality, while enhancing microbial abundance and activities [57,58]. Soil enzyme activities are considered an indicator of the activities of soil microbes. Soil enzyme activity levels can reflect the attributes of soil microbes [59,60]. Our research revealed significant differences in the S-UE, S-SC, S-POD, and S-ACP activities between the application of only chemical fertilizers and mixed organic and chemical fertilizers. This suggested that the combination of chemical and organic fertilizers enhanced soil enzyme activities, which may be attributed to the significant improvement in soil physicochemical properties by organic fertilizers [11]. Long-term soil experiments at four locations in China revealed that the application of organic fertilizers enhanced most soil enzyme activities. They were observed to improve the networks between ion composition and soil enzyme activities in the soil in contrast to chemical fertilizers, which improved its dynamics and stability [61]. Studies by Ning et al. found that the combined application of chemical and organic fertilizers significantly increased SOM, S-POD, and S-UE activities [62]. Furthermore, the carbon content of organic fertilizers can serve as a significant carbon source for soil microorganisms [63]. S-UE, S-SC, S-POD, and S-ACP activities are closely related to soil C, N, and P cycling [64–67], which suggested that organic fertilizers substantially enhanced the activities of enzymes related to soil C, N, and P cycling [68]. In addition, organic fertilizers introduce beneficial microorganisms directly into the soil, and a variety of soil microorganisms can dissolve nutrients such as P, Fe, and Zn, thereby further improving soil enzyme activity.

The manure used in this study was sourced from local farms. This approach to soil fertility management is integrated with local conditions, improves food security and environmental sustainability of agricultural systems, and maximizes crop yields while minimizing the exploitation of soil nutrient reserves and degradation of soil physical and chemical properties. The goals of high and stable yield, quality improvement and sustainable development of hickory orchard were realized.

5. Conclusions

In this study, combined organic and chemical fertilizer treatments significantly enhanced the pecans yield and improved the quality compared with the application of only chemical fertilizers. Moreover, this mixed fertilizer approach significantly increased soil nutrient levels and soil enzyme activities. Based on these findings, we recommend the use of various organic fertilizers in conjunction with chemical fertilizers in orchards to enhance the utilization of nutrients, improve soil quality, and increase crop productivity.

Author Contributions: C.F. and S.F.: conceptualization. Y.T. and Z.W.: writing of the manuscript and preparation of figures; data curation, D.G.; writing—review and editing, X.M. (Xiaoxiang Ma) and X.M. (Xiaomin Ma): visualization, C.H.; project administration, F.Y. All authors have read and agreed to the published version of the manuscript.

Funding: This work was financially supported by Special Major Science and Technology Project of Anhui Province (202103a06020007), the Forestry Science and Technology Promotion Demonstration Project of State (wan 2023TG22), the Science and Technology Project of Anhui Province (GJ2022QN14) and the Huzhou Natural Science Foundation Project (2022YZ19).

Data Availability Statement: The data presented in this study are available upon request from the corresponding author. The data are not publicly available due to compliance with data protection regulations.

Acknowledgments: We thank Xiang Ge (Fuyang harvest agricultural planting professional cooperative), Duxin Gong, Cheng Huang, Xiaomin Ma, Xiaoxiang Ma, Feiyang Yuan, (Anhui Agricultural University), Zhaocheng Wang, (Anhui Agricultural University; Huzhou vocational & technical college) for their support in the collection of field data.

Conflicts of Interest: The authors declare that the research was conducted in the absence of any commercial or financial relationships that could be construed as potential conflicts of interest.

References

1. Zhang, D.; Wang, C.; Li, X.-L.; Yang, X.-S.; Zhao, L.-B.; Xia, S.-J. Correlation of production constraints with the yield gap of apple cropping systems in Luochuan County, China. *J. Integr. Agric.* **2019**, *18*, 1714–1725. [CrossRef]
2. Conijn, J.G.; Bindraban, P.S.; Schröder, J.J.; Jongschaap, R.E.E. Can our global food system meet food demand within planetary boundaries? *Agric. Ecosyst. Environ.* **2018**, *251*, 244–256. [CrossRef]

3. Kopittke, P.M.; Menzies, N.W.; Wang, P.; McKenna, B.A.; Lombi, E. Soil and the intensification of agriculture for global food security. *Environ. Int.* **2019**, *132*, 105078. [CrossRef] [PubMed]
4. Wakelin, S.A.; Condron, L.M.; Gerard, E.; Dignam, B.E.A.; Black, A.; O'Callaghan, M. Long-term P fertilisation of pasture soil did not increase soil organic matter stocks but increased microbial biomass and activity. *Biol. Fertil. Soils* **2017**, *53*, 511–521. [CrossRef]
5. Ge, S.; Zhu, Z.; Jiang, Y. Long-term impact of fertilization on soil pH and fertility in an apple production system. *J. Soil Sci. Plant Nutr.* **2018**, *18*, 282–293. [CrossRef]
6. Wang, N.; Wolf, J.; Zhang, F.-S. Towards sustainable intensification of apple production in China—Yield gaps and nutrient use efficiency in apple farming systems. *J. Integr. Agric.* **2016**, *15*, 716–725. [CrossRef]
7. Gu, X.B.; Cai, H.J.; Du, Y.D.; Li, Y.N. Effects of film mulching and nitrogen fertilization on rhizosphere soil environment, root growth and nutrient uptake of winter oilseed rape in northwest China. *Soil Tillage Res.* **2019**, *187*, 194–203. [CrossRef]
8. Zhang, X.; Sun, N.; Wu, L.; Xu, M.; Bingham, I.J.; Li, Z. Effects of enhancing soil organic carbon sequestration in the topsoil by fertilization on crop productivity and stability: Evidence from long-term experiments with wheat-maize cropping systems in China. *Sci. Total Environ.* **2016**, *562*, 247–259. [CrossRef]
9. Zhao, Z.; Zhang, C.; Wang, H.; Li, F.; Pan, H.; Yang, Q.; Li, J.; Zhang, J. The Effects of Natural Humus Material Amendment on Soil Organic Matter and Integrated Fertility in the Black Soil of Northeast China: Preliminary Results. *Agronomy* **2023**, *13*, 794. [CrossRef]
10. Mohamed, H.; Popov, A.I.; Mohamed, R. Integrated use of bio-organic fertilizers for enhancing soil fertility-plant nutrition, germination status and initial growth of corn (*Zea mays* L.). *Environ. Technol. Innov.* **2021**, *21*, 101329. [CrossRef]
11. Huang, S.; Zhang, W.; Yu, X.; Huang, Q. Effects of long-term fertilization on corn productivity and its sustainability in an Ultisol of southern China. *Agric. Ecosyst. Environ.* **2010**, *138*, 44–50. [CrossRef]
12. Ananthi, T.; Vennila, C. Influence of Organic Manures and Inorganic Fertilizers on Growth and Yield of Fodder Maize (*Zea mays* L.) Grown in North Eastern Zone of Tamil Nadu. *Curr. J. Appl. Sci. Technol.* **2021**, *40*, 70–78. [CrossRef]
13. Ye, G.; Lin, Y.; Liu, D.; Chen, Z.; Luo, J.; Bolan, N.; Fan, J.; Ding, W. Long-term application of manure over plant residues mitigates acidification, builds soil organic carbon and shifts prokaryotic diversity in acidic Ultisols. *Appl. Soil Ecol.* **2018**, *133*, 24–33. [CrossRef]
14. Yue, X.; Zhang, J.; Shi, A.; Yao, S.; Zhang, B. Manure substitution of mineral fertilizers increased functional stability through changing structure and physiology of microbial communities. *Eur. J. Soil Biol.* **2016**, *77*, 34–43. [CrossRef]
15. Zou, C.; Li, Y.; Huang, W.; Zhao, G.; Pu, G.; Su, J.; Coyne, M.S.; Chen, Y.; Wang, L.; Hu, X.; et al. Rotation and manure amendment increase soil macro-aggregates and associated carbon and nitrogen stocks in flue-cured tobacco production. *Geoderma* **2018**, *325*, 49–58. [CrossRef]
16. Shiwakoti, L.D.; Chalise, K.; Dahal, P.; Shiwakoti, R.; Katuwal, N.; Kc, Y. Effect of fertilizer application on tea plant productivity and phytochemicals in prepared black tea. *Cogent Food Agric.* **2023**, *9*, 2184013. [CrossRef]
17. Ye, L.; Zhao, X.; Bao, E.; Li, J.; Zou, Z.; Cao, K. Bio-organic fertilizer with reduced rates of chemical fertilization improves soil fertility and enhances tomato yield and quality. *Sci. Rep.* **2020**, *10*, 177. [CrossRef]
18. Wang, Y.; Heerema, R.J.; Walworth, J.L.; Dungan, B.; VanLeeuwen, D.; Holguin, F.O. Nutraceutical Properties of Pecan Kernels Are Affected by Soil Zinc Fertilizer Application. *Hortscience* **2020**, *55*, 2001–2007. [CrossRef]
19. Zhao, Z.-B.; He, J.-Z.; Quan, Z.; Wu, C.-F.; Sheng, R.; Zhang, L.-M.; Geisen, S. Fertilization changes soil microbiome functioning, especially phagotrophic protists. *Soil Biol. Biochem.* **2020**, *148*, 107863. [CrossRef]
20. Dinca, L.C.; Grenni, P.; Onet, C.; Onet, A. Fertilization and Soil Microbial Community: A Review. *Appl. Sci.* **2022**, *12*, 1198. [CrossRef]
21. Gao, M.; Yang, J.; Liu, C.; Gu, B.; Han, M.; Li, J.; Li, N.; Liu, N.; An, N.; Dai, J.; et al. Effects of long-term biochar and biochar-based fertilizer application on brown earth soil bacterial communities. *Agric. Ecosyst. Environ.* **2021**, *309*, 107285. [CrossRef]
22. Wardle, D.A. A comparative assessment of factors which influence microbial biomass carbon and nitrogen levels in soil. *Biol. Rev.* **1992**, *67*, 321–358. [CrossRef]
23. Lori, M.; Hartmann, M.; Kundel, D.; Mayer, J.; Mueller, R.C.; Mäder, P.; Krause, H.M. Soil microbial communities are sensitive to differences in fertilization intensity in organic and conventional farming systems. *FEMS Microbiol. Ecol.* **2023**, *99*, fiad046. [CrossRef] [PubMed]
24. Kumar, S.; Bauddh, K.; Barman, S.C.; Singh, R.P. Amendments of microbial biofertilizers and organic substances reduces requirement of urea and DAP with enhanced nutrient availability and productivity of wheat (*Triticum aestivum* L.). *Ecol. Eng.* **2014**, *71*, 432–437. [CrossRef]
25. Zhalnina, K.; Louie, K.B.; Hao, Z.; Mansoori, N.; Da Rocha, U.N.; Shi, S.; Cho, H.; Karaoz, U.; Loqué, D.; Bowen, B.P.; et al. Dynamic root exudate chemistry and microbial substrate preferences drive patterns in rhizosphere microbial community assembly. *Nat. Microbiol.* **2018**, *3*, 470–480. [CrossRef] [PubMed]
26. Scharfy, D.; Güsewell, S.; Gessner, M.O.; Venterink, H.O. Invasion of Solidago gigantea in contrasting experimental plant communities: Effects on soil microbes, nutrients and plant-soil feedbacks. *J. Ecol.* **2010**, *98*, 1379–1388. [CrossRef]
27. Jalilian, J.; Modarres-Sanavy, S.A.M.; Saberali, S.F.; Sadat-Asilan, K. Effects of the combination of beneficial microbes and nitrogen on sunflower seed yields and seed quality traits under different irrigation regimes. *Field Crops Res.* **2012**, *127*, 26–34. [CrossRef]

28. Noperı-Mosqueda, L.C.; Soto-Parra, J.M.; Sánchez, E.; Piña-Ramírez, F.J.; Pérez-Leal, R.; Flores-Córdova, M.A.; Salas-Salazar, N.A. Impact of Organic and Mineral Fertilization in Pecan Nut on Production, Quality and Antioxidant Capacity. *Agric. Sci.* **2019**, *10*, 227–240. [CrossRef]
29. Atanasov, A.G.; Sabharanjak, S.M.; Zengin, G.; Mollica, A.; Szostak, A.; Simirgiotis, M.; Huminiecki, L.; Horbanczuk, O.K.; Nabavi, S.M.; Mocan, A. Pecan nuts: A review of reported bioactivities and health effects. *Trends Food Sci. Technol.* **2018**, *71*, 246–257. [CrossRef]
30. Domínguez-Avila, J.A.; Alvarez-Parrilla, E.; López-Díaz, J.A.; Maldonado-Mendoza, I.E.; Gómez-García, M.d.C.; de la Rosa, L.A. The pecan nut (*Carya illinoinensis*) and its oil and polyphenolic fractions differentially modulate lipid metabolism and the antioxidant enzyme activities in rats fed high-fat diets. *Food Chem.* **2015**, *168*, 529–537. [CrossRef]
31. de la Rosa, L.A.; Vazquez-Flores, A.A.; Alvarez-Parrilla, E.; Rodrigo-García, J.; Medina-Campos, O.N.; Ávila-Nava, A.; González-Reyes, S.; Pedraza-Chaverri, J. Content of major classes of polyphenolic compounds, antioxidant, antiproliferative, and cell protective activity of pecan crude extracts and their fractions. *J. Funct. Foods* **2014**, *7*, 219–228. [CrossRef]
32. Reisner, Y.; de Filippi, R.; Herzog, F.; Palma, J. Target regions for silvoarable agroforestry in Europe. *Ecol. Eng.* **2006**, *29*, 401–418. [CrossRef]
33. Milošević, T.; Milošević, N. Soil fertility: Plant nutrition vis-à-vis fruit yield and quality of stone fruits. *Fruit Crops* **2020**, *45*, 583–606. [CrossRef]
34. Zhu, M.; Long, Y.; Chen, Y.; Huang, Y.; Tang, L.; Gan, B.; Yu, Q.; Xie, J. Fast determination of lipid and protein content in green coffee beans from different origins using NIR spectroscopy and chemometrics. *J. Food Compos. Anal.* **2021**, *102*, 104055. [CrossRef]
35. El Riachy, M.; Hamade, A.; Ayoub, R.; Dandachi, F.; Chalak, L. Oil Content, Fatty Acid and Phenolic Profiles of Some Olive Varieties Growing in Lebanon. *Front. Nutr.* **2019**, *6*, 94. [CrossRef] [PubMed]
36. Yang, S.; Feng, C.; Ma, Y.; Wang, W.; Huang, C.; Qi, C.; Fu, S.; Chen, H.Y. Transition from N to P limited soil nutrients over time since restoration in degraded subtropical broadleaved mixed forests. *For. Ecol. Manag.* **2021**, *494*, 119298. [CrossRef]
37. Soltanpour, P.N.; Schwab, A.P. A new soil test for simultaneous extraction of macro- and micro-nutrients in alkaline soils. *Commun. Soil Sci. Plant Anal.* **1977**, *8*, 195–207. [CrossRef]
38. Sun, X.; Zhu, L.; Wang, J.; Wang, J.; Su, B.; Liu, T.; Zhang, C.; Gao, C.; Shao, Y. Toxic effects of ionic liquid 1-octyl-3-methylimidazolium tetrafluoroborate on soil enzyme activity and soil microbial community diversity. *Ecotoxicol. Environ. Saf.* **2017**, *135*, 201–208. [CrossRef]
39. Kandeler, E.; Gerber, H. Short-term assay of soil urease activity using colorimetric determination of ammonium. *Biol. Fertil. Soils* **1988**, *6*, 68–72. [CrossRef]
40. Taylor, J.P.; Wilson, B.; Mills, M.S.; Burns, R.G. Comparison of microbial numbers and enzymatic activities in surface soils and subsoils using various techniques. *Soil Biol. Biochem.* **2002**, *34*, 381–401. [CrossRef]
41. Ai, C.; Liang, G.Q.; Sun, J.W.; Wang, X.B.; Zhou, W. Responses of extracellular enzyme activities and microbial community in both the rhizosphere and bulk soil to long-term fertilization practices in a fluvo-aquic soil. *Geoderma* **2012**, *173*, 330–338. [CrossRef]
42. Somenahally, A.C.; McLawrence, J.; Chaganti, V.N.; Ganjegunte, G.K.; Obayomi, O.; Brady, J.A. Response of soil microbial Communities, inorganic and organic soil carbon pools in arid saline soils to alternative land use practices. *Ecol. Indic.* **2023**, *150*, 110227. [CrossRef]
43. Kumar, T.; Swaminathan, V.; Kumar, S.N. Influence of nitrogen, phosphorus and biofertilizers on growth, yield and essential oil constituents in ratoon crop of davana (*Artemisia pallens* Wall.). *Electron. J. Environ. Agric. Food Chem.* **2009**, *8*, 86–95.
44. Chen, L.; Wang, X.; Zhou, W.; Guo, S.; Zhu, R.; Qin, Y.; Sun, J. Responses of crop yields, soil enzymatic activities, and microbial communities to different long-term organic materials applied with chemical fertilizer in purple soil. *Eur. J. Soil Biol.* **2021**, *105*, 103319. [CrossRef]
45. Moeskops, B.; Sukristiyonubowo; Buchan, D.; Sleutel, S.; Herawaty, L.; Husen, E.; Saraswati, R.; Setyorini, D.; Neve, S.D. Soil microbial communities and activities under intensive organic and conventional vegetable farming in West Java, Indonesia. *Appl. Soil Ecol.* **2010**, *45*, 112–120. [CrossRef]
46. Zheng, H.; Wang, X.; Luo, X.; Wang, Z.; Xing, B. Biochar-induced negative carbon mineralization priming effects in a coastal wetland soil: Roles of soil aggregation and microbial modulation. *Sci. Total Environ.* **2018**, *610–611*, 951–960. [CrossRef] [PubMed]
47. Deljooy-e-Tohidi, T.; Torkashv, A.M.; Hashemabadi, D. The possibility using some organic wastes as growth medium and nutrition method on the growth of English daisy (*Bellis perennis*). *Eur. J. Exp. Biol.* **2013**, *3*, 139–147.
48. Pardon, P.; Mertens, J.; Reubens, B.; Reheul, D.; Coussement, T.; Elsen, A.; Nelissen, V.; Verheyen, K. Juglans regia (walnut) in temperate arable agroforestry systems: Effects on soil characteristics, arthropod diversity and crop yield. *Renew. Agric. Food Syst.* **2019**, *35*, 1–17. [CrossRef]
49. Verardo, V.; Riciputi, Y.; Sorrenti, G.; Ornaghi, P.; Marangoni, B.; Caboni, M.F. Effect of nitrogen fertilisation rates on the content of fatty acids, sterols, tocopherols and phenolic compounds, and on the oxidative stability of walnuts. *LWT Food Sci. Technol.* **2013**, *50*, 732–738. [CrossRef]
50. Du, Y.; Cui, B.; Zhang, Q.; Wang, Z.; Sun, J.; Niu, W. Effects of manure fertilizer on crop yield and soil properties in China: A meta-analysis. *Catena* **2020**, *193*, 104617. [CrossRef]
51. Krzyszczak, J.; Brodowska, M.; Bednarek, W.; Dresler, S.; Tkaczyk, P.; Krzyszczak, J.; Baranowski, P. Content of certain macro- and microelements in orchard soils in relation to agronomic categories and reaction of these soils. *J. Elem.* **2018**, *23*, 1361–1372. [CrossRef]

52. Wang, G.-Y.; Zhang, X.-Z.; Wang, Y.; Xu, X.-F.; Han, Z.-H. Key minerals influencing apple quality in Chinese orchard identified by nutritional diagnosis of leaf and soil analysis. *J. Integr. Agric.* **2015**, *14*, 864–874. [CrossRef]
53. Cai, A.D.; Xu, M.G.; Wang, B.R.; Zhang, W.J.; Liang, G.P.; Hou, E.Q.; Luo, Y.Q. Manure acts as a better fertilizer for increasing crop yields than synthetic fertilizer does by improving soil fertility. *Soil Tillage Res.* **2019**, *189*, 168–175. [CrossRef]
54. Choudhary, M.; Panday, S.C.; Meena, V.S.; Singh, S.; Yadav, R.P.; Mahanta, D.; Mondal, T.; Mishra, P.K.; Bisht, J.K.; Pattanayak, A. Long-term effects of organic manure and inorganic fertilization on sustainability and chemical soil quality indicators of soybean-wheat cropping system in the Indian mid-Himalayas. *Agric. Ecosyst. Environ.* **2018**, *257*, 38–46. [CrossRef]
55. Jones, D.L.; Darrah, P.R. Re-sorption of organic compounds by roots of *Zea mays* L. and its consequences in the rhizosphere: III. Characteristics of sugar influx and efflux. *Plant Soil* **1996**, *178*, 153–160. [CrossRef]
56. Ni, H.; Su, W.; Fan, S.; Chu, H. Effects of intensive management practices on rhizosphere soil properties, root growth, and nutrient uptake in Moso bamboo plantations in subtropical China. *For. Ecol. Manag.* **2021**, *493*, 119083. [CrossRef]
57. Sun, B.; Gao, Y.; Wu, X.; Ma, H.; Zheng, C.; Wang, X.; Zhang, H.; Li, Z.; Yang, H. The relative contributions of pH, organic anions, and phosphatase to rhizosphere soil phosphorus mobilization and crop phosphorus uptake in maize/alfalfa polyculture. *Plant Soi* **2020**, *447*, 117–133. [CrossRef]
58. Guo, Z.; Han, J.; Li, J.; Xu, Y.; Wang, X. Effects of long-term fertilization on soil organic carbon mineralization and microbial community structure. *PLoS ONE* **2019**, *14*, e0211163. [CrossRef]
59. Nannipieri, P.; Ascher, J.; Ceccherini, M.T.; Landi, L.; Pietramellara, G.; Renella, G. Microbial diversity and soil functions. *Eur. J. Soil Sci.* **2003**, *54*, 655–670. [CrossRef]
60. Schloter, M.; Nannipieri, P.; Sørensen, S.J.; van Elsas, J.D. Microbial indicators for soil quality. *Biol. Fertil. Soils* **2018**, *54*, 1–10. [CrossRef]
61. Feng, X.; Ling, N.; Chen, H.; Zhu, C.; Duan, Y.; Peng, C.; Yu, G.; Ran, W.; Shen, Q.; Guo, S. Soil ionomic and enzymatic responses and correlations to fertilizations amended with and without organic fertilizer in long-term experiments. *Sci. Rep.* **2016**, *6*, 24559. [CrossRef] [PubMed]
62. Ning, C.-C.; Gao, P.-D.; Wang, B.-Q.; Lin, W.-P.; Jiang, N.-H.; Cai, K.-Z. Impacts of chemical fertilizer reduction and organic amendments supplementation on soil nutrient, enzyme activity and heavy metal content. *J. Integr. Agric.* **2017**, *16*, 1819–1831. [CrossRef]
63. Cheng, J.; Jin, H.; Zhang, J.; Xu, Z.; Yang, X.; Liu, H.; Xu, X.; Min, D.; Lu, D.; Qin, B. Effects of Allelochemicals, Soil Enzyme Activities, and Environmental Factors on Rhizosphere Soil Microbial Community of *Stellera chamaejasme* L. along a Growth-Coverage Gradient. *Microorganisms* **2022**, *10*, 158. [CrossRef] [PubMed]
64. Li, Y.; Wang, Y.; Gu, C.; Shen, C.; Xu, L.; Zhao, Y.; Yi, S.; Zuo, W.; Shan, Y.; Zhang, Z.; et al. Differential Effects of Organic Ameliorants on the Reassembly of Bacterial Communities in Newly Amended Coastal Mudflat Salt-Affected Soil. *Agronomy* **2022**, *12*, 2525. [CrossRef]
65. Parlak, K.U.; Yilmaz, D.D. Response of antioxidant defences to Zn stress in three duckweed species. *Ecotoxicol. Environ. Saf.* **2012**, *85*, 52–58. [CrossRef]
66. Duff, S.M.G.; Sarath, G.; Plaxton, W.C. The role of acid phosphatases in plant phosphorus metabolism. *Physiol. Plant.* **1994**, *90*, 791–800. [CrossRef]
67. Sannino, F.; Gianfreda, L. Pesticide influence on soil enzymatic activities. *Chemosphere* **2001**, *45*, 417–425. [CrossRef]
68. Robinson, M.D.; McCarthy, D.J.; Smyth, G.K. edgeR: A Bioconductor package for differential expression analysis of digital gene expression data. *Bioinformatics* **2010**, *26*, 139–140. [CrossRef]

horticulturae

MDPI

Article

Effect of Biofertilizers on Broccoli Yield and Soil Quality Indicators

Irene Ollio [1,2,*], Vanesa Santás-Miguel [3,4], Diego Soto Gómez [1,3], Eva Lloret [1,2], Virginia Sánchez-Navarro [1,2], Silvia Martínez-Martínez [1], Catalina Egea-Gilabert [1,2], Juan A. Fernández [1,2], David Fernández Calviño [3,4] and Raúl Zornoza [1,2]

[1] Grupo de Gestión, Aprovechamiento y Recuperación de Suelos y Agua (GARSA), Departamento de Ingeniería Agronómica, Universidad Politécnica de Cartagena, Paseo Alfonso XIII 48, 30203 Cartagena, Spain; diego.soto@upct.es (D.S.G.); eva.lloret@upct.es (E.L.); virginia.sanchez@upct.es (V.S.-N.); silvia.martinez@upct.es (S.M.-M.); catalina.egea@upct.es (C.E.-G.); juan.fernandez@upct.es (J.A.F.); raul.zornoza@upct.es (R.Z.)

[2] Instituto de Biotecnología Vegetal (IBV), Edificio I+D+I, Campus Muralla del Mar, Universidad Politécnica de Cartagena, 30202 Cartagena, Spain

[3] Departamento de Bioloxía Vexetal e Ciencia do Solo, Facultade de Ciencias, Universida de Vigo, As Lagoas s/n, 32004 Ourense, Spain; vsantas@uvigo.es (V.S.-M.); davidfc@uvigo.gal (D.F.C.)

[4] Instituto de Agroecoloxía e Alimentación (IAA), Universidade de Vigo—Campus Auga, 32004 Ourense, Spain

* Correspondence: irene.ollio@upct.es

Abstract: High rates of fertilizer applications potentially have significant environmental consequences, such as soil and water contamination and biodiversity loss. This study aimed to compare the use of biofertilizers and inorganic fertilizers in a broccoli crop to determine their impact on soil microorganism abundance, microbial community structure, functional gene diversity, yield, and greenhouse gas emissions. Four different fertilization treatments were designed: (i) inorganic fertilizers applied at a rate to cover the nutritional demands of the crop (F100); (ii) 50% of the rate of inorganic fertilizers added in F100 (F50); (iii) F50 + the application of a formulation of various bacteria (BA); and (iv) F50 + the application of a formulation of bacteria and non-mycorrhizal fungi (BA + FU). The results showed that reduced fertilization and the addition of both biofertilizer products had no significant effect on soil nutrients, microbial population, microbial activity, or yield when compared to conventional inorganic fertilization. Thus, microbial inoculants were ineffective in enhancing soil microbial abundance and activity, and there were no changes in GHG emissions or crop yields. Nonetheless, crop yield was positively related to total soil N, microbial activity, and CO_2 emissions, confirming the positive effect of soil biodiversity on production. The application of biofertilizers can help reduce mineral fertilization in a broccoli crop with no negative effect on yield.

Keywords: CO_2; N_2O; CH_4; biofertilizers; enzyme activities; PLFAs; *Brassica oleracea* var italica Plenck; nutrients

Citation: Ollio, I.; Santás-Miguel, V.; Gómez, D.S.; Lloret, E.; Sánchez-Navarro, V.; Martínez-Martínez, S.; Egea-Gilabert, C.; Fernández, J.A.; Calviño, D.F.; Zornoza, R. Effect of Biofertilizers on Broccoli Yield and Soil Quality Indicators. *Horticulturae* **2024**, *10*, 42. https://doi.org/10.3390/horticulturae10010042

Academic Editors: Francesco De Mastro, Gennaro Brunetti, Karam Farrag and Huadong Zang

Received: 30 November 2023
Revised: 29 December 2023
Accepted: 30 December 2023
Published: 31 December 2023

1. Introduction

Agricultural soils act both as a source and sink for greenhouse gases (GHGs) such as carbon dioxide (CO_2), methane (CH_4), and nitrous oxide (N_2O) [1,2]. N_2O and CH_4 are potent greenhouse gases, with a global warming potential that is, respectively, 296 times and 23 times greater than that of CO_2 [3]. Soil management practices (such as land use, nutrient application, tillage, and reduction in soil compaction) can indirectly influence these fluxes [4]. Furthermore, emission flux rates largely depend on soil water content, soil temperature, nutrient availability, organic matter quantity and quality, and pH [5]. Inorganic fertilizers play an essential role in enhancing crop productivity and soil fertility. However, N fertilizer can also directly influence GHG emissions [6]. Land use influences GHG emissions, which are increasingly higher in agricultural soils [7,8]. The use of nitrogen fertilization [9–11], irrigation [12–14], soil temperature [12,15], changes in microbial biomass [16], and nutrient availability [2] can explain an important part of the temporal variation of GHG fluxes in agroecosystems. Soil GHG emissions are therefore a key issue for climate research and agricultural management [17].

Soils and plants generate CO_2 fluxes through plant aboveground respiration, root respiration, and anaerobic and aerobic microbial respiration processes, with root respiration accounting for 50% of total soil respiration [2]. In the soils, possible sources of NO and N_2O include nitrification by autotrophic and heterotrophic nitrifiers, denitrification by nitrifiers and denitrifiers, nitrate respiration by fermenters, and chemodenitrification [18]. These processes can occur simultaneously, in different microsites of the soil, with consequent increases in N_2O production [19], although this is dependent on the N rate and type [20]. Understanding the responses of N_2O-producing microorganisms to changes in environmental conditions or input handling is the key to regulating gaseous N_2O losses [21]. The production of CH_4 is the result of the anaerobic activity of different groups of microorganisms, including zymogenic bacteria, acetic acid and hydrogen producers, and methanogens [22]. Normally, when the oxygen supply is adequate, most of the C in decomposing organic matter converts to CO_2. Furthermore, under aerobic conditions, CH_4 that has been produced in anaerobic soil microsites and atmospheric CH_4 can be oxidized, resulting in soils "absorbing" CH_4 [4]. In addition to high GHG emissions, the excessive use of inorganic fertilizers has been associated with soil and water pollution by leaching and runoff [23].

An alternative to inorganic fertilizers in agriculture to reduce pollution and GHG emissions is the use of biofertilizers. These are substances that contain living plant growth-promoting rhizobacteria (PGPR) [24] and plant growth-promoting fungi (PGPF) [25], which, when applied to seed, plant surfaces, or soil, colonize the rhizosphere or the interior of the plant and encourage development by improving the supply or availability of primary nutrients to the host plant, as well as providing indirect biological control of plant diseases [26–28]. According to numerous studies, they improve soil fertility by fixing the atmospheric N [29], solubilizing insoluble phosphates [30] and potassium [31], and producing plant growth-promoting substances in the soil [32]. With plant growth-promoting microorganisms (PGPMs), plant growth can be stimulated, and so more C can be allocated to plant biomass, which is a prominent option for climate change mitigation. Furthermore, by enhancing the production of glomalin in the rhizosphere by increasing mycorrhizal colonization, PGPM creates important reservoirs of C and N in the soil [33]. The net transfer of biologically fixed N directly from the bacteria to the host plant occurs concurrently with a significant transfer of photosynthetically fixed plant carbon to the N-fixing bacteria [34]. According to various studies, the inoculation of PGPM significantly reduces GHG emissions depending on the inoculation dose, the different humidity levels of the growth substrate, and the C and N availability of the soil [35–37]. This could be critical in evaluating and mitigating the environmental impacts of various agricultural management practices. Other compounds responsible for indirect stimulation of plant growth and produced by microorganisms, and which may be enhanced by the use of biofertilizers (PGPR and/or PGPF), include enzymes, nitric oxide, osmolytes, siderophores, organic acids, and antibiotics [38,39]. Moreover, plants exude ethylene into the soil, especially during stressful events, which inhibits oxidation of the methane present in the soil. The biosynthesis of ethylene can be interrupted by the activity of the enzyme ACC deaminase. The use of biofertilizers with microorganisms able to express this enzyme is perhaps the simplest and most effective way to reduce the inhibitory effect of ethylene on CH_4 oxidation and plant growth while reducing the CH_4 emissions [40].

Although the exact mechanisms of plant growth stimulation are vastly complicated, it is known that they differ between fungal and bacterial strains, environmental conditions, crops, and cultivated genotypes, and most certainly depend on the various compounds released by the different microorganisms [38,39]. Soil microbiota and their activity are crucially important and actively involved in soil fertility, sustainability, and crop production [41,42]. Assessing soil microorganism abundance and structure following biofertilizer application is critical for optimizing the sustainability and fertility of both soils and crops, as well as determining how the biofertilizer may affect native microbial communities [43]. Along with microbial abundance, their functionality is also of great importance. Some of the major microbial gene clusters in soil are related to C and N cycles because they are involved in the supply of utilizable nutrients to the crops. Most important genes are related to ammonia-oxidizing enzymes (*amoA*) [44], nitrite reductases encoding for denitrifying process (*nirK*) [45], N2-fixing microbial gene clusters such as *nifH* [46], CO_2-fixing microbial genes such as *cbbL* [47], or cellulose-breaking activity through cellobiohydrolase coding genes (*GH7*) [48]. Related to soil microbial functioning, soil enzymes, mainly produced by the cellular metabolism of soil microorganisms, catalyze processes of decomposition of organic matter and influence the cycle of nutrients [49,50]. Soil enzymes such as β-glucosidase, cellulase, urease, and arylesterase are involved in the C and N cycle [51–53]. The sensitivity of soil microbial indicators to soil management has been reported to be higher than that of soil physicochemical properties, and so is more suited to explaining changes in soil GHG emissions [54,55]. Several studies have demonstrated significant changes in soil indicators such as enzymes and microbial community structure and abundance [56–58], and the abundance of functional genes [59,60], after inoculation with PGPR or PGPF.

We designed an experiment comparing the use of inorganic fertilizers in a broccoli crop with the use of two types of biofertilizers: bacteria and bacteria + fungi, associated with a decrease of 50% in the inorganic fertilization rate. Broccoli is a crop with high demand for soil fertilizers, with potential negative environmental impacts in the regions where it is produced, mostly related to soil and water pollution and low biodiversity [61]. Thus, the partial substitution of mineral fertilizers by a microbial inoculant may contribute to reducing the negative impacts of high fertilization. Thus, we expected that BA and BA + FU treatments in a broccoli crop would increase soil GHG emissions due to increased microbial metabolism. The abundance of soil organisms may increase with the addition of biofertilizers owing to the incorporation of new microbes into the soil, although the release of allelopathic compounds by broccoli, as a Brassicaceae species, may restrain these increases. Furthermore, the microbial community structure might differ due to lower nutrient availability as a consequence of a lower rate of fertilizer application and the addition of external beneficial organisms. The activity of soil microorganisms may enhance crop yield by solubilizing soil nutrients, and increase plant protection, associated with lower CO_2 equivalent emissions per unit of crop production. Furthermore, GHG emissions may be related to higher soil organic carbon content, bioavailable soil nitrogen forms, and microbial activity measured as enzyme activities and functional genes. The objectives of this study were to: (i) assess whether the use of biofertilizers may modify the abundance of soil microorganisms, the soil microbial community structure, and the diversity of functional genes related to the C and N cycles and yield in a broccoli crop, compared to a broccoli crop fertilized only with inorganic fertilizers; (ii) assess if the use of biofertilizers may modify soil GHG emissions owing to a more active soil microbiota; and (iii) elucidate if soil GHG emissions are related to soil chemical and biological properties and crop yield.

2. Materials and Methods

2.1. Study Site and Experimental Design

This study was carried out in Cartagena, southeastern Spain, at the Tomás Ferro Experimental Field of the Polytechnic University of Cartagena (UPCT), Spain (37°41′16.6″ N 0°56′55.6″ W). The climate is semiarid Mediterranean with a total annual precipitation of 275 mm and a mean annual temperature of 18 °C. Annual potential evapotranspiration surpasses 900 mm. Soil is classified as Haplic Calcisol (loamic, hypercalcic) [62], with clay loam texture, organic matter content of 1.80%, and pH of 8. Soil analyses were performed in the laboratories of the research group of GARSA (Management, Use and Recovery of Soils and Water), UPCT. The experiment was performed on a broccoli crop (*Brassica* oleracea var italica Plenck, cultivar Parthenon), (Sakata Seed Iberica—Murcia, Spain) grown from 5 October 2021 to 10 January 2022. Prior to our experimental setup, there was a crop of potato from December 2020 to May 2021. A 1 year field experiment aimed to evaluate the impact of biofertilizers on broccoli, a crop that requires crop rotation to avoid soil and biodiversity issues. As a common practice in this area, farmers tend to rotate crops every year to avoid negative effects of monocultures. The experiment focused on the effectiveness of biofertilizers on Brassicaceae, which release allelopathic compounds. Treatments, including reduced mineral fertilization and fertilizer addition, began in January 2021 with a potato crop. However, no data on microbial abundance, activity, or GHG emissions were collected until the establishment of the broccoli crop in October 2021. Following the local practices, the crop was established under drip irrigation and inorganic fertilization. The separation of seedlings at planting was 100 cm × 20 cm. Thus, the density of the broccoli plants was 50,000 plants ha^{-1}. Four different fertilization treatments were designed: (i) inorganic fertilizers applied at the nutritional demands of the crop (F100); (ii) 50% of the rate of inorganic fertilizers added in F100 (F50); (iii) F50 + the application of a formulation of nitrogen-fixing and phosphorus- and potassium-solubilizing bacteria (BA); and (iv) F50 + the application of a formulation of bacteria and non-mycorrhizal fungi (BA + FU). The formulation of nitrogen-fixing and phosphorus- and potassium-solubilizing bacteria was mostly based on plant growth-promoting rhizobacteria (PGPR) such as *Azospirillum*, *Pseudomonas*, and *Bacillus* (Bactoneco®); the formulation of bacteria and non-mycorrhizal fungi was mostly based on a mix of PGPR and beneficial fungi such as *Bacillus*, *Azotobacter*, and non-mycorrhizal fungi (Nuve®). These products were provided by Fertilizantes y Nutrientes Ecológicos, S.L. (Spain), and the exact compositions were not shared due to the protection of intellectual property rights of the providers. The field experiment was established as a completely randomized design with four replications, and each plot had a size of 700 m^2. All treatments received the same quantity of irrigation (1100 m^3 ha^{-1}). The irrigation was scheduled according to the climatic conditions, crop coefficient, and evapotranspiration rate. Meteorological data were measured using an automatic weather station located on the experimental farm (Figure 1A1). In all treatments, the soil was tilled at a 25 cm depth and a preparatory herbicide application (Metazachlor 50% (SC) p/v) was carried out; subsequently, the crop was kept free from weeds by manual hoeing when necessary. The inorganic fertilization rate in F100 consisted of 158 kg ha^{-1} of N, 68 kg ha^{-1} of P_2O_5, and 255 kg ha^{-1} of K_2O applied by fertigation through the lifespan of the crop [63]. The used fertilizers included ammonium nitrate (34.5% N), monoammonium phosphate (61% P_2O_5, 12% N), and potassium nitrate (46% K_2O, 13% N). Biofertilizers were applied by drip irrigation according to the producer's recommendations. The product dose used in the BA treatment consisted of two applications (30 November 2021; 9 December 2021) of Bactoneco N®, Bactoneco P®, and Bactoneco K® at a total dose of 6 L ha^{-1}. The BA + FU treatment was 30 L ha^{-1}, divided into three applications (30 November 2021; 9 December 2021; 14 December 2021) of Nuve® product. An insecticide (1.5% p/v Lambda-cyhalothrin) and fungicide (25% p/v Azoxistrobin) were applied as a single preventative treatment on 2 November 2021.

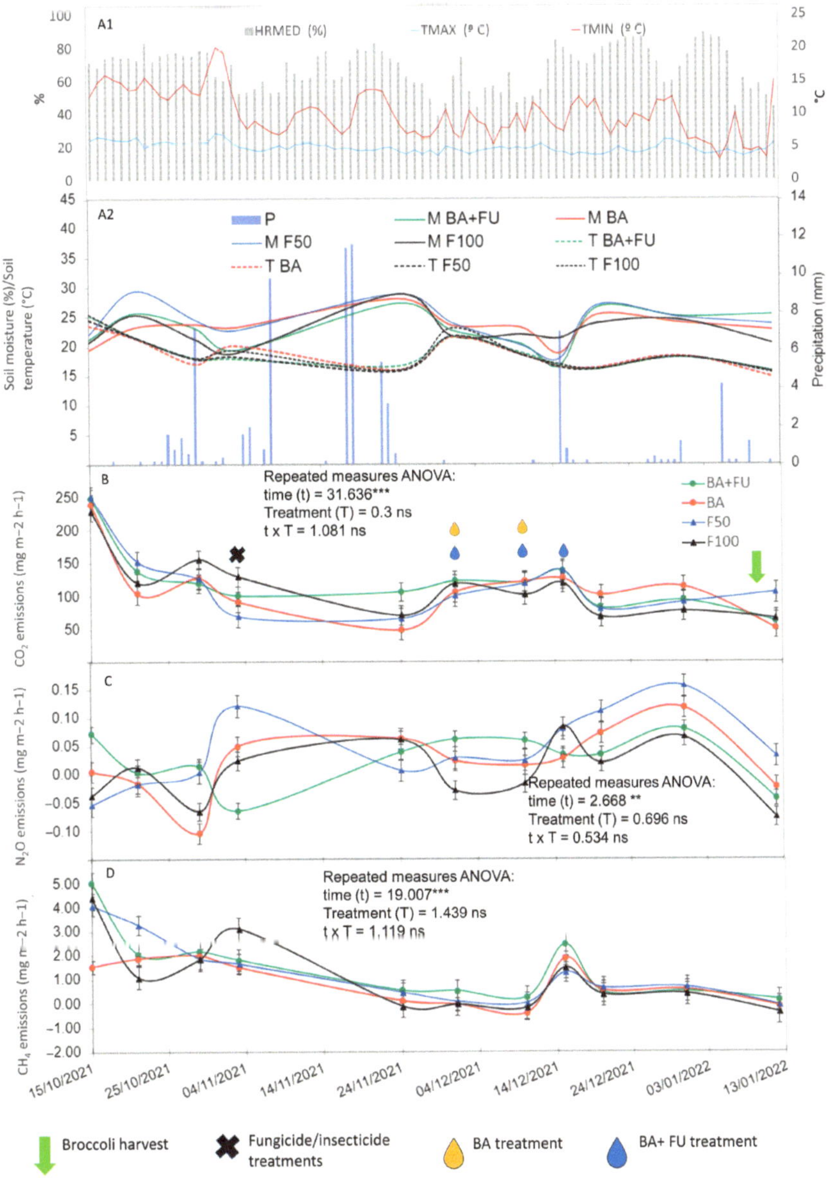

Figure 1. Environmental conditions over the course of the experiment (**A1**,**A2**), soil CO_2 emission rates (**B**), soil N_2O emission rates (**C**), and CH_4 emission rates (**D**) in broccoli cultivation with different fertilization treatments. Vertical bars denote the standard error. P: precipitation; T: soil temperature.; M: soil moisture; TMAX and TMIN: maximum and minimum temperature, respectively; HMED: relative humidity; BA + FU (F50 + the application of a formulation of bacteria and non-mycorrhizal fungi), BA (F50 + the application of a formulation of nitrogen-fixing and phosphorus- and potassium-solubilizing bacteria), F50 (50% of the rate of inorganic fertilizers added in F100), F100 (inorganic fertilizers applied at the rate to cover the nutritional demands of the crop). Blue and yellow icons indicate the days of applications of biofertilizers. Arrows indicate the days of harvest. For repeated measures ANOVA data: significant at *** $p < 0.001$; ** $p < 0.01$; ns: not significant ($p > 0.05$).

2.2. Soil Greenhouse Gas Measurements

Measurements of CO_2, N_2O, and CH_4 were made every 7 days in all replicated treatments from 15 October 2021 to 10 January 2022, between 9:00 and 12:00. Moreover, we measured the GHGs 24 h after a fertigation episode or the addition of biofertilizers. This procedure was established because microorganisms are considered to bioactivate after 24 h of application in soil [64]. The basic experimental procedure used in this study was the dynamic gas chamber technique [65]. The chamber was made of non-oxidizable steel, with a diameter of 7.5 cm and a height of 20 cm, with one inlet and one outlet connected to a photoacoustic infrared spectroscopy multi-gas analyzer with an ultra-sensitive cantilever pressure sensor (Gasera One, Gasera Ltd., Helsinki, Finland). The dynamic system with an inlet and outlet in the chamber permits a continuous flow and avoids pressure fluctuations. The chambers were inserted into the bare soil to a depth of 10 cm, within two broccoli plants. CO_2, N_2O, and CH_4 were quantified every 1 min for a period of 5 min to assess the linear trend. CO_2, N_2O, and CH_4 emission rates were expressed as the difference between the quantification at the end and the beginning of the measurement period divided by the time. CO_2, N_2O, and CH_4 cumulative emissions for each treatment were estimated using numerical integration [66]. GHG emissions were converted into CO_2 equivalent (CO_2e), and then cumulative emission data (g m^{-2}) were also expressed on a production basis (g kg^{-1}) for the experimental period to assess the emissions per product of each treatment. Soil temperature (T) and soil moisture (M) were measured using ProCheck and 5TM sensors (Decagon Devices Inc., Pullman, WA, USA), introduced at a 15 cm depth adjacent to the place where GHG measurements were made.

2.3. Soil and Plant Sampling

Soil sampling was carried out on 17 December 2021, after the end of the fertigation schedule in all treatments. All plots were sampled at a 0–25 cm depth (Ap horizon). One composite soil sample derived from four sampling points per plot was collected, thereby avoiding the border effect. Soil was collected in the crop line between two plants. Each sample was divided into two aliquots in the field. The first aliquot was air dried for 7 days, sieved <2 mm, and stored at room temperature for chemical analyses and enzyme activities [67]. The other aliquot was stored in a cool box with ice to be taken to the lab immediately and stored at 4 °C for nitrate and ammonium analysis, and at −20 °C for molecular and phospholipid fatty acid (PLFA) analysis. Harvesting was performed on 5 January 2022 and 10 January 2022, collecting the heads that were formed with the buds of the head firm and tight. Broccoli crop yield was determined by weighing the heads when they reached the marketable size.

2.4. Soil Chemical and Biochemical Analyses

Soil organic carbon (SOC) and total nitrogen (Nt) were analyzed using an elemental CHN (CHN 628, Leco). The soluble carbon (Csol) and soluble nitrogen (Nsol) were extracted with 0.5 M K_2SO_4 (1:5 ratio w/v) [68] and measured using a CN analyzer for liquid samples (Multi N/C 3100 Analytic Jena). Soil NO^{3-} was extracted with deionized water in a 1:10 ratio (w/v) [69] and measured via ion chromatography (Metrohm 861). The NH_4^+ was extracted with 2 M KCl in a 1:10 ratio (w/v) and colorimetrically measured [70]. The β-glucosidase activity (Glu) was measured based on the determination of p-nitrophenol released after incubation at 37 °C with β-D-glucopyranoside [71]. The arylesterase activity (Aryl) was determined based on the production of p-nitrophenol released after incubation with p-nitrophenyl acetate at 37 °C [67]. The cellulase activity (Cls) was assessed via the determination of gearbox sugars using amorphous cellulose as a substrate [72,73]. The urease activity (Urs) was based on the determination of the ammonium released after incubation of the soil with urea at 37 °C [74]. All analyses are reported on an air-dry weight basis.

2.5. DNA Extraction and Quantitative PCR (qPCR) Gene Analysis

Quantitative PCR (qPCR) analysis was used to quantify the copy number of the microbial functional genes *amoA* (ammonium-oxidizing, nitrifying bacteria), *nirK* (nitrite reductase, denitrifying bacteria), and *nifH* (nitrogenase, N-fixing bacteria) involved in the nitrogen cycle, and *cbbL* red-like (ribulose-1,5-bisphosphate carboxylase/oxygenase in autotrophic bacteria) and *GH7* (cellulose degradation) involved in the carbon cycle. Soil DNA was extracted from 0.25 g (wet weight) of soil using the DNeasy PowerSoil Pro Kit (Qiagen, Hilden, Germany), following the manufacturer's protocol. The DNA was eluted in a final volume of 60 µL. The quantity and quality of the DNA extracts were quantified using a Qubit 3.0 Fluorometer (Invitrogen, Thermo Fisher Scientific, Waltham, MA, USA) and a Nano Photometer N60 (Implen Scientific Inc., Germany), respectively. Subsequently, extracts were purified with magnetic beads (AMPure XP beads (Beckman Coulter, High Wycombe, UK). For the construction of the qPCR standards, the DNA extracted from soil samples as described above served as the template of PCR reactions. Amplification was performed in a MultiGene OptiMax Thermalcycler (Labnet International Inc., New York, NY, USA) and conducted using the primer pairs listed in Supplementary Table S1. The expected size of the PCR products was verified by electrophoresis on a 1.5% (*w*/*v*) agarose gel in 1× Tris-acetate-EDTA (TAE) buffer stained with ethidium bromide. Triplicate amplicons were pooled and purified using the QIAquick PCR Purification Kit (Qiagen, Hilden, Germany), and quantified with a Qubit® 2.0 Fluorometer and Qubit® dsDNA HS Assay Kit (Invitrogen, Merelbeke, Belgium). The purified PCR product was ligated into the pGEM®-T Easy Vector Systems kit (Promega, Madison, WI, USA), and the resulting ligation products were used to transform into *Escherichia coli* JM109 competent cells (Promega, Madison, WI, USA) following the manufacturer's instructions. Transformants were grown on LB plates containing ampicillin (100 µg mL^{-1}), IPTG (0.5 mM), and X-Gal (80 µg mL^{-1}). Individual white colonies were randomly selected and cultured overnight at 37 °C in 5 mL of LB broth medium (Lennox) supplemented with ampicillin (100 µg mL^{-1}), and the plasmids were extracted and purified (QIAprep Spin Miniprep Kit, Qiagen, Hilden, Germany). Cloning screening was performed with reamplification using the vector-specific M13F and M13R primers (Table S1 [46,75–78]) and the PCR products were examined via electrophoresis (1.5% (*w*/*v*) agarose gel in 1× TAE buffer). All plasmid standards were digested with NdeI restriction enzyme (New England BioLabs, Ipswich, MA, USA) with a restriction enzyme reaction composed of 20 µL of plasmid DNA, 5 µL NEBuffer (New England Biolabs, Ipswich, MA, USA), and 1 µL/100 U NdeI in a final volume of 50 µL. The enzymatic reaction was carried out for 1 h at 37 °C and the linearized plasmid DNA was purified (QIAquick PCR Purification Kit, Qiagen, Hilden, Germany), checked on an agarose gel (1.5% (*w*/*v*) 1× TAE), and quantified (Qubit® 2.0 Fluorometer and Qubit® dsDNA HS Assay Kit, Invitrogen, Merelbeke, Belgium). The copy numbers of each of the genes of interest were calculated from the known concentration of the extracted DNA plasmid. Ten-fold serial dilutions of linearized plasmids (107 to 101) containing the gene fragment of interest were run in each qPCR assay in triplicate to generate a standard curve. Gene abundances were determined by qPCR in replicated samples using the Rotor-Gene Q (Qiagen, Hilden, Germany) employing the same primers as for cloning. Each reaction was performed in a 20 µL volume containing 10 µL of PowerUp SYBR Green Master Mix (Applied Biosystems, Waltham, MA, USA), 0.56 µL of bovine serum albumin (0.56 mg mL^{-1}, Invitrogen, Merelbeke, Belgium), 400 nM (*cbbL*, *GH7*, *amoA*, and *nirK* genes) or 500 nM (*nifH* gene) of each primer, 5 µL of DNA template, and nuclease-free water. Amplification conditions are described in Tables S2–S6. A final dissociation stage, melt curve analysis with continuous fluorescence acquisition from 65 to 95 °C at a rate of 0.25 °C per 5 s, was performed to detect nonspecific amplification. qPCR products were also checked via 1.5% (*w*/*v*) agarose gel electrophoresis to check the specificity of the amplification.

2.6. Phospholipid Fatty Acid (PLFA) Analysis

The abundance of microbial groups was estimated by phospholipid fatty acid (PLFA) analysis [79]. Briefly, lipids were extracted from soils by weighing 2 g (dry weight) of soil with a chloroform/methanol/citrate buffer mixture (1:2:0.8 $v/v/v$) and separated into neutral lipids, glycolipids, and phospholipids using a prepacked silica column. Phospholipids were then subjected to a mild alkaline methanolysis, and the resulting fatty acid methyl esters were identified via gas chromatography. A total of 32 different PLFAs were identified and quantified. The PLFAs were designated in terms of the total number of carbon atoms and double bonds, followed by the position of the double bond from the methyl end of the molecule. Furthermore, cis and trans configurations are indicated by "c" and "t", respectively. The prefixes "a" and "i" indicate anteiso- and iso-branching positions, "br" indicates the unknown methyl group branching position, "Me" indicates a methyl group on the tenth carbon atom from the carboxyl end of the molecule, and "cy" refers to cyclopropane fatty acids. The abundance of different microbial groups was estimated following Joergensen [80]: Firmicutes: i14:0, i15:0, i16:0a, i17:0, i18, a15:0, a16:0, a17:0, a18:0, a19:0; Actinobacteria: 10Me16:0, 10Me17:0, 10Me18:0; Gram positive (G+) bacteria: Firmicutes + Actinobacteria; Gram negative (G−) bacteria: cy17:0, cy19:0, 16:1ω7, 16:1ω9, 17:1ω8, 18:1ω7; Bacteria: G+ plus G−; Arbuscular mycorrhiza fungi (AMF): 16:1ω5c; Zygomycota: 18:1ω9c; Ascomycota and Basidiomycota: 18:2ω6c; Unspecific fungal PLFA: 18:3ω6,9,12; Fungi: AMF + Zygomycota + Ascomycota and Basidiomycota + unspecific fungal; Unspecific microbial PLFA: 14:0, 15:0, 16:0, 17:0, 18:0, 20:0, 20:4ω6,9,12,15; Total microbial PLFA: bacterial + fungal + unspecific microbial.

2.7. Statistical Analysis

Data were checked to ensure normal distribution using the Kolmogorov–Smirnov test at $p < 0.05$. GHG emission data were submitted to two-way repeated measures ANOVA, with measurement date as the within-subject factor, and treatment (F100, F50, BA and BA + FU) as the between-subject factor. GHG data were also submitted, independently for each date, to one-way ANOVA and Tukey's post-hoc test ($p < 0.05$) to compare significant differences between treatments. Crop yield, soil chemical and biological properties, and cumulative GHG emission values for the experimental period were submitted to a one-way ANOVA and Tukey's post hoc test ($p < 0.05$) to compare significant differences between treatments. Relationships among properties were studied using Pearson correlations. Multiple linear regression analysis ($Y = m1 \times 1 + m2 \times 2 + \cdots + mnXn + b$) was carried out using the stepwise method, with cumulative values of GHG as independent variables, and soil chemical and biological properties and crop yield as dependent variables. The standardized coefficient (β) and partial correlation values were used for the analysis. The β coefficient compares the intensity of the effect of each independent variable with that of the dependent variable. The higher the absolute value of the beta coefficient, the stronger the effect. The partial correlation measures the correlation between two variables, while controlling for the effect of one or more other variables. The unstandardized coefficients (m) were used to interpret the effect of each independent variable on the outcome of the regression model. Data from all soil properties were subjected to a principal component analysis (PCA) to examine dependency and correlation structures. Statistical analyses were performed with the software IBM SPSS for Windows, Version 26.

3. Results

3.1. Soil Greenhouse Gas Emission Rates

Soil CO_2 emission rates followed the trend of soil temperature, with a significant positive correlation ($R = 0.61$; $p \leq 0.01$) (Figure 1A1–B). On the contrary, the CO_2 emission rates showed a slightly negative correlation to soil moisture ($R = -0.30$; $p \leq 0.01$) (Figure 1A1–B). Hence, the highest CO_2 emission rates were related to the highest soil temperatures and the lowest soil moisture levels. There were no significant differences in terms of soil temperature and moisture regarding treatments (Figure 1A1,A2). Soil CO_2 emission rates showed no significant differences between treatments (Figure 1B). On average, CO_2 emission rates during the crop cycle were 117 mg m^{-2} h^{-1} for all treatments.

Soil N_2O emission rates had a flat trend with small oscillations around 0 mg m^{-2} h^{-1}, and they were not correlated with either soil temperature or soil moisture (Figure 1A1,A2,C). Soil N_2O emission rates were not significantly different between treatments at any given time. On average, N_2O emission rates were 0.029 mg m^{-2} h^{-1} for all treatments during the experimental period.

Soil CH_4 emission rates followed the trend of soil temperature, with a significant positive correlation (R = 0.39; $p \leq 0.01$). However, CH_4 emission rates were not correlated with soil moisture. CH_4 emission rates showed no significant differences between treatments (Figure 1D), with an average value of 1.08 mg m^{-2} h^{-1}.

Thus, for all GHGs, there was no significant effect of fertilization treatment, and emissions were only affected by sampling dates, showing a slight time variability (Figure 1). The interaction between treatment and sampling time was not significant for any GHG.

3.2. Overall Cumulative Soil Emissions

The estimation of cumulative CO_2, N_2O, CH_4, and CO_2e released during the experimental period showed no significant differences between the treatments (Table 1). The cumulative CO_2 emission was positively correlated with crop yield (R = 0.633; $p \leq 0.05$) and Csol (R = 0.560; $p \leq 0.05$) (Table S7). Cumulative N_2O and CH_4 had no significant correlations with soil properties or crop yield.

Table 1. Cumulative values of soil CO_2, N_2O, CH_4, and total CO_2 equivalent emissions, crop yield, and cumulative CO_2 equivalent emission data expressed on a production basis released from the soil in the broccoli crop with different fertilization treatments. The values shown are mean ± standard error (n = 4).

Treatment	Cumulative CO_2	Cumulative N_2O	Cumulative CH_4	CO_2e	Crop Yield	CO_2e
	g m^{-2}				kg ha^{-1}	g kg^{-1} of Crop Yield
BA + FU	261.5 ± 19.7	0.0 ± 0.0	3.0 ± 0.3	351.3 ± 14.4	15,082.3 ± 894.3	225.0 ± 12.6
BA	242.8 ± 24.4	0.1 ± 0.1	2.1 ± 0.4	327.7 ± 36.7	14,928.3 ± 886.4	203.3 ± 18.4
F50	233.8 ± 16.7	0.1 ± 0.1	3.0 ± 0.7	360.0 ± 43.6	15,622.7 ± 1736.2	262.3 ± 27.6
F100	250.3 ± 8.1	0.0 ± 0.0	3.1 ± 0.4	347.2 ± 22.3	16,797.3 ± 650.3	204.3 ± 24.9
F ANOVA	0.744 ns	0.336 ns	0.518 ns	0.903 ns	0.649 ns	0.260 ns

BA + FU (F50 + the application of a formulation of bacteria and non-mycorrhizal fungi), BA (F50 + the application of a formulation of nitrogen-fixing and phosphorus- and potassium-solubilizing bacteria), F50 (50% of the rate of inorganic fertilizers added in F100), F100 (inorganic fertilizers applied at the rate to cover the nutritional demands of the crop), ns: not significant ($p > 0.05$).

3.3. Crop Yield, Soil Enzyme Activities, and Chemical Properties

Broccoli yield showed no significant differences between treatments, with an average value of 16,797 kg ha^{-1} for F100, 15,621 kg ha^{-1} for F50, 14,928 kg ha^{-1} for BA, and 15,082 kg ha^{-1} for BA + FU (Table 1). Enzyme activities and chemical properties showed no significant differences between treatments (Table 2). We found a positive correlation between crop yield and Nt (R = 0.618; $p \leq 0.05$) and Cls (R = 0.620; $p \leq 0.05$) (Table S7). SOC was positively correlated with Nt (R = 0.662, $p \leq 0.01$) and negatively with Csol (R = −0.701; $p \leq 0.01$) (Table S7).

Table 2. Soil enzyme activities and chemical properties in terms of fertilization treatments in the broccoli crop. Values are mean ± standard error (n = 4).

		BA + FU	BA	F50	F100	F-ANOVA
Cls	µmol gearbox sugars g^{-1} h^{-1}	36.3 ± 6.0	20.3 ± 11.1	30.1 ± 10.3	31.0 ± 3.45	0.649 ns
Urs	µmol NH_4^+ g^{-1} h^{-1}	4.9 ± 1.4	4.7 ± 0.8	4.1 ± 0.7	7.2 ± 2.1	1.049 ns
Aryl	µmol PNP g^{-1} h^{-1}	13.3 ± 1.8	15.4 ± 1.9	14.4 ± 1.2	15.6 ± 0.9	0.490 ns
Glu	µmol PNP g^{-1} h^{-1}	0.3 ± 0.0	0.3 ± 0.0	0.3 ± 0.1	0.3 ± 0.0	0.505 ns
M	%	14.4 ± 0.6	15.5 ± 0.9	15.4 ± 0.8	13.6 ± 0.5	0.254 ns
SOC	%	1.0 ± 0.0	1.0 ± 0.1	1.1 ± 0.0	1.0 ± 0.1	0.656 ns
Nt	%	0.1 ± 0.0	0.1 ± 0.0	0.1 ± 0.0	0.1 ± 0.0	0.411 ns
Csol	mg kg^{-1}	1405.8 ± 71.4	1255.5 ± 136.4	1080.5 ± 52.3	1129.5 ± 56.6	0.082 ns
Nsol	mg kg^{-1}	94.6 ± 8.3	78.2 ± 7.2	98.1 ± 16.8	95.0 ± 4.4	0.534 ns
NO_3^-	mg kg^{-1}	47.2 ± 13.6	33.2 ± 5.5	40.4 ± 6.1	79.8 ± 19.0	0.088 ns
NH_4^+	mg kg^{-1}	1.7 ± 0.0	1.8 ± 0.1	1.7 ± 0.0	1.8 ± 0.0	0.293 ns

BA + FU (F50 + the application of a formulation of bacteria and non-mycorrhizal fungi), BA (F50 + the application of a formulation of nitrogen-fixing and phosphorus- and potassium-solubilizing bacteria), F50 (50% of the rate of inorganic fertilizers added in F100), F100 (inorganic fertilizers applied at the rate to cover the nutritional demands of the crop). Cls: Cellulase activity; Urs: Urease activity; Aryl: Arylestrerase activity; Glu: β-glucosidase activity; M: Soil moisture; SOC: soil organic carbon; Nt: Total nitrogen; Csol: Soluble carbon; Nsol: Soluble nitrogen; PNP: *p*-nitrophenol; ns: not significant ($p > 0.05$).

3.4. PLFA Biomarkers and Functional Genes

Microbial biomass, assessed as total PLFAs, and microbial groups based on PLFA biomarkers, showed no significant differences between the fertilization treatments. Total PLFA averaged 15.4 nmol g^{-1}, bacteria and fungi averaged 7.63 nmol g^{-1} and 2.64 nmol g^{-1}, respectively, and G+ were more abundant (4.97 nmol g^{-1}) than G− (2.66 nmol g^{-1}) (Table 3). Total PLFAs were positively correlated with SOC and *amoA* (R = 0.728; $p \leq 0.01$; R = 0.648; $p \leq 0.01$) and negatively with Csol (R = −0.512; $p \leq 0.05$) (Table S7).

Table 3. Microbial biomass and microbial groups based on PLFA biomarkers (nmol g^{-1}). Values are mean and standard error (n = 4).

Microbial Groups	BA + FU	BA	F50	F100	F-ANOVA
			nmol g^{-1}		
Total PLFA	15.6 ± 1.0	15.8 ± 1.1	15.6± 1.1	14.6 ± 0.7	0.834 ns
Firmicutes	3.6 ± 0.2	3.5 ± 0.3	3.4 ± 0.3	3.1 ± 0.2	0.523 ns
Actinobacteria	1.8 ± 0.1	1.7 ± 0.1	1.4 ± 0.1	1.5 ± 0.1	0.097 ns
Gram-positive	5.4 ± 0.3	5.2 ± 0.4	4.7 ± 0.2	4.6 ± 0.3	0.265 ns
Gram-negative	2.5 ± 0.2	2.7 ± 0.2	2.9 ± 0.2	2.6 ± 0.2	0.729 ns
Bacteria	7.9 ± 0.5	7.8 ± 0.6	7.6 ± 0.4	7.2 ± 0.5	0.763 ns
AMF	bdl	bdl	bdl	bdl	
Zygomycota	2.0 ± 0.2	2.1 ± 0.1	2.2 ± 0.2	1.9 ± 0.1	0.558 ns
Ascomycota and Basidiomycota	0.5 ± 0.0	0.6 ± 0.0	0.6 ± 0.0	0.6 ± 0.0	0.194 ns
Total fungi	2.6 ± 0.2	2.7 ± 0.1	2.8 ± 0.3	2.5 ± 0.1	0.530 ns
Unspecific microbial PLFA	5.1 ± 0.4	5.3 ± 0.4	5.1 ± 0.4	4.9 ± 0.2	0.922 ns

ns: not significant; bdl: below detection limit.

None of the C- and N-cycle gene abundances showed significant differences between treatments (Figure 2a–e). The *amoA* was positively correlated with SOC, NH_4^+, Csol, *nifH*, and Csol (R = 0.644, $p < 0.01$; R = 0.603, $p < 0.01$, R = 0.636, $p < 0.05$; R = 0.618, $p < 0.05$ respectively) (Table S7). The abundance of the *nifH* gene was correlated with SOC and NH_4^+ (R = 0.519; $p \leq 0.05$; R = 0.557, $p \leq 0.05$) (Table S7). *cbbL* was correlated with *nirK* and *GH7* (R = 0.713; $p \leq 0.01$; R = 0.567 $p \leq 0.01$, respectively), while no other significant correlations were detected (Table S7).

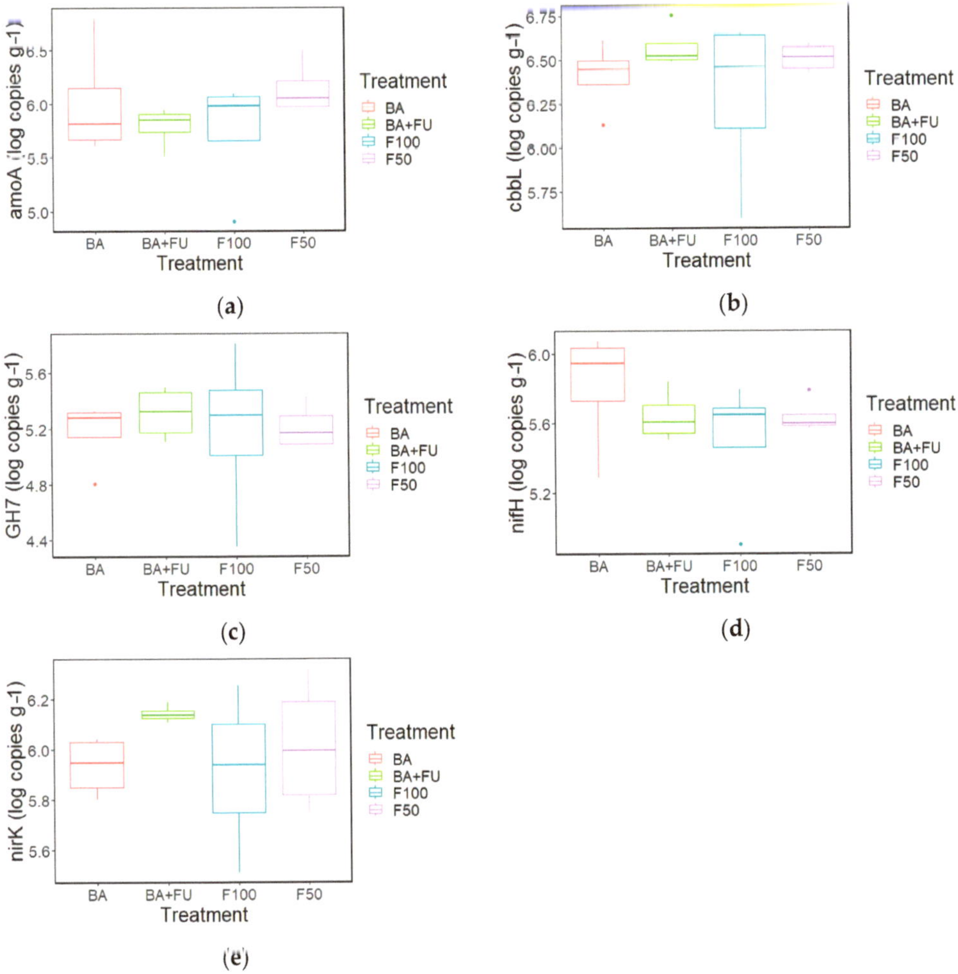

Figure 2. Boxplots show abundance of N- and C-cycle genes: *amoA* (**a**), *nifH* (**b**), *nirK* (**c**). *GH7* (**d**), *cbbL* (**e**) in soils under the different fertilization treatments. BA + FU (F50 + the application of a formulation of bacteria and non-mycorrhizal fungi) in green, BA (F50 + the application of a formulation of nitrogen-fixing and phosphorus- and potassium-solubilizing bacteria) in red, F50 (50% of the rate of inorganic fertilizers added in F100) in purple, F100 (inorganic fertilizers applied at the rate to cover the nutritional demands of the crop) in blue. Values are mean ± standard error (n = 4).

3.5. Interrelationship between GHGs, Soil Properties and Microbial Abundance and Functioning

Multiple linear regression analysis (Table 4) showed that the cumulative CO_2 emissions were positively related to crop yield, and negatively to *nifH* (R2 = 0.70; F = 10.34; $p \leq 0.01$).

Table 4. Multiple linear regression model for cumulative CO_2 emissions in a broccoli crop.

Y	X	m	Partial Correlation	β	R2	R2 adj	F Value
Cumulative CO_2 (mg m^{-2})	Constant	422.86					10.34 ($p < 0.01$)
	nifH (Log copies g^{-1} dry weight)	−57.51	−0.70	−0.56	0.70	0.63	
	Crop yield (kg ha^{-1})	0.01	0.68	0.52			

The PCA performed on soil chemical properties, cumulative values of GHG emissions, enzyme activities, functional genes, and PLFA showed that 64.19% of the total variability of data can be explained by four PCs (Table 5). None of the PC's were able to show separations between treatments, with all samples clustering together. Therefore, the structure of dependence and correlation between soil properties has not been significantly influenced by the fertilization treatments (Figure 3). PC1, which explained 33.98% of the data variability, was associated with most of the PLFA indicators, including total PLFA, bacteria, Firmicutes, Zygomicota, fungi, G− and G+, SOC, *amoA*, and β-glucosidase activity. Thus, most variability in data is related to microbial community structure, which is associated with higher SOC content and higher ammonification and degradation of oligosaccharides (Table 5). PC2, which explained 11.50% of data variability, was associated with cumulative CO_2e, N_2O and CH_4 emissions, and *nirK*, indicating that GHG emissions are the second set of variables contributing to explain data variability, with a negative relationship with denitrification processes. The rest of the properties explained <10% of the data variability.

Table 5. Matrix of PCA obtained with all soil properties, including cumulative values of GHG emissions.

Variance Explained	PC1 (33.98%)	PC2 (11.50%)	PC3 (10.01%)	PC4 (8.70%)
Total PLFA	0.98	0.06	0.01	−0.01
Bacteria	0.95	0.02	0.15	0.03
Unspecific microbial PLFA	0.95	0.16	−0.08	0.05
Firmicutes	0.91	−0.04	0.05	−0.05
Zygomycota	0.90	0.01	−0.20	−0.23
Gram-negative	0.90	0.21	0.15	0.07
Fungi	0.88	−0.01	−0.19	−0.22
Gram-positive	0.85	−0.10	0.13	0.00
Soil organic C	0.78	0.04	0.41	0.16
amoA gene	0.69	0.11	−0.12	0.60
β-glucosidase	0.68	−0.05	0.05	0.34
Soluble C	−0.57	0.01	−0.21	−0.22
CO_2e	0.16	0.89	0.02	−0.17
N_2O	0.27	0.73	0.07	0.05
nirK gene	0.04	−0.71	0.38	0.00
CH_4	0.26	0.59	0.25	−0.08
GH7 gene	0.24	0.08	0.89	−0.15
cbbL gene	0.29	−0.24	0.61	0.15
NO_3^-	0.07	−0.12	0.60	0.06
Urease	−0.42	0.44	0.50	0.15
NH_4^+	0.05	−0.08	−0.12	0.89
nifH gene	0.45	−0.18	0.21	0.68
Actinobacteria	0.43	−0.15	0.23	0.09
Ascomycota and Basidiomycota	0.41	−0.11	−0.14	−0.16
Moisture	0.35	0.09	−0.01	0.11
Total N	0.50	0.04	0.38	0.33
Soluble N	0.02	0.12	−0.18	−0.07
CO_2	−0.19	0.39	−0.23	−0.25
Cellulase	0.23	−0.03	−0.18	−0.54
Arylesterase	0.02	0.18	0.05	0.08

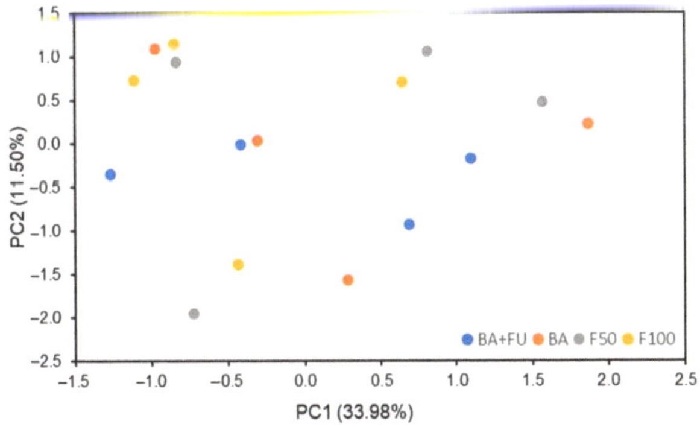

Figure 3. PCA factor scores of variations in soil properties in the broccoli crop submitted to different fertilization strategies. BA + FU (F50 + the application of a formulation of bacteria and non-mycorrhizal fungi), BA (F50 + the application of a formulation of nitrogen-fixing and phosphorus- and potassium-solubilizing bacteria), F50 (50% of the rate of inorganic fertilizers added in F100), F100 (inorganic fertilizers applied at the rate to cover the nutritional demands of the crop).

4. Discussion

Results showed no significant changes in soil properties and GHG emissions when applying two different types of biofertilizers in a broccoli crop, consequently partially rejecting our initial hypotheses: there was no increase in GHG owing to a higher microbial activity caused by the addition of fertilizers, there was no increase in microbial abundances with the addition of microorganisms, and there was no decrease in the emission of CO_2e per unit of product. Nonetheless, we can partially confirm our hypothesis since GHG emissions were related to soluble organic C and microbial activity assessed by functional genes and enzyme activities.

4.1. GHG Emissions

Soil temperature exhibited a positive correlation with soil CO_2 emissions (Figure 1), aligned with previous studies, and confirming that soil temperature is the most important factor controlling CO_2 emissions when water is not the limiting factor [14,81,82]. However, under the irrigation conditions of the broccoli crop, when water is always available for crop development, slight increases in water supply can contribute to decreased CO_2 emissions. This is likely because of the presence of anaerobic microsites in the soil in irrigated systems. When O_2 supply is adequate, most of the organic C is converted to CO_2 by rhizospheric and heterotrophic soil respiration, which is highly affected by temperature [83,84]. Due to the limitation of gas permeability and O_2 availability, the CO_2 fluxes tend to decrease, and organic C is converted anaerobically to CH_4 [84,85]. However, no effect of irrigation or water availability was observed in terms of CH_4 emissions, and so, no overall anaerobic conditions were present in the field. Thus, it is likely that an excessive water supply has reduced the aeration and thus respiration by aerobic CO_2-producing microorganisms and roots [86,87], without reaching the level to increase methanogenesis and methanogenic populations [88]. As observed with CO_2, CH_4 emissions showed a significant positive correlation with soil temperature, confirming again that temperature is the most important factor controlling emissions, as previously reported in other studies [12,89].

In terms of fertilization treatments, the reduction in fertilizer rates or the addition of biofertilizers did not contribute to changes in CO_2 or CH_4 emissions, contrary to our initial hypothesis. This might be attributed to the lack of an effect of fertilization treatments on microbial abundance and activity in the short term [90], implying that fertilization modifications had no direct influence on microbial communities and their functioning [91] during the broccoli crop. Soil type, weather conditions, and seasonal differences, along with soil moisture and pH, are considered the most influential drivers of microbial community structure [92]. The use of microorganisms can either reduce or have no effect on CO_2 emissions, depending on a variety of abiotic (moisture conditions, C and N substrates, SOM, fertilization, etc.) and biotic (microbial strains, soil enzymes, and microbial activity) factors [37,93,94]. Previous studies have also reported no effect of fertilization rates on CH_4 emissions [93,95,96].

The cumulative CO_2 emission is positively correlated with crop yield. This may be explained by the fact that soil respiration is strongly linked to plant metabolism, autotrophic respiration, and photosynthesis, thus promoting plant growth [86]. The respiration of the plant and the metabolism associated with soil microbial biomass can explain this correlation, including a positive correlation with Csol. Csol is a substrate for microorganisms (heterotrophic respiration) that could support CO_2 fluxes, and so an important pool for microbial activity, rather than the total stock of soil organic carbon measured as SOC [97,98]. The negative correlation between cumulative CO_2 and the *nifH* gene may indicate the highest efficiency in C use when there is high biological N fixation, leading to lower CO_2 emissions. This is because diazotrophs use large amounts of soil C and energy to produce bioavailable soil N [99]. Consistent with other studies, the *nifH* gene abundance is generally associated with higher organic carbon levels [54]. PGPR addition can also decrease C-cycling enzyme activity and stimulate N-cycling enzyme activity in the soil, reducing CO_2 emissions from soil [94], although this effect was not observed in our study.

Most research has reported that N fertilization influences N_2O emissions [10,11,89], something also not found in our experiment. Furthermore, most of the increases in N_2O emissions happen after irrigation events in previous studies [13]. However, in our experiment, fertilization was always accompanied by irrigation and soil moisture was always maintained at the appropriate levels to ensure crop development, so moisture was not a limiting factor. This may explain the lack of a relationship between N_2O emissions and soil moisture in our study. Production of N_2O primarily occurs through microbially mediated nitrification and denitrification [100]. Under aerobic conditions, ammonium (NH_4^+) is oxidized in soil to nitrite (NO^{2-}) and nitrate (NO^{3-}), resulting in N_2O production, whereas under anaerobic conditions, NO^{3-} can be reduced to N_2O and/or dinitrogen (N_2) [101]. The fact that changes in fertilization regime or addition of biofertilizers did not affect these processes may be related to the lack of modification of microbial communities with the treatments, as assessed by PFLA analysis, enzyme activities, and functional genes. Huang et al. [36], contrary to our results, found that application of PGPM in a cucumber crop reduced soil N_2O emissions by 22.6–33.5%, depending on the inoculation dose, and was associated with the enrichment of the nitrifier AOB gene and the denitrifiers *nirK* and *nosZ* gene abundances. Wu et al. [60] carried out a biofertilizer treatment in an oil-seed rape crop (*Brassica campestris*) and reported increased relative abundances of bacteria involved in denitrification and increased numbers of nosZ gene copies, which led to the increased reduction of N_2O to N_2.

4.2. Soil Chemical Properties, Microbial Abundance, and Potential Activity

Many studies showed that SOC is critical for maintaining the soil microbial community. SOC concentration, composition, and dissolving rate drive soil respiration, which is significantly correlated with the shifts in bacterial community compositions [97,98]. Thus, the lack of differences in SOC, Csol, or available N under the different treatments may have contributed to maintain the same microbial activity, and therefore the same GHG emissions. The significant correlations between SOC and some functional genes suggest that, in our experiment, SOC is controlling microbial functioning rather than external fertilization or the addition of exogenous microorganisms. In this sense, SOC exerts a strong influence on the abundance and diversity of the N-cycling genes [102,103], as it is considered more limiting to free-living microbial activity than nutrient availability [104]. The *nifH* gene that fixes atmospheric N, plus the mineralization of organic N, leads to the formation of NH_4^+ (ammonification) moving on to nitrification by the *amoA* gene, which is also positively correlated with SOC and microbial abundance [103]. The *cbbL* gene (RubisCO) was also significantly positively correlated with SOC, confirming its key role in CO_2 fixation and sequestration in soils [105]. The *cbbL* gene was also positively correlated with the *GH7* gene, which explains how microorganisms simultaneously contribute to the processes involved in the C cycle [106]. Several authors have identified different bacteria and fungi with cellulolytic and nitrogen-fixing attributes [107–109]. These bacteria and fungi play a crucial role in plant nutrient accessibility, sustaining soil fertility, plant growth, and, proportionally, crop yield [110]. In this line, we observed a positive correlation between crop yield, Nt, and cellulase activity. A wide range of soil enzymes have been identified as being strongly associated with soil organic matter (SOM) decomposition [111]. Urease plays a significant role in N cycling, and this explains the correlation in our study between the urease enzyme and the N-cycle genes (*amoA*, *nifH*).

As confirmed by the correlation and multivariate analyses, the abundance of microorganisms was mostly controlled by SOC and Csol, as highlighted in previous studies [112–114], but not affected by fertilization regimes. SOC and Nt may have promoted the increase in bacterial biomass until a peak was reached, after which some other biotic factors, such as competition, may have prevented bacterial growth even if nutrient levels increased [115]. This may explain the lack of significant differences in microbial biomass in our study, and also how the application of external microorganisms did not affect the stock of microorganisms [115,116]. However, other research described how different N fertilization strategies and the application of PGPM increased the content of various PLFAs markers [56,57] and, in general, microbial abundance and biodiversity [117] and crop yields [118]. These controversial results may suggest that the effect of microbial inoculants is soil-, climate-, crop-, and management-dependent. In this line, conventional practices (tillage and pesticide addition) could weaken the effect of inoculants [119]. Additionally, it is also known that bacterial species decline after inoculation in soil, mainly due to competition with native communities and the hostility of biotic or abiotic interactions [120]. These conditions may have greatly affected the microbial life in our applications. Furthermore, Brassicaceae are known to release allelopathic compounds, such as glucosinolates, that may have contributed to limiting the growth of microbial inoculants, explaining the differences in all the properties measured [121,122].

Increased activity of soil enzymes after the inoculation with PGPR strains has been reported in other studies [56,58,123], contrary to our results. In a greenhouse experiment, the legume *Hedysarum carnosum* was sown using a loamy soil collected from southern Spain, observing that the inoculation with *Bacillus subtilis* increased dehydrogenase, β-glucosidase, urease, and alkaline phosphatase activities compared to the non-inoculated control [124]. However, other studies have also reported no effects on microbial properties after inoculation of soil with PGPR. For instance, Chaudhary et al. [56] reported that reduced fertilization and PGPR application in a peanut field had no significant effect on microbial biomass, β-glucosidase, and urease activities. They consider it a positive result because inoculation with beneficial bacteria, which had a significant effect by influencing and improving the availability of nutrients (N, P, K, and Fe) in the soil, does not perturb the natural microbial community of the soil. Angelina et al. [119] reported in barley cultivation that the use of microbial inoculants (*Bacillus subtilis* and *Pseudomonas fluorescens*) led to an increase in microbial biomass only in the organic system, while no differences were detected in the conventional system, as we have reported. Moreover, the activity of β-glucosidase was not affected, indicating the independence of the carbon cycle on inoculation.

4.3. Crop Yield

It is known that PGPM inoculation could compensate for nutrient deficiency and improve plant development through the production of plant growth regulators, stimulating the development of plant roots and leading to better absorption of water and nutrients from the soil [125]. Various studies consider that the most significant effect of biofertilizers on crops occurs in poor soils or in conditions of stress [125,126]. However, in our study, the reduction in nutrients did not affect the production in the treatment without biofertilizers.

It is possible that nutrients already present in the soil compensated for the lack of NPK fertilizer in each treatment where a reduction in fertilizer was applied (see Supplementary Table S8). Two microbial biofertilizer preparations were applied in an organic arable crop rotation in central Europe, and the results showed no effect on crop yield, soil microbial biomass, activity parameters, substrate turnover, or soil microbial community structure [88]. Nuzzo et al. [127] tested different formulations of plant growth-promoting bacteria (*Lactobacillus*, *Rhizobia*, etc.), yeasts, and mycorrhizal fungi on a tomato crop under greenhouse conditions and found no significant effect on plant growth. Biofertilizers tested in a pot experiment of *Lolium perenne* L. crop did not affect N mineralization or plant growth, but may have suppressive effects on the zymogenic microbial biomass of the soil due to the substrate of the biofertilizer suspensions [128].

Application methodologies, doses, and inoculation times of microorganisms vary between studies, and can thus affect the results; for example, inoculation efficacy depends on the rhizosphere competence of the bacteria for the particular crop type [129]. Thus, the fact that in our study no significant differences were found between treatments may be a result of the type or crop being less compatible with the applied PGPM. Fiorentino et al. [121] showed that different botanical families have different cultural performances when treated with Trichoderma-based biostimulants, with improvements in lettuce (Asteraceae), but not in broccoli (Brassicaceae), consistent with our study. Another explanation could be that Brassicaceae species have an allelopathic effect on soil microbes, bacteria, and fungi due to the production of numerous inhibitory compounds, such as glucosinolates, that are released into the soil of the rhizosphere [121,122]. Due to limited knowledge of the ecological factors that determine the survival of inoculants, such as competition with indigenous bacteria for available growth substrates [129], a general assumption regarding the exogenous microorganism survival applied in the soil cannot be determined in our study. Some variables such as soil type, crop type, natural selection, and agricultural management, including pesticide use, may be among the factors influencing colonization with preselected beneficial microorganisms [38].

In our case, when a 50% reduction in the conventional dose of N fertilizer was applied (158 kg ha^{-1}), we found no significant difference in crop production. A study of broccoli fertilization regime carried out in Italy under a Mediterranean climate suggested that the most effective dose of fertilizer N was 75 kg ha^{-1} and each unit of N above 75 kg ha^{-1} produced about 41% less fresh weight of the head [61]. However, according to Conversa et al. [61], an application rate of 150 kg ha^{-1} of N is the advisable rate to enhance overall production and is currently being used in the study region. According to Mourão and Brito [126], who performed an experiment with different rates of N fertilizers (0, 60, 120, 180, and 240 kg N ha^{-1}) in a broccoli crop, the yield increases were not significant above 120 kg N ha^{-1}. Similarly, Kim et al. [130] conducted a comprehensive simulation-based study of broccoli and concluded that the amount of N fertilizer does not significantly alter the crop yield above 75 kg ha^{-1} N input. Thus, even if the N supply was reduced compared to the dose conventionally applied by farmers in the area, plants did not suffer from lack of nutrients, and the microorganisms applied had no significant impact on crop yield, soil microbial biomass, soil microbial activity parameters, soil microbial community structure, or greenhouse gas emissions.

5. Conclusions

The reduction in fertilization and the addition of biofertilizers did not significantly affect the yield, soil chemical properties, or biological properties, compared to the conventional inorganic fertilization, applied to a broccoli crop in this area. Thus, the addition of microbial inoculants was not effective in increasing soil microbial abundance and activity, and so no changes were observed in emissions of GHGs such as CO_2, N_2O, and CH_4. GHG emissions responded to soil organic C content (mostly the soluble fraction), available N, and microbial activity assessed by enzyme activities and functional genes, which were not different between fertilization regimes. This may be related to the crop type (Brassicaceae, with release of allelopathic compounds), the lack of nutrient limitation, and the conventional management with use of tillage and pesticides. Crop yield responded positively to total N, microbial activity, and CO_2 emissions, suggesting that the active soil microbial community is related to high yields. Nevertheless, these positive correlations were observed under short-term experiments and an overall statistical difference was not observed. In our experiment, it was concluded that, following a 50% reduction in fertilizer compared to conventional practices in a farm location in SE Spain, total crop yield was not affected. Specifically for the location of this study, the crop type and cultivar, the climatic conditions, and the crop management used, it is possible to reduce the amount of fertilizer and still obtain optimal production. This may contribute to reducing production costs and possible water pollution by leaching and runoff, and eventually improve farm profitability. The present study is based on a single inoculation of a product based on microorganisms in a short-term experiment. Therefore, long-term application of biofertilizers under different soil and climate conditions and crop types would be needed to fully assess their effect on GHG emissions and soil microbial communities along with their effect on soil microbial activity and final crop yield.

Supplementary Materials: The following supporting information can be downloaded at: https://www.mdpi.com/article/10.3390/horticulturae10010042/s1. Table S1. Primers used in this study [46,75–78]; Table S2. *amoA* PCR cycling conditions; Table S3. *nirK* PCR cycling conditions; Table S4. *cbbL* PCR cycling conditions; Table S5. *GH7* PCR cycling conditions; Table S6. *nifH* PCR cycling conditions; Table S7. Pearson correlation results of the properties analyzed; Table S8. Main soil characteristics. Values mean ± standard error (n = 4). References [126–130] are cited in Supplementary Materials.

Author Contributions: Conceptualization, R.Z., S.M.-M., C.E.-G. and J.A.F.; methodology, R.Z., D.F.C., S.M.-M., C.E.-G. and J.A.F.; formal analysis, I.O., V.S.-M., D.S.G., E.L. and V.S.-N.; resources, R.Z. and D.F.C.; data curation, I.O.; writing—original draft preparation, I.O.; writing—review and editing, R.Z.; supervision, S.M.-M., C.E.-G., J.A.F., D.F.C. and R.Z.; project administration, D.F.C.; funding acquisition, D.F.C. and R.Z. All authors have read and agreed to the published version of the manuscript.

Funding: This work was supported by the European Commission Horizon 2020 project SoildiverAgro [grant agreement 817819].

Data Availability Statement: Data fully accessible on the SoildiverAgro Community of the repository Zenodo.

Conflicts of Interest: The authors declare no conflicts of interest.

References

1. Cooper, R.N.; Houghton, J.T.; McCarthy, J.J.; Metz, B. Climate Change 2001: The Scientific Basis. *Foreign Aff.* **2002**, *81*, 208. [CrossRef]
2. Oertel, C.; Matschullat, J.; Zurba, K.; Zimmermann, F.; Erasmi, S. Greenhouse Gas Emissions from Soils—A Review. *Geochemistry* **2016**, *76*, 327–352. [CrossRef]
3. Prather, M.; Ehhalt, D.; Dentener, F.; Derwent, R.; Dlugokencky, E.; Holland, E.; Isaksen, I.; Katima, J.; Kirchhoff, V.; Matson, P.; et al. Atmospheric Chemistry and Greenhouse Gases. In *Precise Soil Management as a Tool to Reduce CH_4 and N_2O Emissions from Agricultural Soils*; Houghton, J.T., Ed.; Cambridge University Press: New York, NY, USA, 2001; pp. 239–287.
4. Mosquera, J.; Hilhorst, M.A. Precise Soil Management as a Tool to Reduce N2O and CH4 Emissions from Agricultural Soils: Literature Review. In Proceedings of the 4th International Symposium on non-CO_2 Greenhouse Gases (NCGG-4), Science, Control, Policy and Implementation, Utrecht, The Netherlands, 4–6 July 2005; Millpress Science Publishers: Rotterdam, The Netherlands, 2005; pp. 5–12.
5. Ludwig, J.; Meixner, F.; Vogel, B.; Förstner, J. Soil-Air Exchange of Nitric Oxide: An Overview of Processes, Environmental Factors, and Modeling Studies. *Biogeochemistry* **2001**, *52*, 225–257. [CrossRef]
6. Maraseni, T.N.; Cockfield, G. Including the Costs of Water and Greenhouse Gas Emissions in a Reassessment of the Profitability of Irrigation. *Agric. Water Manag.* **2012**, *103*, 25–32. [CrossRef]
7. Iqbal, J.; Ronggui, H.; Lijun, D.; Lan, L.; Shan, L.; Tao, C.; Leilei, R. Differences in Soil CO_2 Flux between Different Land Use Types in Mid-Subtropical China. *Soil Biol. Biochem.* **2008**, *40*, 2324–2333. [CrossRef]

8. Sainju, U.M.; Stevens, W.B.; Caesar-TonThat, T.; Jabro, J.D. Land Use and Management Practices Impact on Plant Biomass Carbon and Soil Carbon Dioxide Emission. *Soil Sci. Soc. Am. J.* **2010**, *74*, 1613–1622. [CrossRef]
9. Linquist, B.A.; Adviento-Borbe, M.A.; Pittelkow, C.M.; van Kessel, C.; van Groenigen, K.J. Fertilizer Management Practices and Greenhouse Gas Emissions from Rice Systems: A Quantitative Review and Analysis. *Field Crops Res.* **2012**, *135*, 10–21. [CrossRef]
10. Piva, J.T.; Dieckow, J.; Bayer, C.; Zanatta, J.A.; de Moraes, A.; Tomazi, M.; Pauletti, V.; Barth, G.; Piccolo, M.d.C. Soil Gaseous N_2O and CH_4 Emissions and Carbon Pool Due to Integrated Crop-Livestock in a Subtropical Ferralsol. *Agric. Ecosyst. Environ.* **2014**, *190*, 87–93. [CrossRef]
11. Sanz-Cobena, A.; García-Marco, S.; Quemada, M.; Gabriel, J.L.; Almendros, P.; Vallejo, A. Do Cover Crops Enhance N_2O, CO_2 or CH_4 Emissions from Soil in Mediterranean Arable Systems? *Sci. Total Environ.* **2014**, *466–467*, 164–174. [CrossRef]
12. Schaufler, G.; Kitzler, B.; Schindlbacher, A.; Skiba, U.; Sutton, M.A.; Zechmeister-Boltenstern, S. Greenhouse Gas Emissions from European Soils under Different Land Use: Effects of Soil Moisture and Temperature. *Eur. J. Soil Sci.* **2010**, *61*, 683–696. [CrossRef]
13. Yan, H.; Xie, L.; Guo, L.; Fan, J.; Diao, T.; Lin, M.; Zhang, H.; Lin, E. Characteristics of Nitrous Oxide Emissions and the Affecting Factors from Vegetable Fields on the North China Plain. *J. Environ. Manag.* **2014**, *144*, 316–321. [CrossRef] [PubMed]
14. Zornoza, R.; Acosta, J.A.; Gabarrón, M.; Gómez-Garrido, M.; Sánchez-Navarro, V.; Terrero, A.; Martínez-Martínez, S.; Faz, Á.; Pérez-Pastor, A. Greenhouse Gas Emissions and Soil Organic Matter Dynamics in Woody Crop Orchards with Different Irrigation Regimes. *Sci. Total Environ.* **2018**, *644*, 1429–1438. [CrossRef] [PubMed]
15. Hou, C.; Song, C.; Li, Y.; Wang, J.; Song, Y.; Wang, X. Effects of Water Table Changes on Soil CO_2, CH_4 and N_2O Fluxes during the Growing Season in Freshwater Marsh of Northeast China. *Environ. Earth Sci.* **2013**, *69*, 1963–1971. [CrossRef]
16. Ren, C.; Wang, T.; Xu, Y.; Deng, J.; Zhao, F.; Yang, G.; Han, X.; Feng, Y.; Ren, G. Differential Soil Microbial Community Responses to the Linkage of Soil Organic Carbon Fractions with Respiration across Land-Use Changes. *For. Ecol. Manag.* **2018**, *409*, 170–178. [CrossRef]
17. Sanz-Cobena, A.; Lassaletta, L.; Aguilera, E.; del Prado, A.; Garnier, J.; Billen, G.; Iglesias, A.; Sánchez, B.; Guardia, G.; Abalos, D.; et al. Strategies for Greenhouse Gas Emissions Mitigation in Mediterranean Agriculture: A Review. *Agric. Ecosyst. Environ.* **2017**, *238*, 5–24. [CrossRef]
18. Hui, K.; Yuan, Y.; Xi, B.; Tan, W. A Review of the Factors Affecting the Emission of the Ozone Chemical Precursors VOCs and NO_x from the Soil. Environ. Int. Anderson, I.C.; Levine, J.S. Relative Rates of Nitric Oxide and Nitrous Oxide Production by Nitrifiers, Denitrifiers, and Nitrate Respirers. *Environ. Int.* **2023**, *172*, 107799. [PubMed]
19. Abbasi, M.K.; Adams, W.A. Gaseous N Emission during Simultaneous Nitrification–Denitrification Associated with Mineral N Fertilization to a Grassland Soil under Field Conditions. *Soil Biol. Biochem.* **2000**, *32*, 1251–1259. [CrossRef]
20. Wang, X.; Zou, C.; Gao, X.; Guan, X.; Zhang, W.; Zhang, Y.; Shi, X.; Chen, X. Nitrous Oxide Emissions in Chinese Vegetable Systems: A Meta-Analysis. *Environ. Pollut.* **2018**, *239*, 375–383. [CrossRef]
21. Senbayram, M.; Chen, R.; Mühling, K.H.; Dittert, K. Contribution of Nitrification and Denitrification to Nitrous Oxide Emissions from Soils after Application of Biogas Waste and Other Fertilizers. *Rapid Commun. Mass Spectrom.* **2009**, *23*, 2489–2498. [CrossRef]
22. Hou, A.X.; Chen, G.X.; Wang, Z.P.; Van Cleemput, O.; Patrick, W.H., Jr. Methane and Nitrous Oxide Emissions from a Rice Field in Relation to Soil Redox and Microbiological Processes. *Soil Sci. Soc. Am. J.* **2000**, *64*, 2180–2186. [CrossRef]
23. Foley, J.A.; DeFries, R.; Asner, G.P.; Barford, C.; Bonan, G.; Carpenter, S.R.; Chapin, F.S.; Coe, M.T.; Daily, G.C.; Gibbs, H.K.; et al. Global Consequences of Land Use. *Science* **2005**, *309*, 570–574. [CrossRef] [PubMed]
24. Riaz, U.; Murtaza, D.G.; Anum, W.; Samreen, T.; Sarfraz, M.; Nazir, M. Plant Growth-Promoting Rhizobacteria (PGPR) as Biofertilizers and Biopesticides. In *Microbiota and Biofertilizers*; Springer: Cham, Switzerland, 2020; pp. 181–196. ISBN 978-3-030-48770-6.
25. Hossain, M.M.; Sultana, F.; Islam, S. Plant Growth-Promoting Fungi (PGPF): Phytostimulation and Induced Systemic Resistance. In *Plant-Microbe Interactions in Agro-Ecological Perspectives: Volume 2: Microbial Interactions and Agro-Ecological Impacts*; Singh, D.P., Singh, H.B., Prabha, R., Eds.; Springer: Singapore, 2017; pp. 135–191. ISBN 978-981-10-6593-4.
26. El-Saadony, M.T.; Saad, A.M.; Soliman, S.M.; Salem, H.M.; Ahmed, A.I.; Mahmood, M.; El-Tahan, A.M.; Ebrahim, A.A.M.; Abd El-Mageed, T.A.; Negm, S.H.; et al. Plant Growth-Promoting Microorganisms as Biocontrol Agents of Plant Diseases: Mechanisms, Challenges and Future Perspectives. *Front. Plant Sci.* **2022**, *13*, 923880. [CrossRef] [PubMed]
27. Malusá, E.; Sas-Paszt, L.; Ciesielska, J. Technologies for Beneficial Microorganisms Inocula Used as Biofertilizers. *Sci. World J.* **2012**, *2012*, e491206. [CrossRef] [PubMed]
28. Vessey, J.K. Plant Growth Promoting Rhizobacteria as Biofertilizers. *Plant Soil* **2003**, *255*, 571–586. [CrossRef]
29. Xu, J.; Kloepper, J.W.; Huang, P.; McInroy, J.A.; Hu, C.H. Isolation and Characterization of N2-Fixing Bacteria from Giant Reed and Switchgrass for Plant Growth Promotion and Nutrient Uptake. *J. Basic Microbiol.* **2018**, *58*, 459–471. [CrossRef]
30. Khan, M.S.; Zaidi, A.; Wani, P.A. Role of Phosphate-Solubilizing Microorganisms in Sustainable Agriculture—A Review. *Agron. Sustain. Dev.* **2007**, *27*, 29–43. [CrossRef]
31. Ali, A.M.; Awad, M.Y.M.; Hegab, S.A.; Gawad, A.M.A.E.; Eissa, M.A. Effect of Potassium Solubilizing Bacteria (*Bacillus cereus*) on Growth and Yield of Potato. *J. Plant Nutr.* **2021**, *44*, 411–420. [CrossRef]
32. Calvo, P.; Nelson, L.; Kloepper, J.W. Agricultural Uses of Plant Biostimulants. *Plant Soil* **2014**, *383*, 3–41. [CrossRef]
33. Sarfraz, R.; Hussain, A.; Sabir, A.; Ben Fekih, I.; Ditta, A.; Xing, S. Role of Biochar and Plant Growth Promoting Rhizobacteria to Enhance Soil Carbon Sequestration—A Review. *Env. Monit. Assess.* **2019**, *191*, 251. [CrossRef]

34. Pankievicz, V.C.S.; do Amaral, F.P.; Santos, K.F.D.N.; Agtuca, B.; Xu, Y.; Schueller, M.J.; Arisi, A.C.M.; Steffens, M.B.; de Souza, E.M.; Pedrosa, F.O.; et al. Robust Biological Nitrogen Fixation in a Model Grass–Bacterial Association. *Plant J.* **2015**, *81*, 907–919. [CrossRef]
35. Florio, A.; Bréfort, C.; Gervaix, J.; Bérard, A.; Le Roux, X. The Responses of NO_2- and N_2O-Reducing Bacteria to Maize Inoculation by the PGPR Azospirillum Lipoferum CRT1 Depend on Carbon Availability and Determine Soil Gross and Net N_2O Production. *Soil Biol. Biochem.* **2019**, *136*, 107524. [CrossRef]
36. Huang, M.; Zhang, Y.; Wu, J.; Wang, Y.; Xie, Y.; Geng, Y.; Zhang, N.; Michelsen, A.; Li, S.; Zhang, R.; et al. Bacillus Velezensis SQR9 Inhibition to Fungal Denitrification Responsible for Decreased N_2O Emissions from Acidic Soils. *Sci. Total Environ.* **2023**, *885*, 163789. [CrossRef] [PubMed]
37. Paliwoda, D.; Mikiciuk, G.; Mikiciuk, M.; Miller, T.; Kisiel, A.; Sas-Paszt, L.; Kozioł, A.; Brysiewicz, A. The Use of Plant Growth Promoting Rhizobacteria to Reduce Greenhouse Gases in Strawberry Cultivation under Different Soil Moisture Conditions. *Agronomy* **2023**, *13*, 754. [CrossRef]
38. Dimkpa, C.; Weinand, T.; Asch, F. Plant–Rhizobacteria Interactions Alleviate Abiotic Stress Conditions. *Plant Cell Environ.* **2009**, *32*, 1682–1694. [CrossRef] [PubMed]
39. Malgioglio, G.; Rizzo, G.F.; Nigro, S.; Lefebvre du Prey, V.; Herforth-Rahmé, J.; Catara, V.; Branca, F. Plant-Microbe Interaction in Sustainable Agriculture: The Factors That May Influence the Efficacy of PGPM Application. *Sustainability* **2022**, *14*, 2253. [CrossRef]
40. Zhou, X.; Smaill, S.J.; Clinton, P.W. Methane Oxidation Needs Less Stressed Plants. *Trends Plant Sci.* **2013**, *18*, 657–659. [CrossRef]
41. Shah, A.M.; Khan, I.M.; Shah, T.I.; Bangroo, S.A.; Kirmani, N.A.; Nazir, S.; Malik, A.R.; Aezum, A.M.; Mir, Y.H.; Hilal, A.; et al. Soil Microbiome: A Treasure Trove for Soil Health Sustainability under Changing Climate. *Land* **2022**, *11*, 1887. [CrossRef]
42. Ortega Pérez, R.; Nieto García, J.C.; Gallegos-Cedillo, V.M.; Domene Ruiz, M.Á.; Santos Hernández, M.; Nájera, C.; Miralles Mellado, I.; Diánez Martínez, F. Biofertilizers Enriched with PGPB Improve Soil Fertility and the Productivity of an Intensive Tomato Crop. *Agronomy* **2023**, *13*, 2286. [CrossRef]
43. Lori, M.; Symnaczik, S.; Mäder, P.; De Deyn, G.; Gattinger, A. Organic Farming Enhances Soil Microbial Abundance and Activity—A Meta-Analysis and Meta-Regression. *PLoS ONE* **2017**, *12*, e0180442. [CrossRef]
44. Chen, X.-P.; Zhu, Y.-G.; Xia, Y.; Shen, J.-P.; He, J.-Z. Ammonia-Oxidizing Archaea: Important Players in Paddy Rhizosphere Soil? *Environ. Microbiol.* **2008**, *10*, 1978–1987. [CrossRef]
45. Henry, S.; Bru, D.; Stres, B.; Hallet, S.; Philippot, L. Quantitative Detection of the nosZ Gene, Encoding Nitrous Oxide Reductase, and Comparison of the Abundances of 16S rRNA, narG, nirK, and nosZ Genes in Soils. *Appl. Environ. Microbiol.* **2006**, *72*, 5181–5189. [CrossRef]
46. Poly, F.; Ranjard, L.; Nazaret, S.; Gourbière, F.; Monrozier, L.J. Comparison of nifH Gene Pools in Soils and Soil Microenvironments with Contrasting Properties. *Appl. Environ. Microbiol.* **2001**, *67*, 2255–2262. [CrossRef]
47. Yousuf, B.; Keshri, J.; Mishra, A.; Jha, B. Application of Targeted Metagenomics to Explore Abundance and Diversity of CO_2-Fixing Bacterial Community Using cbbL Gene from the Rhizosphere of Arachis Hypogaea. *Gene* **2012**, *506*, 18–24. [CrossRef] [PubMed]
48. Fan, F.; Li, Z.; Wakelin, S.A.; Yu, W.; Liang, Y. Mineral Fertilizer Alters Cellulolytic Community Structure and Suppresses Soil Cellobiohydrolase Activity in a Long-Term Fertilization Experiment. *Soil Biol. Biochem.* **2012**, *55*, 70–77. [CrossRef]
49. Henry, H.A.L.; Juarez, J.D.; Field, C.B.; Vitousek, P.M. Interactive Effects of Elevated CO_2, N Deposition and Climate Change on Extracellular Enzyme Activity and Soil Density Fractionation in a California Annual Grassland. *Glob. Chang. Biol.* **2005**, *11*, 1808–1815. [CrossRef]
50. Sinsabaugh, R.L.; Lauber, C.L.; Weintraub, M.N.; Ahmed, B.; Allison, S.D.; Crenshaw, C.; Contosta, A.R.; Cusack, D.; Frey, S.; Gallo, M.E.; et al. Stoichiometry of Soil Enzyme Activity at Global Scale. *Ecol. Lett.* **2008**, *11*, 1252–1264. [CrossRef]
51. Acosta-Martinez, V.; Cano, A.; Johnson, J. Simultaneous Determination of Multiple Soil Enzyme Activities for Soil Health-Biogeochemical Indices. *Appl. Soil Ecol.* **2018**, *126*, 121–128. [CrossRef]
52. Deng, S.P.; Tabatabai, M.A. Cellulase Activity of Soils. *Soil Biol. Biochem.* **1994**, *26*, 1347–1354. [CrossRef]
53. Krajewska, B.; Ureases, I. Functional, Catalytic and Kinetic Properties: A Review. *J. Mol. Catal. B Enzym.* **2009**, *59*, 9–21. [CrossRef]
54. Morales, S.E.; Cosart, T.; Holben, W.E. Bacterial Gene Abundances as Indicators of Greenhouse Gas Emission in Soils. *ISME J.* **2010**, *4*, 799–808. [CrossRef]
55. Nannipieri, P.; Ascher, J.; Ceccherini, M.T.; Landi, L.; Pietramellara, G.; Renella, G. Microbial Diversity and Soil Functions. *Eur. J. Soil Sci.* **2003**, *54*, 655–670. [CrossRef]
56. Chaudhary, D.R.; Rathore, A.P.; Sharma, S. Effect of Halotolerant Plant Growth Promoting Rhizobacteria Inoculation on Soil Microbial Community Structure and Nutrients. *Appl. Soil Ecol.* **2020**, *150*, 103461. [CrossRef]
57. Siddikee, M.A.; Zereen, M.I.; Li, C.-F.; Dai, C.-C. Endophytic Fungus Phomopsis Liquidambari and Different Doses of N-Fertilizer Alter Microbial Community Structure and Function in Rhizosphere of Rice. *Sci. Rep.* **2016**, *6*, 32270. [CrossRef]
58. Wang, H.-W.; Zhu, Y.-X.; Xu, M.; Cai, X.-Y.; Tian, F. Co-Application of Spent Mushroom Substrate and PGPR Alleviates Tomato Continuous Cropping Obstacle by Regulating Soil Microbial Properties. *Rhizosphere* **2022**, *23*, 100563. [CrossRef]
59. Sun, B.; Gu, L.; Bao, L.; Zhang, S.; Wei, Y.; Bai, Z.; Zhuang, G.; Zhuang, X. Application of Biofertilizer Containing Bacillus Subtilis Reduced the Nitrogen Loss in Agricultural Soil. *Soil Biol. Biochem.* **2020**, *148*, 107911. [CrossRef]
60. Wu, S.; Zhuang, G.; Bai, Z.; Cen, Y.; Xu, S.; Sun, H.; Han, X.; Zhuang, X. Mitigation of Nitrous Oxide Emissions from Acidic Soils by Bacillus Amyloliquefaciens, a Plant Growth-Promoting Bacterium. *Glob. Chang. Biol.* **2018**, *24*, 2352–2365. [CrossRef]

61. Conversa, G.; Lazzizera, C.; Bonasia, A.; Elia, A. Growth, N Uptake and N Critical Dilution Curve in Broccoli Cultivars Grown under Mediterranean Conditions. *Sci. Hortic.* **2019**, *244*, 109–121. [CrossRef]
62. IUSS, WRB World Reference Base for Soil Resources. *International Soil Classification System for Naming Soils and Creating Legends for Soil Maps*, 4th ed.; International Union of Soil Sciences (IUSS): Vienna, Austria, 2022.
63. *CARM Normativa Reguladora, Producción Integrada, Alimentos Sanos y Garantizados [Regulatory Rules, Integrated Production, Healthy Food and Guaranteed]*; Consejería de Agricultura y Agua de La Región de Murcia: Murcia, Spain, 1998.
64. He, Y.; Pantigoso, H.A.; Wu, Z.; Vivanco, J.M. Co-Inoculation of *Bacillus* sp. and Pseudomonas Putida at Different Development Stages Acts as a Biostimulant to Promote Growth, Yield and Nutrient Uptake of Tomato. *J. Appl. Microbiol.* **2019**, *127*, 196–207. [CrossRef]
65. Álvaro-Fuentes, J.; Lóczy, D.; Thiele-Bruhn, S.; Zornoza, R. *Handbook of Plant and Soil Analysis for Agricultural Systems*; Universidad Politécnica de Cartagena: Cartagena, Spain, 2019; ISBN 978-84-16325-86-3.
66. Chen, W.; Wang, Y.; Zhao, Z.; Cui, F.; Gu, J.; Zheng, X. The Effect of Planting Density on Carbon Dioxide, Methane and Nitrous Oxide Emissions from a Cold Paddy Field in the Sanjiang Plain, Northeast China. *Agric. Ecosyst. Environ.* **2013**, *178*, 64–70. [CrossRef]
67. Zornoza, R.; Landi, L.; Nannipieri, P.; Renella, G. A Protocol for the Assay of Arylesterase Activity in Soil. *Soil Biol. Biochem.* **2009**, *41*, 659–662. [CrossRef]
68. Vance, E.D.; Brookes, P.C.; Jenkinson, D.S. An Extraction Method for Measuring Soil Microbial Biomass C. *Soil Biol. Biochem.* **1987**, *19*, 703–707. [CrossRef]
69. Keeney, D.R.; Nelson, D.w. Nitrogen—Inorganic Forms. In *Methods of Soil Analysis*; John Wiley & Sons, Ltd.: Hoboken, NJ, USA, 1983; pp. 643–698. ISBN 978-0-89118-977-0.
70. Kandeler, E.; Gerber, H. Short-Term Assay of Soil Urease Activity Using Colorimetric Determination of Ammonium. *Biol. Fertil. Soils* **1988**, *6*, 68–72. [CrossRef]
71. Ekenler, M.; Tabatabai, M.A. β-Glucosaminidase Activity as an Index of Nitrogen Mineralization in Soils. *Commun. Soil Sci. Plant Anal.* **2004**, *35*, 1081–1094. [CrossRef]
72. Garcia-Alvarez, A.; Ibañez, J.J. Seasonal Fluctuations and Crop Influence on Microbiota and Enzyme Activity in Fully Developed Soils of Central Spain. *Arid Soil Res. Rehabil.* **1994**, *8*, 161–178. [CrossRef]
73. Nelson, N. A Photometric Adaptation of the Somogyi Method for the Determination of Glucose. *J. Biol. Chem.* **1944**, *153*, 375–380. [CrossRef]
74. Nannipieri, P.; Johnson, R.L.; Paul, E.A. Criteria for Measurement of Microbial Growth and Activity in Soil. *Soil Biol. Biochem.* **1978**, *10*, 223–229. [CrossRef]
75. Rotthauwe, J.H.; Witzel, K.P.; Liesack, W. The Ammonia Monooxygenase Structural Gene amoA as a Functional Marker: Molecular Fine-Scale Analysis of Natural Ammonia-Oxidizing Populations. *Appl. Environ. Microbiol.* **1997**, *63*, 4704–4712. [CrossRef] [PubMed]
76. Henry, S.; Baudoin, E.; López-Gutiérrez, J.C.; Martin-Laurent, F.; Brauman, A.; Philippot, L. Quantification of Denitrifying Bacteria in Soils by nirK Gene Targeted Real-Time PCR. *J. Microbiol. Methods* **2004**, *59*, 327–335. [CrossRef]
77. Selesi, D.; Schmid, M.; Hartmann, A. Diversity of Green-Like and Red-Like Ribulose-1,5-Bisphosphate Carboxylase/Oxygenase Large-Subunit Genes (cbbL) in Differently Managed Agricultural Soils. *Appl. Environ. Microbiol.* **2005**, *71*, 175–184. [CrossRef]
78. Tian, X.; Yang, T.; He, J.; Chu, Q.; Jia, X.; Huang, J. Fungal Community and Cellulose-Degrading Genes in the Composting Process of Chinese Medicinal Herbal Residues. *Bioresour. Technol.* **2017**, *241*, 374–383. [CrossRef]
79. Frostegård, Å.; Bååth, E.; Tunlio, A. Shifts in the Structure of Soil Microbial Communities in Limed Forests as Revealed by Phospholipid Fatty Acid Analysis. *Soil Biol. Biochem.* **1993**, *25*, 723–730. [CrossRef]
80. Joergensen, R.G. Phospholipid Fatty Acids in Soil—Drawbacks and Future Prospects. *Biol. Fertil. Soils* **2022**, *58*, 1–6. [CrossRef]
81. Tang, X.; Liu, S.; Zhou, G.; Zhang, D.; Zhou, C. Soil-Atmospheric Exchange of CO_2, CH_4, and N_2O in Three Subtropical Forest Ecosystems in Southern China. *Glob. Chang. Biol.* **2006**, *12*, 546–560. [CrossRef]
82. Iqbal, J.; Lin, S.; Hu, R.; Feng, M. Temporal Variability of Soil-Atmospheric CO_2 and CH_4 Fluxes from Different Land Uses in Mid-Subtropical China. *Atmos. Environ.* **2009**, *43*, 5865–5875. [CrossRef]
83. Curiel Yuste, J.; Baldocchi, D.D.; Gershenson, A.; Goldstein, A.; Misson, L.; Wong, S. Microbial Soil Respiration and Its Dependency on Carbon Inputs, Soil Temperature and Moisture. *Glob. Chang. Biol.* **2007**, *13*, 2018–2035. [CrossRef]
84. Arai, H.; Hadi, A.; Darung, U.; Limin, S.H.; Takahashi, H.; Hatano, R.; Inubushi, K. Land Use Change Affects Microbial Biomass and Fluxes of Carbon Dioxide and Nitrous Oxide in Tropical Peatlands. *Soil Sci. Plant Nutr.* **2014**, *60*, 423–434. [CrossRef]
85. Furukawa, Y.; Inubushi, K.; Ali, M.; Itang, A.M.; Tsuruta, H. Effect of Changing Groundwater Levels Caused by Land-Use Changes on Greenhouse Gas Fluxes from Tropical Peat Lands. *Nutr. Cycl. Agroecosystems* **2005**, *71*, 81–91. [CrossRef]
86. Li, Y.; Niu, W.; Wang, J.; Liu, L.; Zhang, M.; Xu, J. Effects of Artificial Soil Aeration Volume and Frequency on Soil Enzyme Activity and Microbial Abundance When Cultivating Greenhouse Tomato. *Soil Sci. Soc. Am. J.* **2016**, *80*, 1208–1221. [CrossRef]
87. Zhu, Y.; Cai, H.; Song, L.; Chen, H. Aerated Irrigation Promotes Soil Respiration and Microorganism Abundance around Tomato Rhizosphere. *Soil Sci. Soc. Am. J.* **2019**, *83*, 1343–1355. [CrossRef]
88. Mayer, H.P.; Conrad, R. Factors Influencing the Population of Methanogenic Bacteria and the Initiation of Methane Production upon Flooding of Paddy Soil. *FEMS Microbiol. Ecol.* **1990**, *6*, 103–111. [CrossRef]

89. Fan, Y.; Hao, X.; Carswell, A.; Misselbrook, T.; Ding, R.; Li, S.; Kang, S. Inorganic Nitrogen Fertilizer and High N Application Rate Promote N_2O Emission and Suppress CH_4 Uptake in a Rotational Vegetable System. *Soil Tillage Res.* **2021**, *206*, 104848. [CrossRef]
90. Sun, B.; Dong, Z.-X.; Zhang, X.-X.; Li, Y.; Cao, H.; Cui, Z.-L. Rice to Vegetables: Short- Versus Long-Term Impact of Land-Use Change on the Indigenous Soil Microbial Community. *Microb. Ecol.* **2011**, *62*, 474–485. [CrossRef] [PubMed]
91. Stark, C.H.; Condron, L.M.; O'Callaghan, M.; Stewart, A.; Di, H.J. Differences in Soil Enzyme Activities, Microbial Community Structure and Short-Term Nitrogen Mineralisation Resulting from Farm Management History and Organic Matter Amendments. *Soil Biol. Biochem.* **2008**, *40*, 1352–1363. [CrossRef]
92. Lüneberg, K.; Schneider, D.; Siebe, C.; Daniel, R. Drylands Soil Bacterial Community Is Affected by Land Use Change and Different Irrigation Practices in the Mezquital Valley, Mexico. *Sci. Rep.* **2018**, *8*, 1413. [CrossRef] [PubMed]
93. Matysek, M.; Leake, J.; Banwart, S.; Johnson, I.; Page, S.; Kaduk, J.; Smalley, A.; Cumming, A.; Zona, D. Impact of Fertiliser, Water Table, and Warming on Celery Yield and CO_2 and CH_4 Emissions from Fenland Agricultural Peat. *Sci. Total Environ.* **2019**, *667*, 179–190. [CrossRef] [PubMed]
94. Ding, Y.; Li, C.; Li, Z.; Liu, S.; Zou, Y.; Gao, X.; Cai, Y.; Siddique, K.H.M.; Wu, P.; Zhao, X. Greenhouse Gas Emission Responses to Different Soil Amendments on the Loess Plateau, China. *Agric. Ecosyst. Environ.* **2023**, *342*, 108233. [CrossRef]
95. Flessa, H.; Ruser, R.; Dörsch, P.; Kamp, T.; Jimenez, M.A.; Munch, J.C.; Beese, F. Integrated Evaluation of Greenhouse Gas Emissions (CO_2, CH_4, N_2O) from Two Farming Systems in Southern Germany. *Agric. Ecosyst. Environ.* **2002**, *91*, 175–189. [CrossRef]
96. Glatzel, S.; Stahr, K. Methane and Nitrous Oxide Exchange in Differently Fertilised Grassland in Southern Germany. *Plant Soil* **2001**, *231*, 21–35. [CrossRef]
97. Iqbal, J.; Hu, R.; Feng, M.; Lin, S.; Malghani, S.; Ali, I.M. Microbial Biomass, and Dissolved Organic Carbon and Nitrogen Strongly Affect Soil Respiration in Different Land Uses: A Case Study at Three Gorges Reservoir Area, South China. *Agric. Ecosyst. Environ.* **2010**, *137*, 294–307. [CrossRef]
98. Zhang, S.; Fang, Y.; Luo, Y.; Li, Y.; Ge, T.; Wang, Y.; Wang, H.; Yu, B.; Song, X.; Chen, J.; et al. Linking Soil Carbon Availability, Microbial Community Composition and Enzyme Activities to Organic Carbon Mineralization of a Bamboo Forest Soil Amended with Pyrogenic and Fresh Organic Matter. *Sci. Total Environ.* **2021**, *801*, 149717. [CrossRef]
99. Tang, Y.; Yu, G.; Zhang, X.; Wang, Q.; Tian, J.; Niu, S.; Tian, D.; Ge, J. Different Strategies for Regulating Free-Living N2 Fixation in Nutrient-Amended Subtropical and Temperate Forest Soils. *Appl. Soil Ecol.* **2019**, *136*, 21–29. [CrossRef]
100. Duff, A.M.; Forrestal, P.; Ikoyi, I.; Brennan, F. Assessing the Long-Term Impact of Urease and Nitrification Inhibitor Use on Microbial Community Composition, Diversity and Function in Grassland Soil. *Soil Biol. Biochem.* **2022**, *170*, 108709. [CrossRef]
101. Kuypers, M.M.M.; Marchant, H.K.; Kartal, B. The Microbial Nitrogen-Cycling Network. *Nat. Rev. Microbiol.* **2018**, *16*, 263–276. [CrossRef] [PubMed]
102. Wang, C.; Zheng, M.; Song, W.; Wen, S.; Wang, B.; Zhu, C.; Shen, R. Impact of 25 Years of Inorganic Fertilization on Diazotrophic Abundance and Community Structure in an Acidic Soil in Southern China. *Soil Biol. Biochem.* **2017**, *113*, 240–249. [CrossRef]
103. Ouyang, Y.; Evans, S.E.; Friesen, M.L.; Tiemann, L.K. Effect of Nitrogen Fertilization on the Abundance of Nitrogen Cycling Genes in Agricultural Soils: A Meta-Analysis of Field Studies. *Soil Biol. Biochem.* **2018**, *127*, 71–78. [CrossRef]
104. Dynarski, K.A.; Houlton, B.Z. Nutrient Limitation of Terrestrial Free-Living Nitrogen Fixation. *New Phytol.* **2018**, *217*, 1050–1061. [CrossRef]
105. Yuan, H.; Ge, T.; Chen, C.; O'Donnell, A.G.; Wu, J. Significant Role for Microbial Autotrophy in the Sequestration of Soil Carbon. *Appl. Environ. Microbiol.* **2012**, *78*, 2328–2336. [CrossRef]
106. Van Hees, P.A.W.; Jones, D.L.; Finlay, R.; Godbold, D.L.; Lundström, U.S. The Carbon We Do Not See—The Impact of Low Molecular Weight Compounds on Carbon Dynamics and Respiration in Forest Soils: A Review. *Soil Biol. Biochem.* **2005**, *37*, 1–13. [CrossRef]
107. Abdel-Rahman, M.A.; Nour El-Din, M.; Refaat, B.M.; Abdel-Shakour, E.H.; Ewais, E.E.-D.; Alrefaey, H.M.A. Biotechnological Application of Thermotolerant Cellulose-Decomposing Bacteria in Composting of Rice Straw. *Ann. Agric. Sci.* **2016**, *61*, 135–143. [CrossRef]
108. Ahmad, B.; Nigar, S.; Malik, N.A.; Bashir, S.; Ali, J.; Yousaf, S.; Bangash, J.A.; Jan, I. Isolation and Characterization of Cellulolytic Nitrogen Fixing Azotobacter Species from Wheat Rhizosphere of Khyber Pakhtunkhwa. *World Appl. Sci. J.* **2013**, *27*, 51–60.
109. Emtiazi, G.; Etemadifar, Z.; Tavassoli, M. A Novel Nitrogen-Fixing Cellulytic Bacterium Associated with Root of Corn Is a Candidate for Production of Single Cell Protein. *Biomass Bioenergy* **2003**, *25*, 423–426. [CrossRef]
110. Harindintwali, J.D.; Zhou, J.; Yu, X. Lignocellulosic Crop Residue Composting by Cellulolytic Nitrogen-Fixing Bacteria: A Novel Tool for Environmental Sustainability. *Sci. Total Environ.* **2020**, *715*, 136912. [CrossRef]
111. Miao, F.; Li, Y.; Cui, S.; Jagadamma, S.; Yang, G.; Zhang, Q. Soil Extracellular Enzyme Activities under Long-Term Fertilization Management in the Croplands of China: A Meta-Analysis. *Nutr. Cycl. Agroecosystems* **2019**, *114*, 125–138. [CrossRef]
112. García-Orenes, F.; Morugán-Coronado, A.; Zornoza, R.; Scow, K. Changes in Soil Microbial Community Structure Influenced by Agricultural Management Practices in a Mediterranean Agro-Ecosystem. *PLoS ONE* **2013**, *8*, e80522. [CrossRef] [PubMed]
113. Siles, J.A.; Vera, A.; Díaz-López, M.; García, C.; van den Hoogen, J.; Crowther, T.W.; Eisenhauer, N.; Guerra, C.; Jones, A.; Orgiazzi, A.; et al. Land-Use- and Climate-Mediated Variations in Soil Bacterial and Fungal Biomass across Europe and Their Driving Factors. *Geoderma* **2023**, *434*, 116474. [CrossRef]

114. Yang, T.; Li, X.; Hu, B.; Wei, D.; Wang, Z.; Bao, W. Soil Microbial Biomass and Community Composition along a Latitudinal Gradient in the Arid Valleys of Southwest China. *Geoderma* **2022**, *413*, 115750. [CrossRef]
115. Bastida, F.; Eldridge, D.J.; García, C.; Kenny Png, G.; Bardgett, R.D.; Delgado-Baquerizo, M. Soil Microbial Diversity–Biomass Relationships Are Driven by Soil Carbon Content across Global Biomes. *ISME J.* **2021**, *15*, 2081–2091. [CrossRef]
116. Wang, H.; Liu, S.; Li, H.; Tao, X.; Wang, H.; Qi, J.; Zhang, Z. Large-Scale Homogenization of Soil Bacterial Communities in Response to Agricultural Practices in Paddy Fields, China. *Soil Biol. Biochem.* **2022**, *164*, 108490. [CrossRef]
117. Di Salvo, L.P.; Cellucci, G.C.; Carlino, M.E.; García de Salamone, I.E. Plant Growth-Promoting Rhizobacteria Inoculation and Nitrogen Fertilization Increase Maize (*Zea mays* L.) Grain Yield and Modified Rhizosphere Microbial Communities. *Appl. Soil Ecol.* **2018**, *126*, 113–120. [CrossRef]
118. Sabier Saeed, K.; Abdulla Ahmed, S.; Ahmaed Hassan, I.; Hamed Ahmed, P. Effect of Bio-Fertilizer and Chemical Fertilizer on Growth and Yield in Cucumber (*Cucumis sativus*) in Green House Condition. *Pak. J. Biol. Sci.* **2015**, *18*, 129–134. [CrossRef]
119. Angelina, E.; Papatheodorou, E.M.; Demirtzoglou, T.; Monokrousos, N. Effects of Bacillus Subtilis and Pseudomonas Fluorescens Inoculation on Attributes of the Lettuce (*Lactuca sativa* L.) Soil Rhizosphere Microbial Community: The Role of the Management System. *Agronomy* **2020**, *10*, 1428. [CrossRef]
120. Qiu, Z.; Egidi, E.; Liu, H.; Kaur, S.; Singh, B.K. New Frontiers in Agriculture Productivity: Optimised Microbial Inoculants and in Situ Microbiome Engineering. *Biotechnol. Adv.* **2019**, *37*, 107371. [CrossRef] [PubMed]
121. Fiorentino, N.; Ventorino, V.; Woo, S.L.; Pepe, O.; De Rosa, A.; Gioia, L.; Romano, I.; Lombardi, N.; Napolitano, M.; Colla, G.; et al. Trichoderma-Based Biostimulants Modulate Rhizosphere Microbial Populations and Improve N Uptake Efficiency, Yield, and Nutritional Quality of Leafy Vegetables. *Front. Plant Sci.* **2018**, *9*, 743. [CrossRef] [PubMed]
122. Siebers, M.; Rohr, T.; Ventura, M.; Schütz, V.; Thies, S.; Kovacic, F.; Jaeger, K.-E.; Berg, M.; Dörmann, P.; Schulz, M. Disruption of Microbial Community Composition and Identification of Plant Growth Promoting Microorganisms after Exposure of Soil to Rapeseed-Derived Glucosinolates. *PLoS ONE* **2018**, *13*, e0200160. [CrossRef]
123. Ju, W.; Liu, L.; Fang, L.; Cui, Y.; Duan, C.; Wu, H. Impact of Co-Inoculation with Plant-Growth-Promoting Rhizobacteria and Rhizobium on the Biochemical Responses of Alfalfa-Soil System in Copper Contaminated Soil. *Ecotoxicol. Environ. Saf.* **2019**, *167*, 218–226. [CrossRef] [PubMed]
124. Hidri, R.; Metoui-Ben Mahmoud, O.; Debez, A.; Abdelly, C.; Barea, J.-M.; Azcon, R. Modulation of C:N:P Stoichiometry Is Involved in the Effectiveness of a PGPR and AM Fungus in Increasing Salt Stress Tolerance of Sulla Carnosa Tunisian Provenances. *Appl. Soil Ecol.* **2019**, *143*, 161–172. [CrossRef]
125. Egamberdiyeva, D. The Effect of Plant Growth Promoting Bacteria on Growth and Nutrient Uptake of Maize in Two Different Soils. *Appl. Soil Ecol.* **2007**, *36*, 184–189. [CrossRef]
126. Mourão, I.; Brito, M. Effects of Direct Film Crop Cover and Top Dress Nitrogen on Earliness and Yield of Broccoli Crop (*Brassica oleracea* Var. Italica Plenk). In Proceedings of the International Conference on Environmental Problems Associated with Nitrogen Fertilisation of Field Grown Vegetable Crops 563, Postdam, Germany, 30 August–1 September 1999; pp. 103–109.
127. Nuzzo, A.; Satpute, A.; Albrecht, U.; Strauss, S.L. Impact of Soil Microbial Amendments on Tomato Rhizosphere Microbiome and Plant Growth in Field Soil. *Microb. Ecol.* **2020**, *80*, 398–409. [CrossRef]
128. Schenck zu Schweinsberg-Mickan, M.; Müller, T. Impact of Effective Microorganisms and Other Biofertilizers on Soil Microbial Characteristics, Organic-Matter Decomposition, and Plant Growth. *J. Plant Nutr. Soil Sci.* **2009**, *172*, 704–712. [CrossRef]
129. Martínez-Viveros, O.; Jorquera, M.A.; Crowley, D.E.; Gajardo, G.; Mora, M.L. Mechanisms and practical considerations involved in plant growth promotion by rhizobacteria. *J. Soil Sci. Plant Nutr.* **2010**, *10*, 293–319. [CrossRef]
130. Kim, S.; Kim, S.; Kiniry, J.R.; Ku, K.-M. A Hybrid Decision Tool for Optimizing Broccoli Production in a Changing Climate. *Hortic. Environ. Biotechnol.* **2021**, *62*, 299–312. [CrossRef]

horticulturae

Article

Exploring the Feasibility of Integrating Weed and Nitrogen Management with Seed Meals in Organic Vegetables

Allison Butterfield and Sam E. Wortman *

Department of Agronomy and Horticulture, University of Nebraska—Lincoln, 278 Plant Sciences Hall, Lincoln, NE 68583, USA
* Correspondence: swortman@unl.edu

Abstract: Corn gluten meal (CGM) and soybean meal (SBM) have demonstrated value as bioherbicides and organic fertilizers, but suggested application rates usually target either weed suppression or crop nutrition, not both. The objective of this study was to explore the feasibility of integrating weed and nitrogen management by evaluating effects of increasing seed meal rates within planting holes of plastic mulch film on weed density, soil nitrogen availability, and crop yield in tomato (*Solanum lycopersicum*) and broccoli (*Brassica oleracea*). CGM (10% N) or SBM (7% N) were applied at rates of 0.5, 1, 2, 3.5, or 5 g planting hole^{-1} N (depending on crop and year) after crops were transplanted, and 40 weed seeds per planting hole were seeded. Weed density decreased with increasing seed meal rate, regardless of type, and velvetleaf (*Abutilon theophrasti*) was more susceptible than the grass weeds tested. Velvetleaf suppression at the 5 g planting hole^{-1} N rate ranged from 66% to 97%, relative to the weedy control. Soil nitrogen availability increased with the application rate, but ammonium mineralized from seed meals applied at the highest rates were likely phytotoxic to weeds and crops. Seed meals never increased the crop yield and reduced the tomato yield in 2018 by 39% to 64%, relative to the weed-free control. The results suggest that integrating the management of weeds and nitrogen with seed meals in plastic mulch planting holes is not feasible because application rates required for consistent weed suppression are also toxic to crops.

Keywords: organic farming; non-chemical weed management; integrated weed management; organic fertilizer; biobased inputs

Citation: Butterfield, A.; Wortman, S.E. Exploring the Feasibility of Integrating Weed and Nitrogen Management with Seed Meals in Organic Vegetables. *Horticulturae* **2024**, *10*, 75. https://doi.org/10.3390/horticulturae10010075

Academic Editors: Francesco De Mastro, Gennaro Brunetti, Karam Farrag, Huadong Zang and Moreno Toselli

Received: 9 December 2023
Revised: 7 January 2024
Accepted: 9 January 2024
Published: 11 January 2024

1. Introduction

Managing weeds and soil fertility are among the top four production challenges identified by organic farmers [1]. Weed management on organic farms often includes manual hand weeding, and organic farmers identified accessing labor as the top non-production related challenge [1]. Thus, research is needed to improve weed management and soil fertility and reduce labor needs in organic farms. Integrating the management of nitrogen and weed management using seed meals could potentially reduce weeds, improve crop nutrition, and reduce labor needed for hand weeding. Depending on their application method and rate, seed meals have the potential to control weeds through physical or chemical modes of action [2,3]. Seed meals can be used to physically abrade emerged weed seedlings when applied with compressed air [2] or applied to soil as an herbicide to prevent weed seed germination or stunt seedling growth [3].

Plastic mulch films are a critically important weed management tool for organic specialty crop growers [4], yet weeds can still emerge and compete with crops through planting holes (i.e., the hole created in the mulch to allow for crop planting and growth), along the edges of the mulch, or through the mulch itself. Unmanaged weeds growing through plastic mulch film planting holes have been shown to reduce pepper (*Capsicum annuum*) and tomato (*Solanum lycopersicum*) yield by 29% to 45% [5]. Seed meals applied directly into the planting hole offer a unique opportunity for integrated weed and nitrogen management because many seed meals have potential value as organic fertilizers and bioherbicides [6].

Corn gluten meal (CGM) is one of the most studied seed meals for its herbicidal potential and is commercially available as a natural preemergent herbicide. Bingaman and Christians [7] applied CGM rates of 0, 324, 649, and 973 g m^{-2} to surface soils in greenhouse pots and found that the growth and survival of many weed species was reduced by the lowest rate of 324 g m^{-2}, but efficacy varied greatly among species. Green foxtail (*Setaria viridis*) survival was reduced by 37% at 324 g m^{-2} and 100% at the 973 g m^{-2} rate, whereas shattercane (*Sorghum bicolor*) survival was reduced only 42% to 51% across all rates. Velvetleaf (*Abutilon theophrasti*) was among the least affected weed species, as survival was 100% at the 324 g m^{-2} rate and was only reduced by 35% at the highest rate of 973 g m^{-2} [7]. By contrast, Wortman [2] found that lower rates of 54 to 269 g m^{-2} CGM stimulated weed germination and growth instead of inhibiting them. This highlights the fertilizer potential of CGM, which contains approximately 10% nitrogen by weight. Under field conditions, McDade and Christians [3] tested CGM at rates between 100 and 400 g m^{-2} in direct seeded vegetables and found similar reductions in weed growth (50% to 82%), but the CGM had similarly negative effects on crop seedling survival, which was reduced 41% to 73%. Webber III et al. [8] observed similar weed control of 72% after an application of 400 g m^{-2} CGM in transplanted onion, and transplanting seemed to mitigate herbicidal effects on the crop.

Soybean meal (SBM) is another widely available seed meal but is more commonly marketed as an organic fertilizer, and less is known about its potential as a biobased herbicide. Shrestha et al. [9] found that rates of 124 to 448 g m^{-2} SBM had limited herbicidal potential, particularly beyond two months after application, but increasing SBM rates did increase spinach yield. The results from Wortman [2] in the greenhouse were similar, as rates between 54 and 269 g m^{-2} SBM had no effect on weed emergence and growth. However, An et al. [10] applied 200 g m^{-2} SBM in transplanted rice paddies, and weed density was reduced by 40% to 100% (depending on the species), and the fresh weight of rice increased by 53%. Similarly, El-Metwally et al. [11] found that 150 g m^{-2} SBM reduced the weed biomass of four different broadleaf weeds and two grass weeds by 80% to 84% compared to weedy controls. SBM was applied and concentrated around the crops to a depth of 4–6 cm, which may have acted more as a mulch than as an herbicide. Similar rates applied within planting holes of plastic mulch films may provide these mulching benefits while concurrently delivering nitrogen to the plant.

Given the potential integrated value of seed meals as biobased herbicides, mulch, and fertilizers, our objective was to evaluate the effects of CGM and SBM on weed density, soil nitrogen availability, and crop yield when applied at different rates in the planting holes of plastic mulch film. The results will help to inform optimum application rates and the potential tradeoffs for integrating the use of seed meals as bioherbicides and organic fertilizers.

2. Materials and Methods

To accomplish our study objectives, five field trials were conducted between May 2017 and November 2018 at two locations on the University of Nebraska East Campus Research Farm (lat. 40°50′12″ N, long. 96°39′48″ W). Three trials were conducted in tomato (*Solanum lycopersicum*) (one trial in 2017 and two trials in 2018 in fields separated by approximately 1 km) and two trials in broccoli (*Brassica oleracea*) (one trial in 2017 and 2018 each). The soil type at the farm (in both locations) is a Zook silty clay loam, and pre-plant soil analyses (Ward Laboratories, Keaney, NE, USA) for each field location and year are included in Table 1. Pre-plant soil samples were analyzed for pH (1:1 soil/water dilution), soil organic matter (loss of weight on ignition), nitrate-N (KCl extraction), phosphorus (Mehlich 3 extraction), and potassium (ammonium acetate extraction) [12].

Table 1. Pre-plant analysis of soils collected to a depth of 0 to 20 cm in 2017 and 2018 from two different experimental fields (north and south) at the University of Nebraska—Lincoln East Campus Research Farm.

	pH	Organic Matter ($g\ kg^{-1}$)T	NO_3-N ($mg\ kg^{-1}$)	Mehlich-P ($mg\ kg^{-1}$)	K ($mg\ kg^{-1}$)
2017					
North field (tomato and broccoli)	6.2	4.0	10.7	90	410
2018					
North field (tomato and broccoli)	6.1	4.2	9.6	95	422
South field (tomato)	6.7	3.8	9.5	86	683

2.1. Experimental Design and Treatments

Each experiment was a randomized complete block design with two treatment factors. The treatment factors included seed meal type—CGM (Preen Natural Weed Preventer, Lebanon Seaboard Corporation, Lebanon, PA, USA) or SBM (Phyta-Grow Leafy Green Special Fertilizer, California Organic Fertilizers, Inc., Hanford, CA, USA)—and rate. The rates were standardized by an estimated N analysis of the seed meals (10% for CGM based on Webber III et al. [13] and 7% for SBM based on the labeled guaranteed analysis) and included 0.5, 1, 2, or 5 g planting hole^{-1} N in 2017; and 2, 3.5, or 5 g planting hole^{-1} N in 2018 (Figure 1). The rates were modified in 2018 based on results from 2017 to better identify an optimum rate for integrated N and weed management. All the treatment combinations were compared to weed-free and weedy controls (0 g planting hole^{-1} N). Assuming 46 cm in row spacing and 1.8 m between row centers (typical for vegetables in the U.S.), a seed meal rate of 0.5 g planting hole^{-1} N resulted in the addition of approximately 6 kg ha^{-1} N, and 5 g planting hole^{-1} N contributed approximately 60 kg ha^{-1} N. Experimental plots were 4.6 m long on 1.2 m wide raised beds. The plants were spaced 0.46 m apart within rows for a total of 10 plants plot^{-1}.

Figure 1. Seed meals applied in planting holes of tomato (**left**) and broccoli (**right**).

Tomato (cv. 'Defiant') and broccoli (cv. 'Arctic') seeds (Johnny's Selected Seeds, Winslow, Maine, USA) were planted 6 to 8 weeks prior to field transplanting in 72-cell greenhouse flats filled with a general use soilless media mix in the greenhouse. The seedlings were watered daily and fertilized as needed with 250 mg dm^{-3} of a water soluble 20-20-20 N-P-K fertilizer (Jack's Professional General Purpose, J.R. Peter's, Inc., Allentown, PA, USA). The fields were prepared 1 to 7 days prior to transplanting by roto-tilling the soil and then concurrently shaping the raised beds (RB-448; Nolt's Produce Supplies; Leola, PA, USA) and laying a single drip tape beneath a black or white-on-black biodegradable plastic mulch film (BIO360; Dubois Agrinovation, Saint-Rémi, QC, Canada). The tomatoes were transplanted on 31 May 2017 and 15 May 2018, and broccoli was transplanted on 18 August 2017 and 10 August 2018. Crop seedlings were transplanted 46 cm apart in single rows on each bed top into punched holes in the mulch film. The crops were drip-irrigated between precipitation events to maintain volumetric soil moisture at or above 0.15 cm^3 cm^{-3} in the top 20 cm of the soil.

Weed seeds were planted in each mulch film planting hole within 2 days of transplanting. In 2017, 20 velvetleaf (*Abutilon theophrasti*) and 20 green foxtail (*Setaria viridis*) seeds were used as model broadleaf and grass weeds, respectively. Forty total seeds were placed in each planting hole and covered with 50 g of topsoil. In 2018, 20 shattercane (*Sorghum bicolor*) seeds were used instead of green foxtail due to poor germination in the green foxtail seed lot. Prior to seeding, velvetleaf seeds were submerged in a 70 °C water bath for 1 min to improve germination [14]. Seed meal treatments were applied within planting holes (above the buried weed seeds) immediately after seeding weeds, resulting in a layer of seed meal that covered what is typically bare soil between the crop stem and mulch film (Figure 1). The plots were uniformly irrigated 2 days after seed meal application to ensure weed seed germination.

2.2. Data Collection

Emerged weed seedlings were identified by the species seeded (velvetleaf, green foxtail, or shattercane) and counted in 5 random planting holes per plot between 16 and 33 days after transplanting tomato or broccoli.

In 2018, pairs of cation and anion Plant Root Simulator (PRS) probes (Western Ag Innovations, Saskatoon, SK, Canada) were used to measure changes in total plant-available soil nitrate and ammonium within a subset of treatments in tomato and broccoli planting holes [15]. The units used by Western Ag Innovations are μg nutrient 10 cm^{-2} ion-exchange membrane surface area $time^{-1}$ of burial (2 weeks in this study). In tomato, data were collected for the 2 g and 5 g rates of CGM and the weedy and weed-free controls across both locations. In broccoli, data were collected for the 2 g and 5 g rates for both seed meals, along with the weedy and weed-free controls. Probes remained in the soil for a 2-week incubation, beginning 2 days after seed meal application. Upon removal from the soil, PRS probe pairs were rinsed with deionized water and stored at 4 °C before shipping, extraction, and analysis at Western Ag Innovations. The probes were eluted for 1 h in 17.5 mL of 0.5 mol/L HCl, and nitrate and ammonium in the eluant were quantified colorimetrically using automated flow injection analysis (Skalar San++ Analyzer; Skalar Inc.; Breda, Noord-Brabant, The Netherlands).

The tomato fruits were harvested when ripe every 5–7 days and weighed fresh to determine the total yield. In 2017, there were six tomato harvests, and, in 2018, there were five. Broccoli was harvested by cutting 15 cm below the head and weighed fresh to determine the total yield.

2.3. Data Analysis

Yield, weed density, and soil nitrogen data were analyzed with a generalized linear mixed model analysis of variance (proc GLIMMIX; SAS 9.4; SAS Institute Inc., Cary, NC, USA). Crops and years were analyzed separately due to differences in seed meal rates between years. Fixed effects in the model included seed meal type, application rate, and their interaction. The replicate block was a random effect along with location (in 2018, tomato only). Least square (LS) means and standard errors were calculated for all significant fixed effects and compared using Tukey's honestly significant difference test at $\alpha = 0.05$.

3. Results

3.1. Tomato

3.1.1. Weed Density

Velvetleaf ($p < 0.0001$) and green foxtail ($p = 0.02$) weed density, 33 days after transplanting in 2017, were both influenced by seed meal rate but not by type or their interaction.

Velvetleaf density generally declined with increasing rates of seed meals to a maximum of 5 g planting hole^{-1} N (Figure 2). The trends were similar for green foxtail, but overall emergence was much lower, and tested rates were not different from the weedy control.

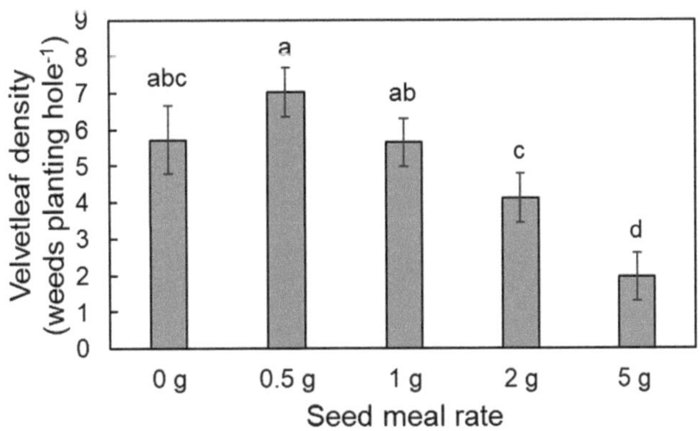

Figure 2. Velvetleaf density in 2017 tomato planting holes 33 days after transplanting as influenced by seed meal rate (g planting hole^{-1} N) pooled across seed meal types (corn gluten meal and soybean meal). Error bars represent ± one standard error of the mean. Different letters above each bar indicate differences among treatments as determined by the Tukey's honestly significant difference test at $\alpha = 0.05$.

In 2018, velvetleaf density was influenced by the seed meal rate ($p < 0.0001$) and type ($p = 0.0003$) but not their interaction. Consistent with 2017, velvetleaf density decreased with increasing seed meal rate (regardless of type) up to 5 g planting hole^{-1} N (Figure 3). However, in 2018, SBM provided slightly better velvetleaf suppression (0.54 ± 0.13 weeds planting hole^{-1}) compared to CGM (1.39 ± 0.13 weeds planting hole^{-1}) across all the tested rates. Shattercane density was not affected by the seed meal type, rate, or their interaction but was overall much lower than velvetleaf.

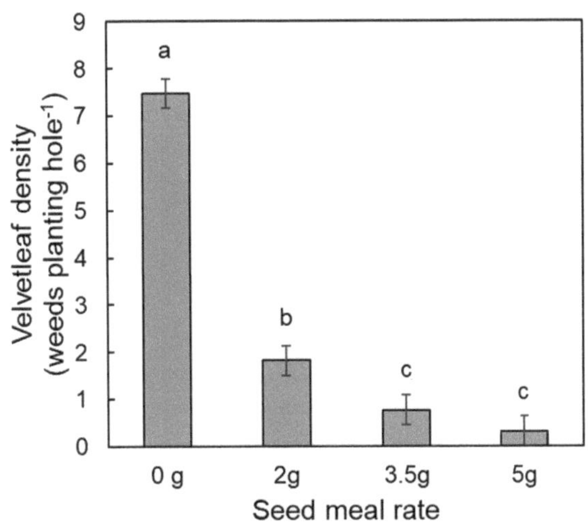

Figure 3. Velvetleaf density in 2018 tomato planting holes 16 days after transplanting as influenced by seed meal rate (g planting hole^{-1} N) pooled across seed meal types (corn gluten meal and soybean meal) and two field locations. Error bars represent ± one standard error of the mean. Different letters above each bar indicate differences among treatments as determined by the Tukey's honestly significant difference test at $\alpha = 0.05$.

3.1.2. Nitrogen Availability

Nitrate availability in tomato planting holes between 2 and 16 days after seed meal application was influenced by the CGM rate ($p < 0.0001$). Nitrate was greatest in the 5 g treatment, followed by 2 g, and lowest in the weedy and weed-free controls that did not receive any CGM (Figure 4). Ammonium availability was also influenced by the CGM rate ($p = 0.02$) and was 3.3× greater in the 5 g treatment compared to the 2 g (Figure 4).

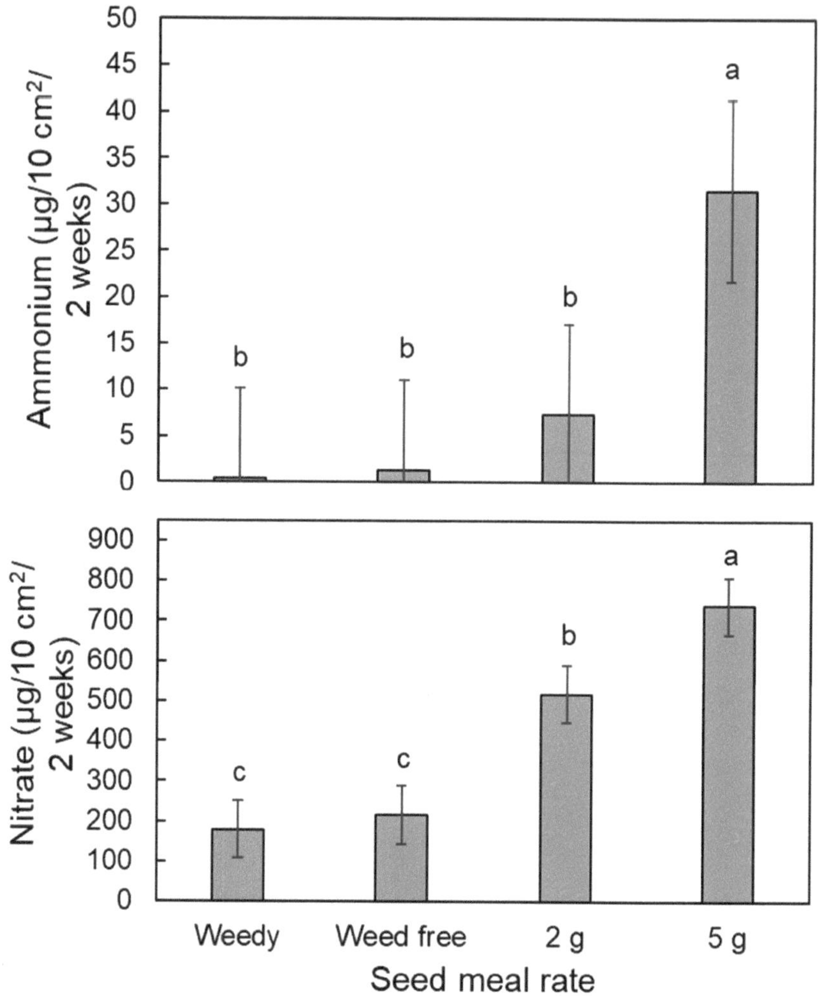

Figure 4. Ammonium (**top**) and nitrate (**bottom**) availability (µg 10 cm^{-2} 2 $weeks^{-1}$; measured by PRS Probes) in 2018 tomato planting holes (between 2 and 16 days after seed meal application) as influenced by corn gluten meal rate (g planting $hole^{-1}$ N) relative to weedy and weed-free controls. Data are pooled across two field locations. Error bars represent ± one standard error of the mean. Different letters above each bar indicate differences among treatments as determined by the Tukey's honestly significant difference test at $\alpha = 0.05$.

3.1.3. Yield

The tomato yield in 2017 was not influenced by the seed meal type ($p = 0.54$), but the effect of the rate was approaching significance ($p = 0.056$) because of reduced yield in the 5 g treatment (103.5 ± 6.1 kg 10 m row^{-1}) compared to the weed-free control (123.0 ± 8.7 kg

10 m row^{-1}). In 2018, the tomato yield was influenced by the seed meal type (p = 0.03) but not the rate (p = 0.13) because seed meals reduced the tomato yield by 39% to 64%, regardless of rate (Figure 5). Overall, the tomato yield was much lower in 2018 (3.7 kg 10 m row^{-1}) compared to 2017 (117.4 kg 10 m row^{-1}) because of disease and delayed fruit maturation throughout the field.

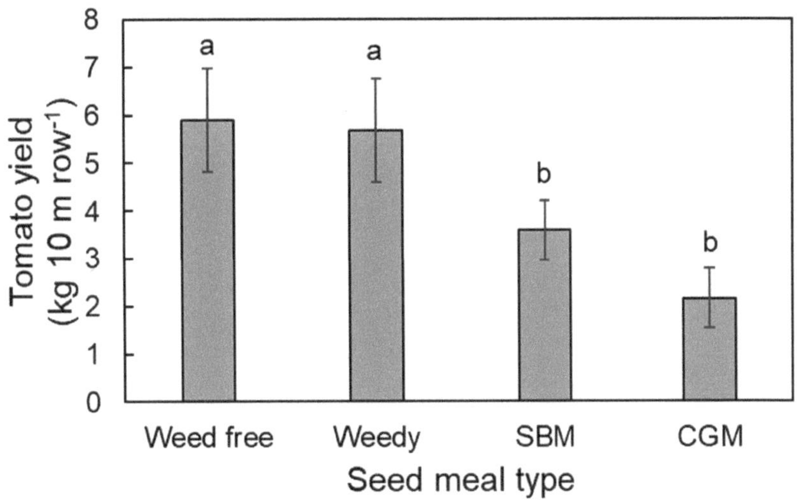

Figure 5. Tomato yield (kg 10 m row^{-1}) in 2018 as influenced by seed meal type (SBM = soybean meal; CGM = corn gluten meal), pooled across rates, compared to weed-free and weedy controls. Data are pooled across seed meal rates and two field locations. Error bars represent ± one standard error of the mean. Different letters above each bar indicate differences among treatments as determined by the Tukey's honestly significant difference test at α = 0.05.

3.2. Broccoli

3.2.1. Weed Density

Velvetleaf weed density, 27 days after transplanting in 2017, was influenced by the seed meal rate (p = 0.0002) but not by type or the interaction with the rate. Velvetleaf density generally declined with increasing rates of seed meal to a maximum of 5 g planting hole^{-1} N (Figure 6). Green foxtail density was not affected by the seed meal rate, type, or their interaction (p > 0.05), and overall emergence was generally low (<2 weeds planting hole^{-1}). In 2018, velvetleaf (p = 0.03) and shattercane (p = 0.0005) density were influenced by the seed meal rate (but not by type or their interaction). Weed density was lowest in the 5 g planting hole^{-1} N rate, followed by the 3.5 g rate (Figure 7).

3.2.2. Nitrogen Availability

Nitrate (p = 0.03) and ammonium (p = 0.03) availability in broccoli planting holes between 2 and 16 days after seed meal application were influenced by the seed meal rate but not by type or their interaction. Both nitrate and ammonium were greater in the 5 g treatment compared to 2 g and the weed-free control, but the weedy control was not different from any other treatment (Figure 8).

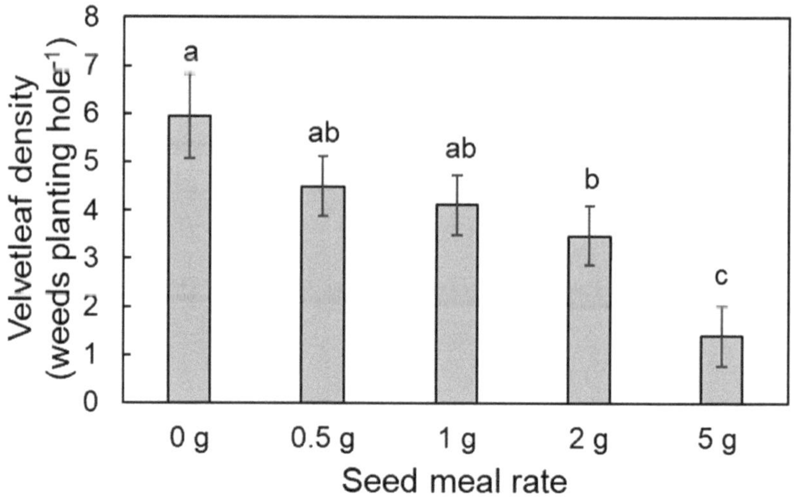

Figure 6. Velvetleaf density in 2017 broccoli planting holes 27 days after transplanting, as influenced by seed meal rate (g planting hole^{-1} N) pooled across seed meal types (CGM and SBM). Error bars represent ± one standard error of the mean. Different letters above each bar indicate differences among treatments as determined by the Tukey's honestly significant difference test at α = 0.05.

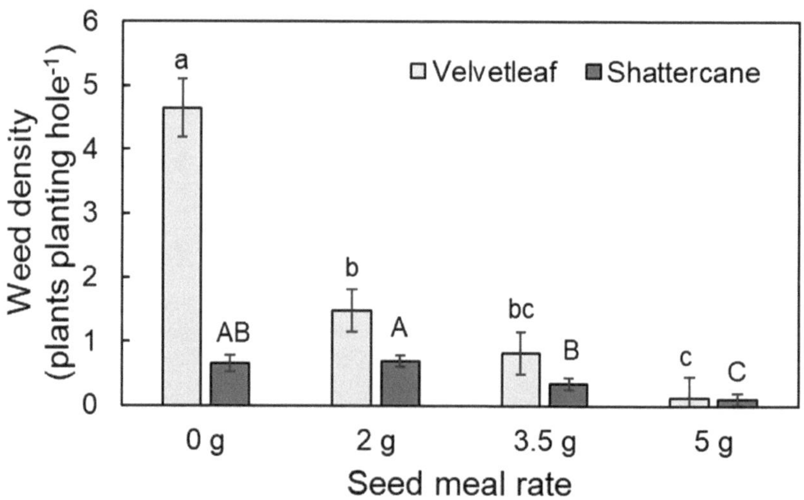

Figure 7. Velvetleaf and shattercane density in 2018 broccoli planting holes 27 days after transplanting as influenced by seed meal rate (g planting hole^{-1} N) pooled across seed meal types (corn gluten meal and soybean meal). Error bars represent ± one standard error of the mean. Different letters above each bar (within each weed species) indicate differences among treatments as determined by the Tukey's honestly significant difference test at α = 0.05.

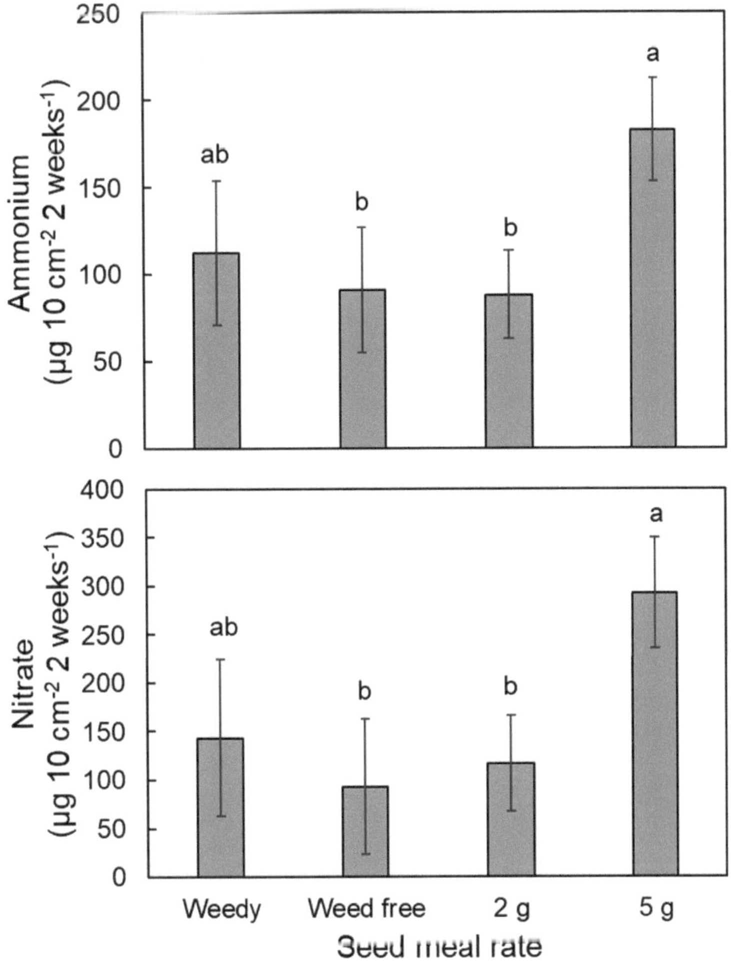

Figure 8. Ammonium (**top**) and nitrate (**bottom**) availability (µg 10 cm^{-2} 2 weeks^{-1}; measured by PRS Probes) in 2018 broccoli planting holes (between 2 and 16 days after seed meal application) as influenced by seed meal rate (g planting hole^{-1} N) relative to weedy and weed-free controls. Data are pooled across seed meal type (corn gluten meal and soybean meal). Error bars represent ± one standard error of the mean. Different letters above each bar indicate differences among treatments as determined by the Tukey's honestly significant difference test at $\alpha = 0.05$.

3.2.3. Yield

The broccoli yield was not influenced by seed meal type, rate, or their interaction in 2017 or 2018 ($p > 0.05$). The mean yield in 2017 was 3.88 kg 10 m row^{-1} and, in 2018, was 0.59 kg 10 m row^{-1}. The yield was relatively low in 2018 because broccoli was harvested before the heads were fully developed (due to an early forecasted killing frost).

4. Discussion

Seed meal effects on weed density in this study were driven by increasing application rates, regardless of seed meal type, and these effects were more commonly observed on velvetleaf (broadleaf) than green foxtail or shattercane (grass). The highest seed meal rate (5 g planting hole^{-1} N) reduced velvetleaf density by 66% to 97% compared to the weedy control, depending on the crop and year. This level of weed suppression is consistent with previous studies, though the rates required to observe the effect were much greater in this study. The 5 g planting hole^{-1} N rate was approximately equivalent to 4840 g m^{-2} CGM or 6880 g m^{-2} SBM concentrated within the planting hole, whereas other studies have observed similar benefits of CGM as low as 224 to 400 g m^{-2} [8,16].

Bingaman and Christians [7] found that velvetleaf was particularly difficult to control, with CGM at the highest rate (973 g m^{-2}) reducing velvetleaf survival by only 35%. At a similar rate tested in this study in 2017 (968 g m^{-2}; 1 g planting hole^{-1} N), there was no observed reduction in velvetleaf density. In most cases, at least 2 g planting hole^{-1} N (1936 g m^{-2} CGM or 2752 g m^{-2} SBM) was required to achieve significant reductions in weed density in this study. At these relatively high rates, it is possible that the weed suppressive mechanism included a mulching effect whereby the thick layer of seed meal (Figure 1) occluded light prior to decomposition. The level of weed suppression observed at the highest rates are consistent with those reported by El-Metwally et al. [11] (80–84%) who used SBM to create a mulch around crop plants.

Green foxtail density was rarely affected by seed meal rate or type, and shattercane density was reduced by 85% in the 5 g planting hole^{-1} N treatment only in 2018 broccoli. Bingaman and Christians [7] observed a better control of green foxtail with CGM as mortality reached 100% at 973 g m^{-2}. However, shattercane proved difficult to control in both studies. The highest rate of 973 g m^{-2} provided only a 51% reduction in shattercane survival in Bingaman and Christians [7], and shattercane was only affected by seed meals in this study in 2018 broccoli. A rate of 3.5 g planting hole^{-1} N (approximately 3388 g m^{-2} CGM or 4816 g m^{-2} SBM) was required to achieve 46% reduction in shattercane density—far greater than the rates required in Bingaman and Christians [7].

Nitrogen availability between 2 and 16 days after seed meal application was generally proportional to the seed meal rate (Figures 4 and 8). Based on the typical guaranteed analyses, we would expect 10% N from CGM and 7% N from SBM by weight. Once fully mineralized, the seed meal rates applied in this study could provide between 0.5 and 5 g/plant N or approximately 6 to 60 kg ha^{-1} N (assuming 1.83 m between-row spacing and 46 cm in-row plant spacing). However, depending on soil conditions, several months is often required before even 50% of organic fertilizer N is mineralized [17].

Of particular interest in this study were the relatively high rates of ammonium observed following the application of the highest seed meal rate (5 g planting hole^{-1} N). While increased nitrate near the crop is potentially beneficial, excessive ammonium has been shown to reduce weed seed germination and is also potentially harmful to the crop [18–20]. The elevated ammonium and potential for toxicity to germinating weed seedlings seems to be the most likely explanation for the weed suppression consistently observed at the 5 g planting hole^{-1} N rate in this study. Where ammonium levels decreased at the 2 g planting hole^{-1} N rate (Figures 4 and 8), weed suppression was often reduced proportionately (Figures 3, 6 and 7) or not detectable (Figure 2).

The potentially toxic effects of elevated ammonium were perhaps most evident in the yield data because, despite improved weed control, the seed meals either had no effect (broccoli in both years and tomato in 2017) or reduced yield (tomato in 2018; Figure 5). While other studies have reported yield benefits from seed meal applications [9–11], none required the high rates to achieve weed control benefits that were observed in this study (>1936 g m^{-2} CGM or >2752 g m^{-2} SBM). McDade and Christians [3] demonstrated the potentially phytotoxic effects of CGM on direct seeded crops at rates of only 100 to 400 g m^{-2}. Transplanting vegetables was suggested as an alternative strategy for avoiding these phytotoxic conditions, but the results of this study suggest that broccoli and tomato transplants may be susceptible to ammonium toxicity even if the plants are not killed by the seed meals.

5. Conclusions

The concept of integrating the use of seed meals as organic fertilizers and biobased herbicides is attractive because it has the potential to mitigate the greatest challenges facing organic farmers, including weed management, soil fertility, and labor availability. However, the results of this study in tomato and broccoli suggest that the seed meal application rates required to achieve consistent weed suppression are potentially toxic to crops. Despite improved weed control and greater nitrogen availability at the highest application rates, the seed meals either had no effect on or reduced crop yields. Ammonium toxicity was the most likely mechanism of toxicity to weeds and crops, and this limits the potential for integrating weed and nitrogen management with seed meals.

Author Contributions: Conceptualization, S.E.W.; methodology, S.E.W.; formal analysis, S.E.W.; investigation, A.B. and S.E.W.; resources, S.E.W.; writing—original draft preparation, A.B. and S.E.W.; writing—review and editing, S.E.W.; visualization, A.B. and S.E.W.; supervision, S.E.W.; project administration, S.E.W.; funding acquisition, S.E.W. All authors have read and agreed to the published version of the manuscript.

Funding: This research was funded by the Nebraska Agricultural Experiment Station with funding from the Hatch Act (accession 1014303) through the USDA National Institute of Food and Agriculture (NIFA), the USDA Agricultural Marketing Service (AMS) Specialty Crop Block Grant Program, and the Nebraska Department of Agriculture.

Data Availability Statement: Data are contained within the article. Additional data can be obtained by contacting the corresponding author of the article.

Conflicts of Interest: The authors declare no conflicts of interest. The funders had no role in the design of the study; in the collection, analyses, or interpretation of the data; in the writing of the manuscript; or in the decision to publish the results.

References

1. Synder, L.; Schonbeck, M.; Velez, T. 2022 National Organic Research Agenda: Outcomes and Recommendations from the 2020 National Organic & Transitioning Farmer Surveys and Focus Groups. Available online: https://ofrf.org/research/nora/ (accessed on 8 December 2023).
2. Wortman, S.E. Integrating Weed and Vegetable Crop Management with Multifunctional Air-Propelled Abrasive Grits. *Weed Technol.* **2014**, *28*, 243–252. [CrossRef]
3. McDade, M.C.; Christians, N.E. Corn Gluten Meal—A Natural Preemergence Herbicide: Effect on Vegetable Seedling Survival and Weed Cover. *Am. J. Altern. Agric.* **2000**, *15*, 189–191. [CrossRef]
4. Miles, C.; DeVetter, L.; Ghimire, S.; Hayes, D.G. Suitability of Biodegradable Plastic Mulches for Organic and Sustainable Agricultural Production Systems. *HortScience* **2017**, *52*, 10–15. [CrossRef]
5. Wortman, S.E. Air-Propelled Abrasive Grits Reduce Weed Abundance and Increase Yields in Organic Vegetable Production. *Crop Prot.* **2015**, *77*, 157–162. [CrossRef]
6. Luong, T.K.N.; Forcella, F.; Clay, S.A.; Douglass, M.S.; Wortman, S.E. Abrasive Weeding as a Vehicle for Precision Fertilizer Management in Organic Vegetable Production. *HortTechnology* **2020**, *31*, 136–143. [CrossRef]
7. Bingaman, B.R.; Christians, N.E. Greenhouse Screening of Corn Gluten Meal as a Natural Control Product for Broadleaf and Grass Weeds. *HortScience* **1995**, *30*, 1256–1259. [CrossRef]
8. Webber III, C.L.; Shrefler, J.W.; Taylor, M.J. Corn Gluten Meal as an Alternative Weed Control Option for Spring-Transplanted Onions. *Int. J. Veg. Sci.* **2008**, *13*, 17–33. [CrossRef]
9. Shrestha, A.; Rodriguez, A.; Pasakdee, S.; Bañuelos, G. Comparative Efficacy of White Mustard (*Sinapis alba* L.) and Soybean (*Glycine max* L. Merr.) Seed Meals as Bioherbicides in Organic Broccoli (*Brassica oleracea* var. Botrytis) and Spinach (*Spinacea oleracea*) Production. *Commun. Soil Sci. Plan.* **2015**, *46*, 33–46. [CrossRef]
10. An, X.-H.; Lee, S.B.; Im, I.B.; Kim, S.; Ahn, S.H.; Kim, J.D. Composition of Phenolic Compounds in Soybean Meal Extracts and Their Allelopathic Effects on Paddy Weeds. *Korean J. Weed Sci.* **2008**, *28*, 236–241.
11. El-Metwally, I.M.; Saudy, H.S.; Elewa, T.A. Natural Plant By-Products and Mulching Materials to Suppress Weeds and Improve Sugar Beet (*Beta vulgaris* L.) Yield and Quality. *J. Soil Sci. Plant Nutr.* **2022**, *22*, 5217–5230. [CrossRef]
12. Ward, R.; Ward Laboratories, Inc. Soil Test Methods. Available online: https://www.wardlab.com/soil-test-methods/ (accessed on 2 January 2024).
13. Webber, C.L., III; Shrefler, J.W.; Taylor, M.J. Influence of Corn Gluten Meal on Squash Plant Survival and Yields. *HortTechnology* **2010**, *20*, 696–699. [CrossRef]
14. Warwick, S.I.; Black, L.D. The Biology of Canadian Weeds: 90. *Abutilon theophrasti*. *Can. J. Plant Sci.* **1988**, *68*, 1069–1085. [CrossRef]
15. Harrison, D.J.; Maynard, D.G. Nitrogen Mineralization Assessment Using PRS™ Probes (Ion-Exchange Membranes) and Soil Extractions in Fertilized and Unfertilized Pine and Spruce Soils. *Can. J. Soil. Sci.* **2014**, *94*, 21–34. [CrossRef]
16. Yu, J.; Morishita, D.W. Response of Seven Weed Species to Corn Gluten Meal and White Mustard (*Sinapis alba*) Seed Meal Rates. *Weed Technol.* **2014**, *28*, 259–265. [CrossRef]

17. Stadler, C.; von Tucher, S.; Schmidhalter, U.; Gutser, R.; Heuwinkel, H. Nitrogen Release from Plant-Derived and Industrially Processed Organic Fertilizers Used in Organic Horticulture. *Z. Pflanzenernähr. Bodenk.* **2006**, *169*, 549–556. [CrossRef]
18. Britto, D.T.; Kronzucker, H.J. NH_4^+ Toxicity in Higher Plants: A Critical Review. *J. Plant Physiol.* **2002**, *159*, 567–584. [CrossRef]
19. Ells, J.E.; MeSay, A.E.; Workman, S.M. Toxic Effects of Manure, Alfalfa, and Ammonia on Emergence and Growth of Cucumber Seedlings. *HortScience* **1991**, *26*, 380–383. [CrossRef]
20. Yamagata, M.; Ae, N. Nitrogen Uptake Response of Crops to Organic Nitrogen. *Soil Sci. Plant Nutr.* **1996**, *42*, 389–394. [CrossRef]

horticulturae

Article

Castor Meal and Ground Hydrothermalized Phonolite Optimize Sweet Potato Nutrition, Yield, and Quality

Renan J. Parecido [1,2], Rogério P. Soratto [1,2,*], Adalton M. Fernandes [1,2,*], Mayara C. Blanes [1], Luis G. Fidelis [1], Harun I. Gitari [3] and Sérgio G. Dutra [4]

1 College of Agricultural Sciences, São Paulo State University (UNESP), Av. Universitária, 3780, Botucatu 18610-034, SP, Brazil; renanjparecido@hotmail.com (R.J.P.); mayara.c.blanes@unesp.br (M.C.B.); guilherme.fidelis@unesp.br (L.G.F.)

2 Center of Tropical Roots and Starches, São Paulo State University (UNESP), Botucatu 18610-034, SP, Brazil

3 Department of Agricultural Science and Technology, School of Agriculture and Environmental Sciences, Kenyatta University, Nairobi 43844-00100, Kenya; harun.gitari@ku.ac.ke

4 Mato Grosso Cotton Institute, Cuiabá 78050-970, MT, Brazil; sgdutra.sc@gmail.com

* Correspondence: rogerio.soratto@unesp.br (R.P.S.); adalton.fernandes@unesp.br (A.M.F.)

Abstract: To assess the effect of pure castor meal and a mixture of castor meal with ground hydrothermalized phonolite rock (CM+HP mixture) in providing nutrients, particularly N and K, and optimizing yield and quality of sweet potato, a field experiment was conducted using a randomized block design. Treatments were the absence and presence of synthetic N and K fertilizers (ammonium nitrate and KCl) combined with rates of organic fertilizers (1.2 and 2.4 Mg ha^{-1} of castor meal, 2.25 and 4.5 Mg ha^{-1} of CM+HP mixture, plus a treatment without organic fertilizers). The CM+HP mixture maintained adequate N and K status in plant leaves. Organic fertilizers increased the number of storage roots per plant and the sweetness of the storage roots, while synthetic fertilizers increased the storage root mean weight. Castor meal combined with synthetic fertilizers improved soil health (increased organic matter and enzyme activity in the soil). The combined application of synthetic fertilizers with 2.4 Mg ha^{-1} of castor meal or 4.5 Mg ha^{-1} of CM+HP mixture had the greatest benefit on storage root yield, with an average increase of 128% (10.9 Mg ha^{-1}) on marketable storage root yield, and the nutrient removal compared with the sole application of organic fertilizers.

Keywords: *Ipomoea batatas; Ricinus communis*; hydrothermalized phonolite; organic fertilization; mineral nutrition; storage root yield

Citation: Parecido, R.J.; Soratto, R.P.; Fernandes, A.M.; Blanes, M.C.; Fidelis, L.G.; Gitari, H.I.; Dutra, S.G.. Castor Meal and Ground Hydrothermalized Phonolite Optimize Sweet Potato Nutrition, Yield, and Quality. *Horticulturae* **2024**, *10*, 775. https://doi.org/10.3390/horticulturae10080775

Academic Editors: Francesco De Mastro, Gennaro Brunetti, Karam Farrag and Huadong Zang

Received: 26 June 2024
Revised: 18 July 2024
Accepted: 22 July 2024
Published: 23 July 2024

1. Introduction

Sweet potatoes (*Ipomoea batatas* L. [Lam.]) are a crucial food source in many nations [1,2], helping to overcome nutrient deficiencies and reduce child malnutrition in some of the world's most deprived regions [1,3,4]. Beyond human consumption, sweet potatoes are also used as animal feed and industrial feed for starch extraction and alcohol production [5,6]. This versatility highlights sweet potatoes as vital raw materials in a global context of constant population growth [7,8].

In South America, Brazil is the largest producer of sweet potatoes, with a cultivated area of 55,200 ha and an annual production of approximately 850,000 Mg [9]. However, sweet potatoes are also cultivated in many other countries. Globally, the leading producers are China and Nigeria, with cultivated areas of 2.2 and 1.5 million ha, respectively.

Sweet potatoes are considered a high-yield crop with good yield even in nutrient-poor tropical soils. However, they still respond to fertilization [1,10–14]. K and N are essential elements for sweet potatoes, being absorbed and utilized in large quantities [7,10,11,15–17]. Deficiencies in these nutrients can substantially reduce productivity by decreasing the size of marketable storage roots and limiting starch accumulation in reserve tissues, leading to changes in important market characteristics, such as the texture and firmness of the storage roots [13,18–20]. A balanced supply of N and K often enhances the response to both nutrients [21], with some studies indicating an elevated response to N in the presence of K [12].

Organic residues release nutrients to plants more gradually and consistently than synthetic fertilizers, providing chemical, physical, and biological benefits to the soil, such as improved structure, aeration, drainage, water retention, and microbial activity [22–26]. In this context, castor meal, a by-product obtained from the extraction of castor seeds (*Ricinus communis* L.), is considered a high-quality organic fertilizer [24,27–30]. Castor meal contains 4.2–7.5% N, 0.7–1.0% K, along with other nutrients, and has a C/N ratio of approximately 12:1 [31,32]. When applied to the soil, castor meal is rapidly mineralized, releasing N for plant uptake [27]. However, due to its relatively low K content, it does not meet the high K demand of crops, such as sweet potatoes, which require large amounts of K [11,17].

To balance the K/N ratio in castor meal, an alternative is to mix it with a gradual K-release natural source, such as finely ground hydrothermalized phonolite rock. Finely ground hydrothermalized phonolite rocks have been demonstrated as viable alternatives for supplying K to grain crops [33–35] and for the Arabica coffee crops (*Coffea arabica* L.) [36]. In addition to K, ground phonolite rocks contain other essential or beneficial elements, such as Si, Mg, and Ca [37–40]. This amendment, free of Cl, can be a valuable option for organic agriculture, where the use of KCl is prohibited [41], improving crop performance and enriching soil quality.

It has been demonstrated that sweet potatoes respond well to organic fertilization with green manure [13,17,42] or cattle manure [43]. Therefore, the balance of nutrients derived from using castor meal and hydrothermalized phonolite rock produced in Brazil could be a viable alternative source of essential nutrients for sweet potato cultivation. Additionally, the constant increase in Brazilian agricultural production and its reliance on imported fertilizers justify the use of ground rocks as sources of K in agriculture. Therefore, these alternative sources (castor meal and ground potassium rocks) could also enhance the economic and environmental sustainability of Brazilian agriculture and serve as a viable option for organic production systems where synthetic and highly soluble sources are not allowed.

In the present study, we aimed to evaluate the effect of alternative non-synthetic fertilizers, specifically pure castor meal and a mixture of castor meal and finely ground hydrothermalized phonolite rock (CM+HP mixture), in providing nutrients, particularly K and N, and optimizing the storage root yield and quality of sweet potato crops.

2. Materials and Methods

2.1. Site Characteristics and Climate

A field experiment was conducted from January to June 2023 in São Manuel, São Paulo State, southeastern Brazil (22°46′22″ S; 48°34′11″ W; 762 m asl). According to the Köppen classification system, this tropical region experiences a Cwa climate with hot and rainy summers and dry winters. During the experimental period, temperatures and rainfall were recorded daily, as shown in Figure 1.

The soil is classified as a sand-textured Ferralsol [44]. Prior to the experiment, the area was previously cultivated with okra [*Abelmoschus esculentus* (L.) Moench] for two years, followed by a 12-month fallow period. The chemical properties of the soil were determined following the procedures described by van Raij et al. [45], while the soil texture was determined using the pipette method [46]. The topsoil layer (0–0.2 m) had a pH($CaCl_2$) of 4.8, organic matter of 16 g kg^{-1}, SO_4-S of 4 mg kg^{-1} and $P_{resin\text{-}extractable}$ of 8 mg kg^{-1}. In addition, it had 1.1, 3.4, 9.5, 16.4, and 30.2 $mmol_c$ kg^{-1} of exchangeable K, Mg, Ca, H+Al, and cation exchange capacity, respectively. The base saturation was 46%, and the silt, sand, and clay contents were 35, 824, and 141 g kg^{-1}, respectively.

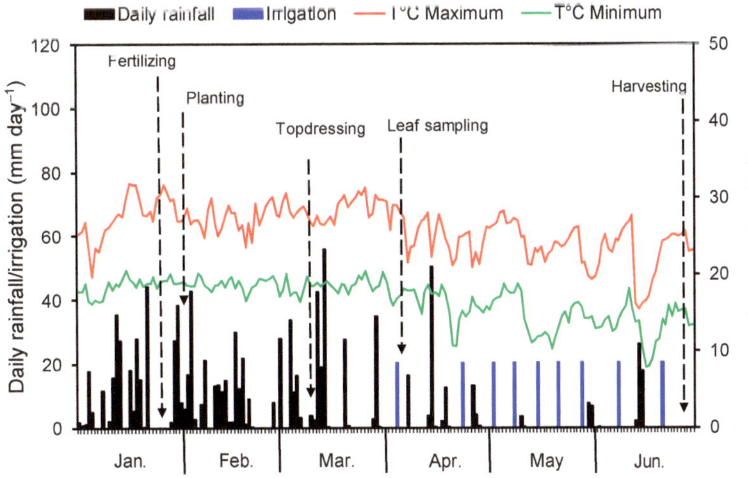

Figure 1. Daily rainfall, irrigation, and maximum and minimum temperatures during the sweet potato growing from January to June 2023 in São Manuel-SP.

2.2. Experimental Design, Treatments, and Crop Management

The experimental design was a randomized complete block in a 2 × 5 factorial scheme with four replicates. The test crop was the sweet potato cultivar Canadense, the most widely grown cultivar in the São Paulo State. Treatments with synthetic fertilizers included the absence or presence of application of recommended N and K fertilizations, such as ammonium nitrate (32% N) and KCl (60% K_2O) fertilizers, respectively. Organic fertilizer treatments consisted of a control (without organic fertilizer) and applications of 1.2 and 2.4 Mg ha^{-1} of pure castor meal, as well as 2.25 and 4.5 Mg ha^{-1} of CM+HP mixture. The castor meal contained 5.0% N, 1.2% K_2O, 0.7% Ca, 1.5% P_2O_5, 0.4% S, 0.5% Mg, and traces of all micronutrients. The hydrothermalized phonolite contained 12.0% total K_2O, 3.0% soluble K_2O in 5% tartaric acid + 0.5% sodium fluoride solution, 25% Si, 0.06% Ca, 0.10% Mg, and traces of Mn, Zn, and Co. Pure castor meal rates were calculated to provide the recommended and twice the recommended N rate (i.e., 60 and 120 kg N ha^{-1}) [47]. The CM+HP mixture rates were calculated to provide 60 or 120 kg N ha^{-1} and 140 or 280 kg K_2O ha^{-1}. Thus, the CM+HP mixture was calibrated to form an N–K_2O formulation of 2.7–6.2 (i.e., balancing the K concentration). Each plot measured 5.2 by 4 m (20.8 m^2), and the sweet potato spacing was 1.30 × 0.35 m. The dimensions of the usable area were 2.6 by 2 m (5.2 m^2) because the outermost rows and 1.0 m at each border were excluded from the evaluation.

Eighteen days before planting, 700 kg ha^{-1} of dolomitic limestone (40% CaO, 28% MgO, and 70% effective $CaCO_3$ equivalence) was applied to raise the base saturation to 60%. The soil was tilled uniformly to a depth of 0–0.20 m with disk harrows. On 26 January 2023, ridges of approximately 0.30 m high were mechanically raised using a tractor-mounted ridger for sweet potatoes. The pure castor meal and CM+HP mixture were applied immediately before raising the ridges. On the same day, all plots received phosphate fertilizer at a rate of 160 kg P_2O_5 ha^{-1} (triple superphosphate, 41% P_2O_5 and 10% Ca) and micronutrients (1 kg B ha^{-1}, 5 kg Zn ha^{-1}, 0.4 kg Cu ha^{-1}, 1.1 kg Mn ha^{-1}, 1.7 kg Fe ha^{-1}, and 0.06 kg Mo ha^{-1}, as fritted trace elements), according to the recommendations of Feltran et al. [47]. In treatments with synthetic N and K fertilizers, 20 kg N ha^{-1} and 70 kg K_2O ha^{-1} were also applied. Sweet potato cuttings, 40 cm long, were planted on 30 January 2023. Topdressing fertilization with synthetic fertilization at 40 kg N ha^{-1} and 70 kg K_2O ha^{-1} was carried out on 10 March 2023 [39 days after planting (DAP)], according to the treatments.

2.3. Sampling and Analyses

2.3.1. Leaf Nutrient Concentrations

At 65 DAP, 15 of the most recently mature leaves were collected from the sampling area of each plot [47]. The leaves were rinsed with deionized water and dried in an oven under forced-air circulation at 65 °C for 72 h. The samples were then ground and sieved using a 40-mesh stainless steel screen. The nutrient (N, P, K, Ca, Mg, S, B, Cu, Fe, Mn, and Zn) concentrations were determined for each sample [48].

2.3.2. Soil Health Indices

Before sweet potato harvesting (27 June 2023), soil samples were collected from the ridges. Three simple samples were taken from the usable area of each plot at a depth of 0–0.20 m using a tubular soil probe to form a composite sample. These samples were analyzed for soil organic matter content [45], and the activities of soil arylsulfatase and β-glucosidase enzymes [49] were determined.

2.3.3. Storage Root Yield and Sorting

Harvesting was performed on 27 June 2023 (148 DAP). At harvest, the storage roots from two 2.0 m-long ridges were collected from the usable area of each plot. The harvested storage roots were washed and counted to obtain yield data. Smooth storage roots with an elongated, uniform shape and weighing between 80 and 800 g were considered marketable [50].

2.3.4. Storage Root Dry Matter Content

A representative subsample of marketable storage roots (proportional to all sizes) was randomly collected from each plot, weighed (fresh weight), sliced, dried in an oven under forced-air circulation at 65 °C for 96 h, and reweighed to obtain the dry weight for computing dry matter (DM) content.

2.3.5. Nutrient Concentration and Removal in the Storage Roots

The samples of dried storage roots were subsequently ground and sieved using a 40-mesh stainless steel screen. Thereafter, the nutrient (N, P, K, Ca, Mg, S, B, Cu, Fe, Mn, and Zn) concentrations were determined [48]. Nutrient removal was calculated by multiplying the amount of DM accumulated in the storage roots per treatment with the concentration of each nutrient. The values were converted accordingly into kg ha^{-1} or g ha^{-1}.

2.3.6. Texture Properties and Soluble Solids of Storage Roots

Another sample of three marketable storage roots was collected from each plot to evaluate quality parameters. Firmness was determined using the TA.XTPlus texturometer (Stable Micro Systems, Surrey, UK) by penetrating 18 mm deep into the pulp of the storage root at a speed of 2.0 mm s^{-1} using a texture probe equipped with a TA 9/1000 tip. Soluble solids extracted from crushed slices of storage root pulp were measured by placing drops of the solute in the prism of a portable refractometer RTD-95 (Instrutherm Instrumentos de Medição Ltd., São Paulo, SP, Brazil), and the results were expressed in °Brix.

2.3.7. Starch, Reducing Sugar, Total Sugar, and Crude Fiber Contents in the Storage Roots

Starch, reducing sugar, and total sugar contents were determined in storage root samples (the same ones used to determine DM content and nutrient concentration) according to Somogyi's procedures [51]. The absorbance of the samples was recorded at 535 nm using a spectrophotometer. Starch, reducing sugar, and total sugar content were then calculated based on a standard curve and expressed as a percentage of fresh weight. In addition, the crude fiber content was assessed following the methodology described by the Association of Official Analytical Chemists [52] and expressed as a percentage of fresh weight.

2.4. Statistical Analyses

Data were processed using a two-way analysis of variance (ANOVA) with the SISVAR statistical software package. Synthetic and organic fertilizers were considered as fixed main factors. The blocks and all of the block interactions were considered random effects. If ANOVA identified a significant effect ($F \leq 0.05$) of main factors or their interaction, the treatment means were separated at the 0.05 probability level using Fisher's protected least significant difference test. When the interaction was significant, a two-way mean separation was performed.

3. Results

3.1. Leaf Nutrient Concentrations

The combined application of synthetic NK and organic fertilizers significantly influenced only the concentrations of Cu and Zn in sweet potato leaves; however, both types of fertilizers affected the concentrations of the other nutrients as well (Table 1). The application of CM+HP mixture at a rate of 4.5 Mg ha^{-1} resulted in increased concentrations of N, Mg, and B in the sweet potato leaves compared to those in other treatments (Table 1). By contrast, the application of synthetic NK fertilizer reduced the leaf concentrations of P, Ca, B, Mg, and Mn. Specifically, in treatments where synthetic NK fertilizers were applied along with a CM+HP mixture at the rate of 4.5 Mg ha^{-1}, there was an increase in leaf Cu concentration compared to that in the other treatments (Figure 2a). Without synthetic NK fertilizers, the application of 1.2 Mg ha^{-1} of castor meal resulted in increased Cu concentration in the sweet potato leaves. The highest leaf Zn concentration occurred in the treatment with 4.5 Mg ha^{-1} of CM+HP mixture combined with synthetic NK fertilizers (Figure 2b).

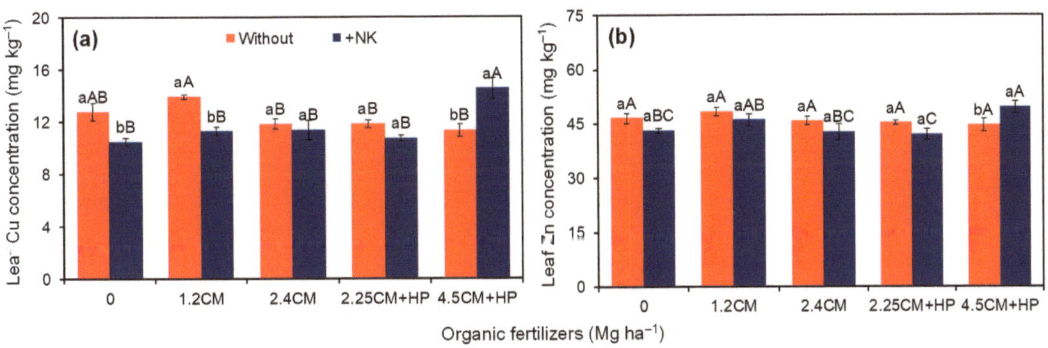

Figure 2. Concentrations of Cu (**a**) and Zn (**b**) in the leaves of sweet potato crop as affected by synthetic NK fertilizers and rates of organic fertilizers composed of castor meal (CM) or castor meal plus hydrothermalized phonolite mixture (CM+HP). The red and blue bars indicate treatments without and with recommended synthetic NK fertilization, respectively. Bars indicate the standard error. Different lowercase letters indicate significant differences by synthetic fertilizer level within each organic fertilizer level, while different uppercase letters indicate significant differences by organic fertilizer level within each synthetic fertilizer level, at $p \leq 0.05$, according to the LSD test.

3.2. Storage Root Yield and Chemical Composition

The treatments that received organic fertilizers (pure castor meal or the CM+HP mixture) had a greater total number of storage roots per plant than those in the treatment without organic fertilizers (Table 2). In the presence of synthetic NK fertilizers, the application of castor meal resulted in the highest number of marketable roots per plant; however, the application of 4.5 Mg ha^{-1} of CM+HP mixture provided the highest value, differing from the application of the same rate of CM+HP mixture alone (Figure 3a). The application of synthetic NK fertilizers increased the mean storage root weight compared to that in the treatments without synthetic NK fertilizers, regardless of organic fertilizer application (Table 2).

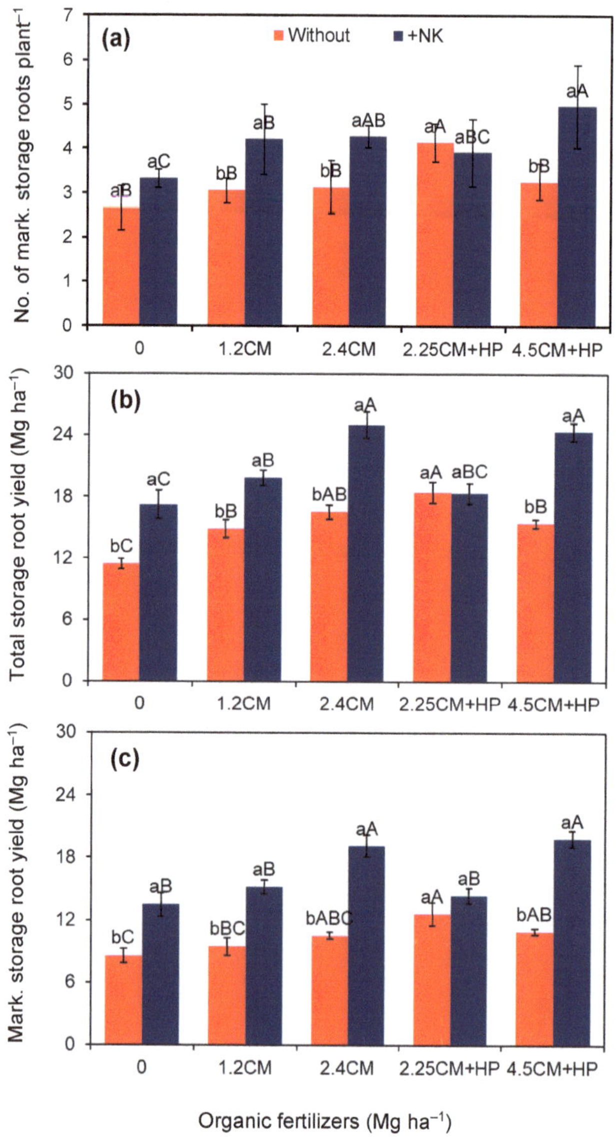

Figure 3. Number of marketable storage roots per plant (**a**), total storage root yield (**b**), and marketable storage root yield (**c**) of sweet potato crop as affected by synthetic NK fertilizers and rates of organic fertilizers composed of castor meal (CM) or castor meal plus hydrothermalized phonolite mixture (CM+HP). The red and blue bars indicate treatments without and with recommended synthetic NK fertilization, respectively. Bars indicate the standard error. Different lowercase letters indicate significant differences by synthetic fertilizer level within each organic fertilizer level, while different uppercase letters indicate significant differences by organic fertilizer level within each synthetic fertilizer level, at $p \leq 0.05$, according to the LSD test.

Table 1. Leaf nutrient (N, P, K, Ca, Mg, S, B, Cu, Fe, Mn, and Zn) concentrations of sweet potato crop as affected by synthetic NK fertilizers (SF) and rates of organic fertilizers (OF) composed of castor meal (CM) or castor meal plus hydrothermalized phonolite mixture (CM+HP). Mean ± standard error.

Variable	Synthetic Fertilizers		Organic Fertilizers (Mg ha^{-1})					ANOVA ($p > F$)			CV (%)
	Without	+NK	0	1.2CM	2.4CM	2.25CM+HP	4.5CM+HP	SF	OF	SF × OF	
N (g kg^{-1})	37.1 ± 0.6 a	36.3 ± 0.5 a	35.3 ± 0.4 b	36.4 ± 0.6 b	35.8 ± 0.9 b	36.9 ± 0.7 ab	38.9 ± 0.9 a	0.22	0.02	0.36	5.6
P (g kg^{-1})	4.2 ± 0.10 a	3.6 ± 0.08 b	3.7 ± 0.15 a	4.1 ± 0.23 a	3.8 ± 0.18 a	3.8 ± 0.17 a	4.2 ± 0.17 a	<0.01	0.09	0.93	10.3
K (g kg^{-1})	38.5 ± 0.8 a	38.4 ± 0.5 a	37.8 ± 0.7 a	38.3 ± 0.8 a	39.4 ± 1.2 a	37.8 ± 1.0 a	39.0 ± 1.7 a	0.87	0.82	0.77	8.3
Ca (g kg^{-1})	7.9 ± 0.14 a	7.4 ± 0.11 b	7.4 ± 0.21 a	7.7 ± 0.17 a	7.6 ± 0.16 a	7.3 ± 0.18 a	8.0 ± 0.29 a	<0.01	0.13	0.32	6.3
Mg (g kg^{-1})	4.1 ± 0.07 a	3.7 ± 0.08 b	3.8 ± 0.13 b	3.9 ± 0.10 b	3.8 ± 0.16 b	3.8 ± 0.10 b	4.3 ± 0.13 a	<0.01	<0.01	0.12	6.7
S (g kg^{-1})	4.3 ± 0.05 a	4.2 ± 0.07 a	4.2 ± 0.09 a	4.3 ± 0.08 a	4.3 ± 0.14 a	4.1 ± 0.06 a	4.1 ± 0.08 a	0.29	0.41	0.12	6.1
B (mg kg^{-1})	51.2 ± 1.2 a	47.8 ± 1.2 b	46.7 ± 1.6 b	47.7 ± 1.5 b	48.5 ± 0.9 b	47.0 ± 1.8 b	57.9 ± 1.1 a	<0.01	<0.01	0.39	6.7
Cu (mg kg^{-1})	12.4 ± 0.3 a	11.7 ± 0.4 b	11.7 ± 0.6 bc	12.6 ± 0.5 ab	11.6 ± 0.4 bc	11.3 ± 0.3 c	13.0 ± 0.7 a	0.04	0.01	<0.01	8.4
Fe (mg kg^{-1})	147.2 ± 7.7 a	154.7 ± 7.2 a	165.9 ± 10.1 a	151.4 ± 8.8 a	128.1 ± 12.2 a	165.5 ± 14.1 a	143.9 ± 9.6 a	0.42	0.08	0.19	19.2
Mn (mg kg^{-1})	119.0 ± 4.4 a	107.9 ± 4.1 b	134.0 ± 8.5 a	111.8 ± 6.8 b	101.6 ± 2.3 b	105.3 ± 1.7 b	114.5 ± 6.8 b	0.04	<0.01	0.80	14.5
Zn (mg kg^{-1})	46.1 ± 0.6 a	44.7 ± 0.9 a	44.9 ± 1.0 ab	47.2 ± 1.0 a	44.3 ± 1.3 b	43.6 ± 0.9 b	47.0 ± 1.4 a	0.12	0.03	0.02	5.8

Values followed by the same letter in the row within each factor are not significantly different at $p < 0.05$ according to the LSD test.

Table 2. Total number of storage roots per plant, number of marketable storage roots per plant, storage root mean weight, total and marketable storage root yield, and chemical composition of storage roots of sweet potato crop as affected by synthetic NK fertilizers (SF) and rates of organic fertilizers (OF) composed of castor meal (CM) or castor meal plus hydrothermalized phonolite mixture (CM+HP). Mean ± standard error.

Variable	Synthetic Fertilizers		Organic Fertilizers (Mg ha^{-1})					ANOVA ($p > F$)			CV (%)
	Without	+NK	0	1.2CM	2.4CM	2.25CM+HP	4.5CM+HP	SF	OF	SF × OF	
Roots plant^{-1}	7.8 ± 0.30 a	8.1 ± 0.36 a	6.0 ± 0.26 b	8.0 ± 0.39 a	8.6 ± 0.30 a	8.5 ± 0.42 a	8.7 ± 0.56 a	0.37	<0.01	0.36	13.6
Mark. roots plant^{-1}	3.3 ± 0.15 b	4.1 ± 0.17 a	3.0 ± 0.19 b	3.6 ± 0.30 a	3.7 ± 0.24 a	4.0 ± 0.19 a	4.1 ± 0.37 a	<0.01	<0.01	0.01	13.9
Root mean weight (g)	98.9 ± 1.8 b	129.6 ± 3.8 a	118.4 ± 8.9 a	111.0 ± 8.1 a	118.0 ± 8.3 a	108.1 ± 3.0 a	115.7 ± 7.3 a	<0.01	0.38	0.07	10.7
Total yield (Mg ha^{-1})	15.3 ± 0.60 b	21.0± 0.85 a	14.3 ± 1.19 d	17.4 ± 1.10 c	20.8 ± 1.74 a	18.4 ± 0.69 bc	19.9 ± 1.80 ab	<0.01	<0.01	<0.01	9.5
Mark. yield (Mg ha^{-1})	10.4 ± 0.40 b	16.4 ± 0.72 a	11.0 ± 1.06 d	12.3 ± 0.20 cd	14.8± 1.72 ab	13.5 ± 0.67 bc	15.4 ± 1.76 a	<0.01	<0.01	<0.01	11.9
Root dry matter (%)	22.2 ± 0.4 a	22.2 ± 0.8 a	21.5 ± 0.6 a	21.8 ± 0.8 a	21.9 ± 0.8 a	21.6 ± 0.7 a	24.3 ± 1.4 a	0.98	0.26	0.84	12.7
Firmness (N)	30.1 ± 0.4 a	30.0 ± 0.4 a	29.0 ± 0.4 a	29.8 ± 0.9 a	30.6 ± 0.6 a	29.9 ± 0.8 a	30.9 ± 0.5 a	0.87	0.27	0.28	5.9
Total sugar (%)	3.6 ± 0.16 a	3.7 ± 0.26 a	3.2 ± 0.15 a	3.3 ± 0.22 a	3.3 ± 0.12 a	4.1 ± 0.43 a	4.3 ± 0.44 a	0.81	0.06	0.36	26.1
Reducing sugars (%)	2.3 ± 0.07 a	2.3 ± 0.17 a	2.2 ± 0.10 a	2.4 ± 0.14 a	2.5 ± 0.15 a	2.0 ± 0.13 a	2.6 ± 0.35 a	0.99	0.24	0.13	23.1
Starch (%)	19.4 ± 0.6 a	18.4 ± 0.9 a	17.5 ± 0.8 a	17.7 ± 1.1 a	19.5 ± 1.2 a	18.9 ± 1.0 a	20.7 ± 1.4 a	0.36	0.31	0.31	17.9
Soluble solids (°Brix)	5.7 ± 0.09 a	5.4 ± 0.12 a	5.2 ± 0.14 b	5.5 ± 0.18 ab	5.7 ± 0.16 a	5.9 ± 0.15 a	5.6 ± 0.17 ab	0.07	0.04	0.14	7.9
Crude fiber (%)	7.0 ± 0.24 a	7.4 ± 0.38 a	7.3 ± 0.47 a	7.4 ± 0.52 a	6.9 ± 0.54 a	6.6 ± 0.21 a	7.7 ± 0.69 a	0.38	0.65	0.78	21.2

Values followed by the same letter in the row within each factor are not significantly different at $p < 0.05$ according to the LSD test.

The application of synthetic NK fertilizers, as well as the isolated application of castor meal or the CM+HP mixture, increased the yield of total and marketable storage roots (Figure 3b,c). However, in the presence of synthetic NK fertilizers, treatments with 2.4 Mg ha^{-1} of castor meal or 4.5 Mg ha^{-1} of CM+HP mixture yielded increased total and marketable storage roots compared to those in the other treatments.

Minimal effects of the treatments were observed on most of the quality traits of sweet potato storage roots, such as firmness and DM, sugar, starch, and fiber contents (Table 2). Nevertheless, the application of 2.4 Mg ha^{-1} of castor meal or 2.25 Mg ha^{-1} of CM+HP mixture increased the soluble solids content in storage roots compared to that in treatments without organic fertilizer, regardless of synthetic NK fertilizer application.

3.3. *Concentration and Removal of Nutrients in the Storage Roots*

Nutrient concentrations and removal by sweet potato storage roots were influenced by the application of organic and synthetic fertilizers, as well as their interaction (Table 3). In the presence of synthetic NK fertilizers, applying the highest rate of castor meal (2.4 Mg ha^{-1}) increased the concentrations of N, P, Fe, K, and Mn in sweet potato storage roots compared to those in the other treatments (Figure 4a–e). When synthetic NK fertilizer was not applied, the application of 2.4 Mg ha^{-1} of castor meal and 2.25 Mg ha^{-1} of CM+HP mixture yielded the highest N concentrations in storage roots (Figure 4a). Conversely, without the application of synthetic NK fertilizer, applying 2.4 Mg ha^{-1} of castor meal resulted in reduced Fe concentrations. A similar case was noted when the CM+HP mixture was applied at an equivalent rate, resulting in reduced Mn concentrations in storage roots (Figure 4d,e). Organic fertilizers had no significant effect on storage root P and K concentrations in the absence of synthetic NK fertilizer (Figure 4b,c). The application of synthetic NK fertilizer increased root Ca concentrations compared to that in the control treatment (Table 3). However, the application of organic fertilizers (pure castor meal or the CM+HP mixture) reduced B concentrations in storage roots, irrespective of the application of synthetic NK fertilizers. Storage roots from treatments with 2.4 Mg ha^{-1} of castor meal or 2.25 Mg ha^{-1} of CM+HP mixture exhibited high Cu concentrations, irrespective of the application of synthetic NK fertilizers. Furthermore, the application of synthetic NK fertilizer increased the Cu and Zn concentrations in storage roots compared to those in the treatments without synthetic NK fertilizer.

In the absence of synthetic NK fertilizer, the application rates of 2.4 Mg ha^{-1} of castor meal and 2.25 Mg ha^{-1} of CM+HP mixture resulted in higher N removal by the roots than with treatments without fertilization (Figure 5a). However, the highest N removal by the sweet potato storage roots was achieved upon the application of 2.4 Mg ha^{-1} of castor meal combined with synthetic NK fertilizers.

Table 3. Nutrient (N, P, K, Ca, Mg, S, B, Cu, Fe, Mn, and Zn) concentration and removal in the sweet potato storage roots as affected by synthetic NK fertilizers (SF) and rates of organic fertilizers (OF) composed of castor meal (CM) or castor meal plus hydrothermalized phonolite mixture (CM+HP). Mean ± standard error.

Variable	Synthetic Fertilizers		Organic Fertilizers (Mg ha^{-1})					ANOVA ($p > F$)			CV (%)
	Without	+NK	0	1.2CM	2.4CM	2.25CM+HP	4.5CM+HP	SF	OF	SF × OF	
N conc. (g kg^{-1})	10.0 ± 0.2 a	10.3 ± 0.4 a	10.3 ± 0.2 b	10.3 ± 0.2 b	11.6 ± 0.8 a	9.6 ± 0.4 bc	9.0 ± 0.2 c	0.29	<0.01	<0.01	9.5
P conc. (g kg^{-1})	1.8 ± 0.07 a	1.7 ± 0.07 a	1.7 ± 0.10 a	1.8 ± 0.07 a	1.9 ± 0.14 a	1.6 ± 0.15 a	1.7 ± 0.07 a	0.13	0.24	<0.01	12.1
K conc. (g kg^{-1})	14.8 ± 0.3 b	15.9 ± 0.3 a	15.4 ± 0.3 a	15.3 ± 0.5 a	16.0 ± 0.9 a	15.6 ± 0.6 a	14.4 ± 0.3 a	<0.01	0.17	0.04	8.4
Ca conc. (g kg^{-1})	1.5 ± 0.08 b	1.8 ± 0.06 a	1.6 ± 0.12 a	1.6 ± 0.07 a	1.5 ± 0.16 a	1.6 ± 0.13 a	1.8 ± 0.12 a	<0.01	0.24	0.56	17.5
Mg conc. (g kg^{-1})	0.6 ± 0.015 a	0.7 ± 0.017 a	0.6 ± 0.018 a	0.6 ± 0.024 a	0.7 ± 0.034 a	0.7 ± 0.027 a	0.7 ± 0.027 a	0.11	0.76	0.15	9.5
S conc. (g kg^{-1})	0.7 ± 0.011 a	0.7 ± 0.010 a	0.7 ± 0.015 a	0.7 ± 0.011 a	0.7 ± 0.023 a	0.7 ± 0.021 a	0.7 ± 0.014 a	0.94	0.67	0.08	6.2
B conc. (mg kg^{-1})	14.4 ± 0.25 a	14.2 ± 0.27 a	15.[illegible] ± 0.17 a	14.2 ± 0.40 b	14.1 ± 0.33 b	14.3 ± 0.38 b	13.4 ± 0.42 b	0.46	<0.01	0.40	6.2
Cu conc. (mg kg^{-1})	4.6 ± 0.10 b	5.0 ± 0.13 a	4.4 ± 0.10 c	4.6 ± 0.15 bc	5.1 ± 0.31 a	5.1 ± 0.16 a	5.0 ± 0.10 ab	<0.01	0.02	0.50	9.7
Fe conc. (mg kg^{-1})	72.9 ± 2.6 a	69.5 ± 3.1 a	74.3 ± 4.0 a	68.2 ± 3.4 a	75.3 ± 6.5 a	71.3 ± 3.7 a	66.9 ± 4.9 a	0.25	0.30	<0.01	12.9
Mn conc. (mg kg^{-1})	23.0 ± 0.8 a	21.6 ± 0.9 a	21.9 ± 1.9 b	22.5 ± 1.1 ab	25.3 ± 1.2 a	19.7 ± 0.7 b	22.1 ± 0.9 b	0.13	0.02	0.02	13.4
Zn conc. (mg kg^{-1})	6.2 ± 0.16 b	7.2 ± 0.15 a	6.7 ± 0.33 a	7.1 ± 0.28 a	6.7 ± 0.33 a	6.7 ± 0.24 a	6.4 ± 0.31 a	<0.01	0.44	0.57	10.8
N rem. (kg ha^{-1})	34.2 ± 1.6 b	48.7 ± 3.8 a	31.9 ± 3.3 c	38.9 ± 2.8 bc	53.8 ± 7.1 a	38.2 ± 2.3 bc	44.5 ± 6.4 ab	<0.01	<0.01	<0.01	21.9
P rem. (kg ha^{-1})	6.1 ± 0.34 b	7.9 ± 0.51 a	5.1 ± 0.51 c	6.7 ± 0.30 b	8.4 ± 0.0.97 a	6.4 ± 0.54 bc	8.2 ± 1.01 a	<0.01	<0.01	<0.01	19.6
K rem. (kg ha^{-1})	50.3 ± 2.5 b	74.0 ± 4.3 a	47.[illegible] ± 4.7 d	57.7 ± 4.4 cd	73.9 ± 9.0 a	61.6 ± 3.2 bc	70.0 ± 8.4 ab	<0.01	<0.01	<0.01	16.7
Ca rem. (kg ha^{-1})	5.1 ± 0.42 b	8.4 ± 0.65 a	5.1 ± 0.66 c	5.9 ± 0.49 bc	7.3 ± 1.22 ab	6.3 ± 0.60 bc	9.2 ± 1.41 a	<0.01	<0.01	0.16	29.1
Mg rem. (kg ha^{-1})	2.2 ± 0.11 b	3.1 ± 0.20 a	2.0 ± 0.22 d	2.3 ± 0.14 cd	3.0 ± 0.36 ab	2.6 ± 0.16 bc	3.2 ± 0.40 a	<0.01	<0.01	0.01	19.2
S rem. (kg ha^{-1})	2.4 ± 0.12 b	3.3 ± 0.21 a	2.2 ± 0.23 d	2.7 ± 0.18 cd	3.3 ± 0.23 ab	2.8 ± 0.16 bc	3.5 ± 0.46 a	<0.01	<0.01	0.02	13.7
B rem. (g ha^{-1})	48.8 ± 2.0 b	65.4 ± 3.2 a	47.5 ± 4.3 b	53.1 ± 2.7 b	63.6 ± 5.2 a	56.4 ± 2.4 ab	64.9 ± 7.6 a	<0.01	<0.01	0.01	16.0
Cu rem. (g ha^{-1})	15.9 ± 0.9 b	23.7 ± 1.6 a	13.7 ± 1.5 d	17.4 ± 1.5 c	23.6 ± 2.9 ab	19.9 ± 0.7 bc	24.3 ± 3.1 a	<0.01	<0.01	<0.01	18.5
Fe rem. (g ha^{-1})	247.8 ± 13.4 b	324.7 ± 24.5 a	227.4 ± 22.8 c	253.3 ± 14.0 c	354.3 ± 55.4 a	282.2 ± 19.2 bc	314.1 ± 27.8 ab	<0.01	<0.01	<0.01	20.7
Mn rem. (g ha^{-1})	78.1 ± 3.9 b	103.5 ± 9.7 a	64.8 ± 4.2 b	84.2 ± 6.0 b	116.8 ± 13.4 a	77.5 ± 2.6 b	110.5 ± 18.5 a	<0.01	<0.01	0.01	26.9
Zn rem. (g ha^{-1})	21.4 ± 1.2 b	33.6 ± 2.3 a	20.7 ± 2.4 b	26.9 ± 2.1 ab	31.0 ± 3.9 a	26.4 ± 1.2 ab	32.7 ± 5.9 a	<0.01	0.02	0.04	25.5

Values followed by the same letter in the row within each factor are not significantly different at $p < 0.05$ according to the LSD test.

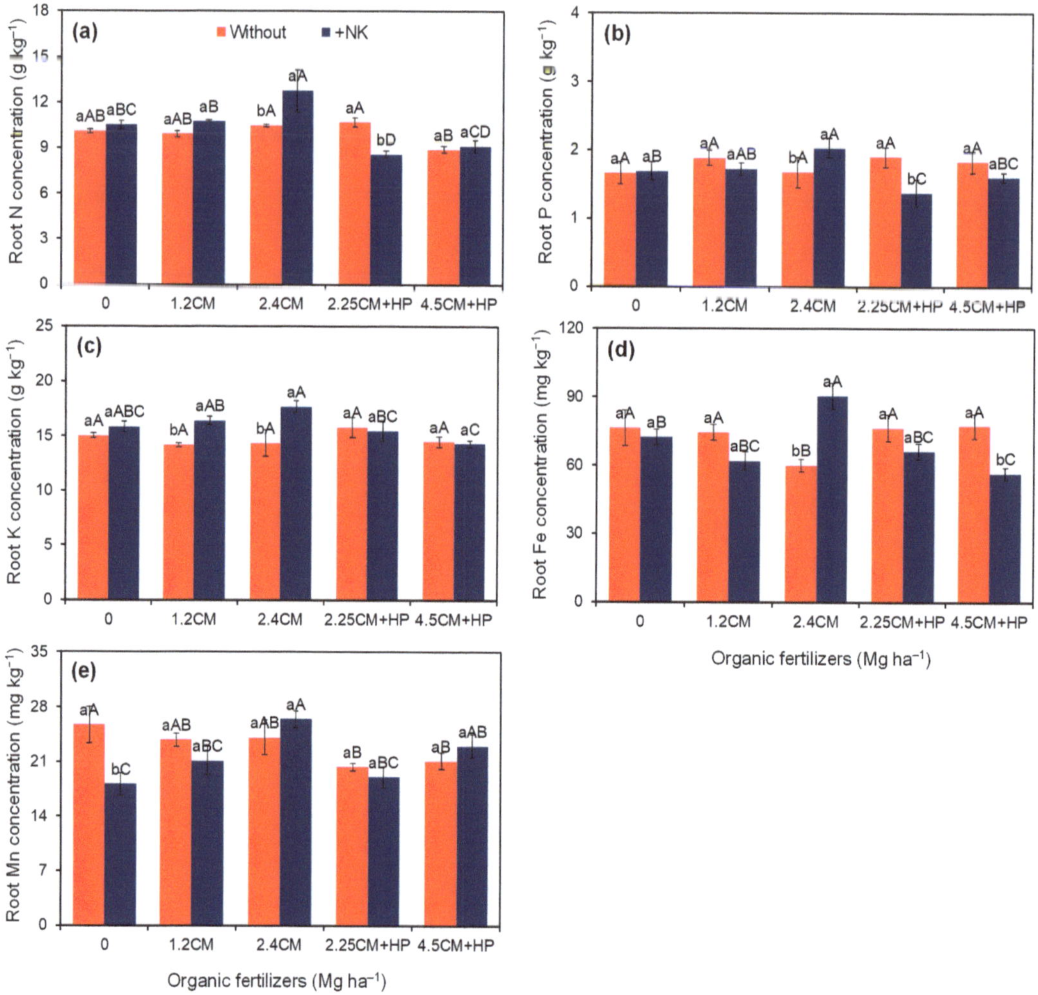

Figure 4. Concentration of N (**a**), P (**b**), K (**c**), Fe (**d**), and Mn (**e**) in the storage roots of sweet potato crop as affected by synthetic NK fertilizers and rates of organic fertilizers composed of castor meal (CM) or castor meal plus hydrothermalized phonolite mixture (CM+HP). The red and blue bars indicate treatments without and with recommended synthetic NK fertilization, respectively. Bars indicate the standard error. Different lowercase letters indicate significant differences by synthetic fertilizer level within each organic fertilizer level, while different uppercase letters indicate significant differences by organic fertilizer level within each synthetic fertilizer level, at $p \leq 0.05$, according to the LSD test.

Similarly, the highest removal of P, K, and Mg was observed with the integrated use of synthetic NK fertilizers along with the highest rates of castor meal or the CM+HP mixture (Figure 5b–d). In the absence of synthetic NK fertilizer, all treatments with organic fertilizers increased P removal by storage roots. The application of 2.25 Mg ha^{-1} of CM+HP mixture resulted in the highest K and Mg removals. Conversely, the lowest removals of N, P, K, and Mg were observed in the absolute control (Figure 5a–d). Ca removal was enhanced in the treatment with synthetic NK fertilizer compared to that in the treatment without synthetic fertilizer, regardless of organic fertilizer usage (Table 3). The application of the highest rate of the CM+HP mixture increased Ca removal, while the lowest removal was observed in the treatment without organic fertilizers, regardless of synthetic NK fertilizers. In the absence of synthetic NK fertilizer, treatments with the highest rate of castor meal and both rates of the CM+HP mixture provided greater S removal than treatments without organic fertilizer (Figure 5e). However, S removal was higher in the treatment with 4.5 Mg ha^{-1} of CM+HP mixture combined with synthetic NK fertilizers than in the other treatments.

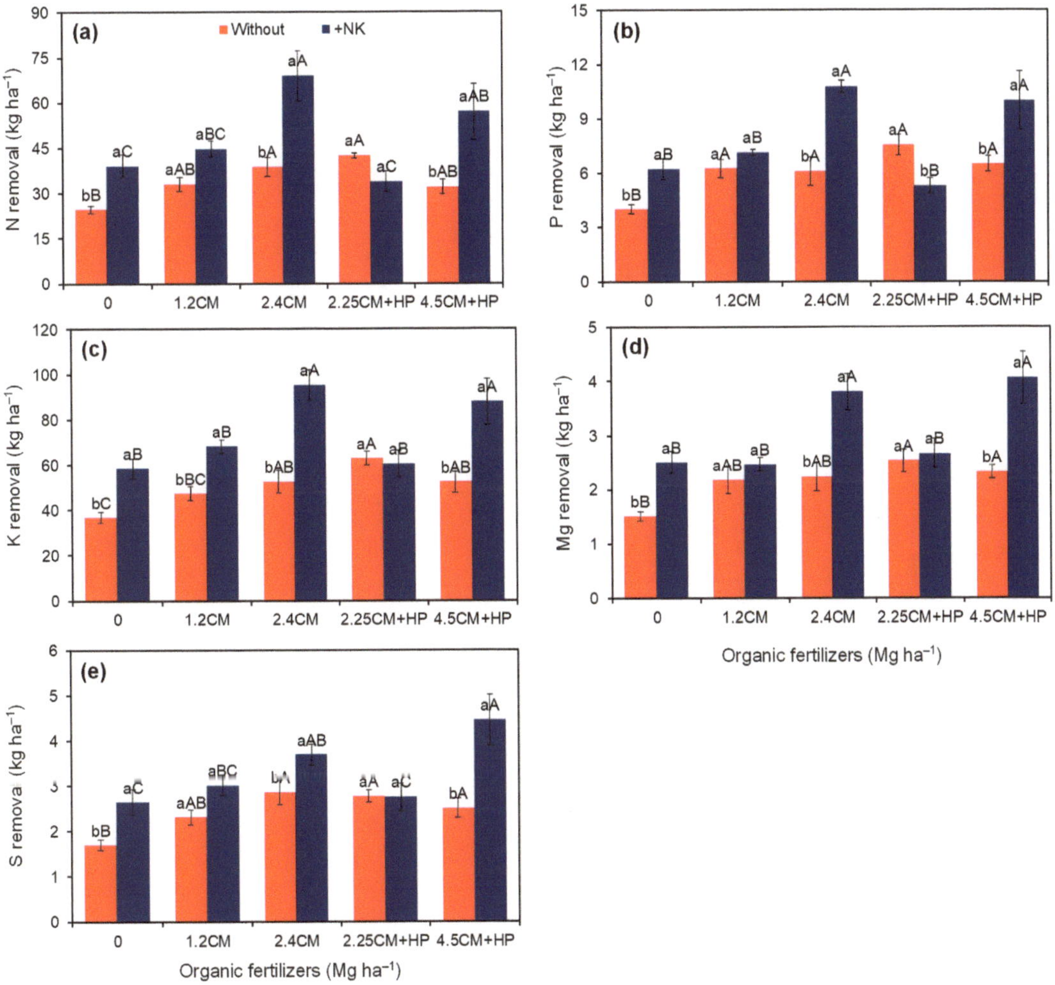

Figure 5. Removal of N (**a**), P (**b**), K (**c**), Mg (**d**), and S (**e**) in the storage roots of sweet potato crop as affected by synthetic NK fertilizers and rates of organic fertilizers composed of castor meal (CM) or castor meal plus hydrothermalized phonolite mixture (CM+HP). The red and blue bars indicate treatments without and with recommended synthetic NK fertilization, respectively. Bars indicate the standard error. Different lowercase letters indicate significant differences by synthetic fertilizer level within each organic fertilizer level, while different uppercase letters indicate significant differences by organic fertilizer level within each synthetic fertilizer level, at $p \leq 0.05$, according to the LSD test.

The application of the highest rates of castor meal and the CM+HP mixture combined with synthetic NK fertilizers resulted in increased removal of B, Cu, and Mn by sweet potato storage roots compared to that in the other treatments (Figure 6a,b,d). In treatments where synthetic NK fertilizers were applied, 2.4 Mg ha^{-1} of castor meal increased Fe removal, whereas the application of 4.5 Mg ha^{-1} of CM+HP mixture resulted in the highest Zn removal (Figure 6c,e). When no synthetic NK fertilizer was used, the application of 2.4 Mg ha^{-1} of castor meal and 2.25 Mg ha^{-1} of CM+HP mixture increased B removal. Additionally, with the application of the highest rate of castor meal and both rates of CM+HP mixture, an increase in Cu removal was observed compared to that in the absolute control. Treatment with high rates of CM+HP mixture demonstrated increased Fe removal, whereas low rates of CM+HP mixture resulted in enhanced Zn removal compared to that in the absolute control when synthetic NK fertilizer was not applied.

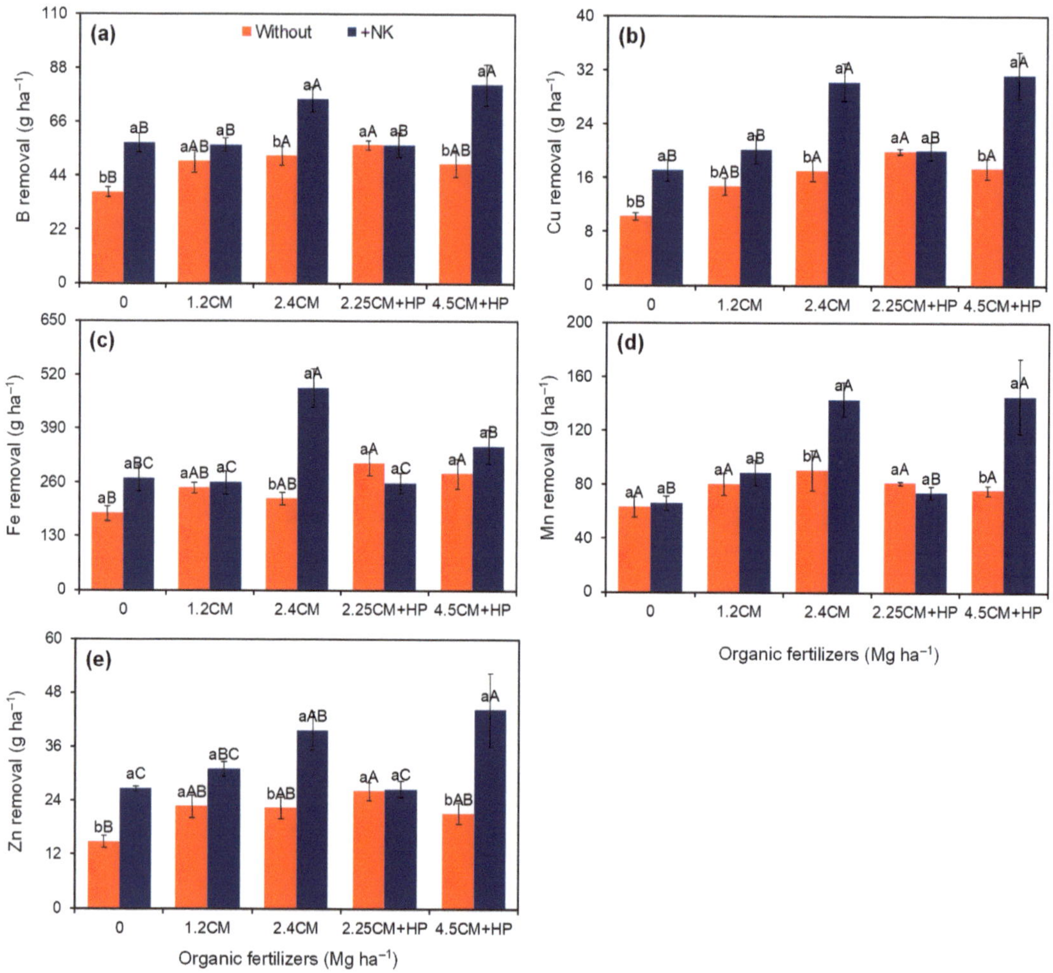

Figure 6. Removal of B (**a**), Cu (**b**), Fe (**c**), Mn (**d**), and Zn (**e**) in the storage roots of sweet potato crop as affected by synthetic NK fertilizers and rates of organic fertilizers composed of castor meal (CM) or castor meal plus hydrothermalized phonolite mixture (CM+HP). The red and blue bars indicate treatments without and with recommended synthetic NK fertilization, respectively. Bars indicate the standard error. Different lowercase letters indicate significant differences by synthetic fertilizer level within each organic fertilizer level, while different uppercase letters indicate significant differences by organic fertilizer level within each synthetic fertilizer level, at $p \leq 0.05$, according to the LSD test.

3.1. Soil Health Indices

An interaction between the studied factors and soil health indices was observed (Figure 7). The application of 2.4 Mg ha^{-1} of castor meal without synthetic NK fertilizer increased soil organic matter content compared to the absolute control (Figure 7a). When synthetic NK fertilizers were used, the highest soil organic matter content was observed with the application of 1.2 Mg ha^{-1} of castor meal. In the absence of synthetic NK fertilizer, the activity of the arylsulfatase enzyme in the soil was elevated upon application of the highest rate of castor meal (Figure 7b). Conversely, in the presence of synthetic NK fertilizer, application of the lowest rate of castor meal resulted in higher arylsulfatase enzyme activity in the soil than in the other treatments. The activity of the β-glucosidase enzyme in the soil was higher in the treatment with synthetic NK fertilizers combined with 1.2 Mg ha^{-1} of castor meal than in the other treatments with organic fertilizers (Figure 7c). When synthetic NK fertilizers were not applied, the activity of the β-glucosidase enzyme in the soil of the absolute control was higher than that in the treatments with organic fertilizers.

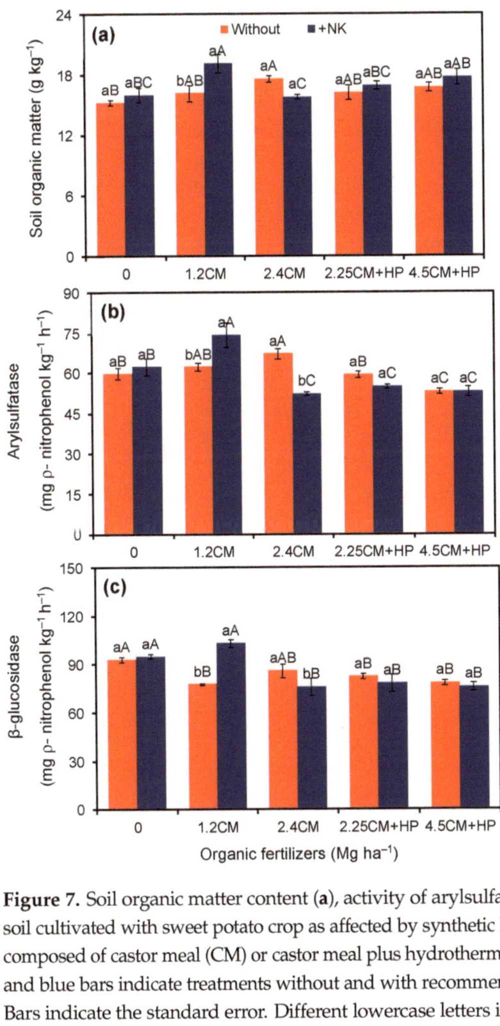

Figure 7. Soil organic matter content (**a**), activity of arylsulfatase (**b**), and β-glucosidase (**c**) enzymes in soil cultivated with sweet potato crop as affected by synthetic NK fertilizers and rates of organic fertilizers composed of castor meal (CM) or castor meal plus hydrothermalized phonolite mixture (CM+HP). The red and blue bars indicate treatments without and with recommended synthetic NK fertilization, respectively. Bars indicate the standard error. Different lowercase letters indicate significant differences by synthetic fertilizer level within each organic fertilizer level, while different uppercase letters indicate significant differences by organic fertilizer level within each synthetic fertilizer level, at $p \leq 0.05$, according to the LSD test.

4. Discussion

Although sweet potatoes can grow and be cultivated in soils with low nutrient availability [1,14], the findings of the present study indicate that organic fertilization (using pure castor meal or castor meal mixed with a K source based on hydrothermalized phonolite rock) improves the nutrition and storage root yield of this root crop, especially when combined with synthetic NK fertilization. The absence of notable interaction between synthetic and organic fertilizers on leaf N and K concentrations indicates that, regardless of the addition of synthetic NK fertilizers, the use of CM+HP mixture proved to be an efficient source of N and K for sweet potatoes, maintaining leaf concentrations at sufficient levels (33–50 g N kg^{-1} and 30–60 g K kg^{-1}) [47]. In sandy, K-deficient soils like the one in the present study (exchangeable K concentration <1.6 mmol$_c$ kg^{-1}) [47], failing to provide K results in a deficiency in sweet potato plants, particularly under heavy rainfall during the growing season [53]. The present study demonstrates that the CM+HP mixture effectively met the K requirements of sweet potato crops.

The application of 4.5 Mg ha^{-1} of organic fertilizer composed of CM+HP mixture increased leaf Mg and B concentrations, even when these were already at sufficient levels (3.0–12.0 g Mg kg^{-1} and 45–75 mg B kg^{-1}) [47]. This indicates that the organic fertilizer either supplied or improved the acquisition of Mg and B from the soil by the sweet potato crop. Conversely, the addition of synthetic NK fertilizers promoted plant development (as evidenced by the increased yield of storage roots), but this growth resulted in the dilution of leaf P, Ca, Mg, B, and Mn concentrations (Table 2). However, this reduction did not lead to nutritional deficiencies in plants, as the concentrations of P, Ca, Mg, B, and Mn remained within the ranges suitable for the growth of sweet potato crops (2.3–5.0 g P kg^{-1}, 7.0–12.0 g Ca kg^{-1}, and 40–250 mg Mn kg^{-1}) [47]. Despite the beneficial effects of the combined application of synthetic NK fertilizers and organic fertilizers on Zn and Cu concentrations in sweet potato leaves, these nutrients did not reach levels of deficiency or toxicity, remaining within the ranges (20–50 mg Zn kg^{-1} and 10–20 mg Cu kg^{-1}) considered suitable by Feltran et al. [47].

The combined application of synthetic NK fertilizers with castor meal or CM+HP mixture may have provided a more balanced nutrient supply, favoring the yield of sweet potato storage roots. For instance, in a study conducted in Brazil, top-dressing with 100 kg N ha^{-1} combined with 120 kg K_2O ha^{-1} resulted in increased sweet potato yields compared to the yields obtained by applying isolated N rates [12]. Organic fertilizers (castor meal or CM+HP mixture), in addition to providing nutrients, can also act as soil conditioners, improving soil structure and enhancing the retention of nutrients and water [54–57]. Furthermore, organic fertilizers may have promoted soil microbial activity and positively influenced plant physiology and metabolism, thereby supporting growth, nutrient uptake, and storage root yield [57–60]. Our results demonstrated that sweet potato crops responded favorably to organic fertilization, with the application of 2.25 Mg ha^{-1} of the CM+HP mixture increasing the total storage root yield by 52%. Organic fertilization, such as using green manure, can have effects comparable to those by NPK fertilization for sweet potato crops or may even help reduce the rate of synthetic N fertilization [13,17,42].

Despite the benefits of using organic fertilizers, synthetic NK fertilizers also proved to be important for the growth of sweet potato storage roots, increasing their weight by an average of 31% irrespective of the organic fertilizer used. The positive impact of the application of synthetic NK fertilizers on the mean storage root weight of sweet potatoes has been particularly observed in K-deficient soils [53,61]. Nitrogen enhances the distribution of DM in plants and, when supplied in adequate quantities, promotes the allocation of DM to storage roots [62,63]. Potassium increases the cambial activity of storage roots by enhancing the translocation of photoassimilates to the roots, thereby increasing root size [64,65]. Therefore, the amount of N supplied via synthetic NK fertilizer was sufficient to promote storage root growth without causing excessive growth of the aboveground portion at the expense of the growth of storage roots, as reported in other studies [17,66,67].

Similarly, K supplied by synthetic fertilizer was essential for increasing the biomass of storage roots, as verified in previous studies [7,53].

The increased P and K concentrations in sweet potato storage roots, observed with the highest application rate of castor meal (2.4 Mg ha^{-1}) combined with synthetic NK fertilizers, can be attributed to the N stimulus in promoting the growth of absorbent roots, thereby enhancing their capacity to uptake and translocate P [68]. Additionally, N uptake can facilitate K uptake due to the synergistic interaction between these two nutrients [21,69]. Baligar et al. [70] highlighted this synergistic effect in plant tissues, noting that a lower supply of N resulted in low K concentrations in plants.

The variation in storage root yields (Table 2 and Figure 3) influenced nutrient removal by sweet potato storage roots (Table 3 and Figures 5 and 6), as the nutrient levels in the roots showed little variation (Figure 4). The nutrient uptake by sweet potato crops is directly correlated with increased storage root yield [71]. Plant biomass and nutrient concentrations in plant parts were also positively correlated with the amount of nutrients taken up [17]. For most tuber crops, K is the nutrient required in the greatest quantities, followed by N and Ca [10,15,16,72]. In the present study, macronutrient removal by sweet potato storage roots followed the order K > N > P > Ca > S > Mg > Fe > Mn > B > Zn > Cu.

Soil organic matter is closely related to improvements in soil quality [73–76]. In the present study, the use of organic fertilizers increased the levels of organic matter and the activity of the enzymes β-glucosidase and arylsulfatase in the soil, indicating an improvement in soil quality. The activities of these enzymes in soil are influenced by several factors. The complexity of interactions can be affected by the amount of organic matter and plant residues, microbial activity, nutrient availability, pH, temperature, and salinity [49,77–80].

In the present study, castor meal proved to be an effective organic fertilizer for sweet potato cultivation since the isolated application of 2.4 Mg ha^{-1} of castor meal increased the total and marketable yields of storage roots by 45% and 23%, respectively. These increases were slightly smaller than those provided by the isolated use of the recommended synthetic NK fertilizer. The use of green manure as an organic fertilizer has also yielded results comparable to those of synthetic NPK fertilizers in sweet potato cultivation [42]. In the absence of synthetic NK fertilizer, the CM+HP mixture did not outperform pure castor meal. However, the combined use of synthetic NK fertilizers with the highest rates of organic fertilizers (castor meal or CM+HP mixture) was particularly effective, resulting in average increases of 116% and 128% in the total and marketable yields of storage roots, respectively, compared to those in the absolute control (without any fertilization). Under these conditions, the highest rates of organic fertilizers also increased the total and marketable yields of storage roots by an average of 60% and 67%, respectively, compared to those in treatments with synthetic and organic fertilizers alone. In the presence of synthetic NK fertilizers, the application of castor meal or a CM+HP mixture, especially at high rates, contributed to a more balanced supply of nutrients, resulting in improvements in the storage root yield and, consequently, greater nutrient removal.

However, neither synthetic nor organic fertilizers decreased the internal quality of the storage roots. On the contrary, organic fertilizers increased the soluble solid content in the storage roots. Other studies have shown that synthetic N fertilizer only improves sweet potato quality parameters, such as starch content if the application is not combined with organic fertilization using green manures [61], which was not observed in this study. Therefore, organic fertilization with the CM+HP mixture is a promising management technique for sweet potato production systems.

5. Conclusions

Regardless of the application of synthetic NK fertilizers, the CM+HP mixture has been shown to be an excellent source of N and K for sweet potatoes. This mixture is particularly effective in maintaining adequate leaf N and K levels. The highest rate (4.5 Mg ha^{-1}) of the CM+HP mixture also increased Mg and B concentrations in sweet potato leaves, irrespective of the use of synthetic NK fertilizers. The application of organic fertilizers increased the number of storage roots per plant and enhanced the soluble solid content in the storage roots, while synthetic NK fertilizers increased the mean weight of the storage root. The combined application of synthetic NK fertilizers with 2.4 Mg ha^{-1} of castor meal or 4.5 Mg ha^{-1} of the CM+HP mixture yielded greater benefits in terms of storage root yield (total and marketable) and nutrient removal than organic fertilizers alone. Moreover, fertilizing with castor meal at a rate of 1.2 Mg ha^{-1} in combination with synthetic NK fertilizers improved soil health, as indicated by the increased levels of organic matter and activity of arylsulfatase and β-glucosidase enzymes in the soil.

Author Contributions: Conceptualization, R.J.P., R.P.S., A.M.F. and S.G.D.; methodology and data collection, R.J.P., M.C.B. and L.G.F.; funding acquisition and project administration, R.P.S.; data curation, R.J.P. and R.P.S.; writing—original draft preparation, R.J.P., M.C.B. and L.G.F.; writing—review and editing, R.P.S., A.M.F., H.I.G. and S.G.D.; supervision, R.P.S. All authors have read and agreed to the published version of the manuscript.

Funding: This research was partially co-funded by A. Azevedo Indústria e Comércio de Óleos Ltda and Mineração Curimbaba Ltda. Additional funding was received from the Dean of Research of the São Paulo State University (Call PROPe 13/2022).

Data Availability Statement: The raw data supporting the conclusions of this article will be made available by the authors on request.

Acknowledgments: Thanks to the Dean of Research of the São Paulo State University (PROPe Call 13/2022) for providing a post-doctoral scholarship to the first author and the National Council for Scientific and Technological Development (CNPq) for providing an award for excellence in research to the second and third authors.

Conflicts of Interest: The authors declare no conflict of interest.

References

1. Truong, V.D.; Avula, R.Y.; Pecota, K.V.; Yencho, G.C. Sweetpotato Production, Processing, and Nutritional Quality. In *Handbook of Vegetables and Vegetable Processing*, 2nd ed.; Siddiq, M., Uebersax, M.A., Eds.; Vol. II. John Wiley & Sons Ltd.: Hoboken, NJ, USA, 2018.
2. Kakabouki, I.; Tataridas, A.; Mavroeidis, A.; Kousta, A.; Roussis, I.; Katsenios, N.; Efthimiadou, A.; Papastylianou, P. Introduction of alternative crops in the Mediterranean to satisfy EU Green Deal goals. A review. *Agron. Sustain. Dev.* **2021**, *41*, 71. [CrossRef]
3. Bechoff, A.; Sindi, K.; Low, J.; Ndyetabula, D. *Everything You Ever Wanted to Know about Sweetpotato: Reaching Agents of Change tot Manual;* International potato Center: Nairobi, Kenya, 2013.
4. CIP—International Potato Center. Facts and Figures about Sweetpotato. 2010. Available online: https://cipotato.org/crops/sweetpotato/ (accessed on 15 January 2024).
5. Ahn, Y.O.; Sun, H.K.; Kim, C.Y.; Lee, J.S. Exogenous sucrose utilization and starch biosynthesis among sweetpotato cultivars. *Carbohyd. Res.* **2010**, *345*, 55–60. [CrossRef]
6. Gonçalves Neto, A.C.; Maluf, W.R.; Gomes, L.A.A.; Gonçalves, R.J.S.; Silva, V.F.; Lasmar, A. Aptidões de genótipos de batata-doce para consumo humano, produção de etanol e alimentação animal. *Pesqu. Agropec. Bras.* **2011**, *46*, 1513–1520. [CrossRef]
7. Geng, J.; Zhao, Q.; Li, Z.; Yang, X.; Lei, S.; Zhang, Q.; Li, H.; Lang, Y.; Huo, X.; Liu, Q. Effects of various potassium fertilizer dosages on agronomic and economic assessment of sweet potato fields. *Horticulturae* **2024**, *10*, 44. [CrossRef]
8. Thiele, G.; Khan, A.; Heider, B.; Kroschel, J.; Harahagazwe, D.; Andrade, M.; Bonierbale, M.; Friedmann, M.; Gemenet, D.; Cherinet, M.; et al. Roots, tubers and bananas: Planning and research for climate resilience. *Open Agric.* **2017**, *2*, 350–361. [CrossRef]
9. FAO—Food and Agriculture Organization of the United Nations. *Food and Agriculture Data. Production-Crops;* FAO: Rome, Italy, 2024; Available online: https://www.fao.org/faostat/en/#data/QCL (accessed on 15 January 2024).
10. Filgueira, F.A.R. *Novo Manual de Olericultura: Agrotecnologia Moderna na Produção e Comercialização de Hortaliças;* UFV: Viçosa, Brazil, 2008.
11. Echer, F.R.; Dominato, J.C.; Creste, J.E. Absorção de nutrientes e distribuição da massa fresca e seca entre órgãos de batata-doce. *Hortic. Bras.* **2009**, *27*, 176–182. [CrossRef]
12. Foloni, J.S.S.; Corte, A.J.; Corte, J.R.N.; Echer, F.R.; Tiritan, C.S. Adubação de cobertura na batata-doce com doses combinadas de nitrogênio e potássio. *Semin.-Cienc. Agrar.* **2013**, *34*, 117–126. [CrossRef]
13. Fernandes, A.M.; Campos, L.G.; Senna, M.S.; Silva, C.L.; Assunção, N.S. Yield and nitrogen use efficiency of sweet potato in response to cover crop and nitrogen management. *Agron. J.* **2018**, *110*, 2004–2015. [CrossRef]

14. Minemba, D.; Gleeson, D.B.; Veneklaas, E.; Ryan, M.H. Variation in morphological and physiological root traits and organic acid exudation of three sweet potato (*Ipomoea batatas*) cultivars under seven phosphorus levels. *Sci. Hort.* **2019**, *256*, 108572. [CrossRef]
15. Brito, C.H.D.; Oliveira, A.P.D.U.; Alves, A.U.; Dorneles, C.S.; Santos, J.F.; Nóbrega, J.P. Produtividade da batata-doce em função das doses de K_2O em solo arenoso. *Hort. Bras.* **2006**, *24*, 320–323. [CrossRef]
16. Thumé, M.A.; Dias, L.E.; Silveira, M.A.D.; Assis, I.R.D. Níveis críticos foliares de nutrientes de três cultivares de batata-doce, selecionados para a produção de etanol. *Rev. Ceres* **2013**, *60*, 863–875. [CrossRef]
17. Fernandes, A.M.; Assunção, N.S.; Ribeiro, N.P.; Gazola, B.; Silva, R.M. Nutrient uptake and removal by sweet potato fertilized with green manure and nitrogen on sandy soil. *Rev. Bras. Cienc. Solo* **2020**, *44*, e0190127. [CrossRef]
18. Silva, J.B.C.; Lopes, C.A.; Magalhães, J.S. Cultura da batata-doce. In *Agricultura: Tuberosas amiláceas Latino Americanas*; Cereda, M.P., Ed.; Fundação Cargill: São Paulo, Brazil, 2002; Volume 2, pp. 448–504.
19. Duan, W.; Zhang, H.; Xie, B.; Wang, B.; Zhang, L. Impacts of nitrogen fertilization rate on the root yield, starch yield and starch physicochemical properties of the sweet potato cultivar Jishu 25. *PLoS ONE* **2019**, *14*, e0221351. [CrossRef]
20. Dong, H.T.; Li, Y.; Henderson, C.; Brown, P.; Xu, C.-Y. Optimum nitrogen application promotes sweetpotato storage root initiation. *Horticulturae* **2022**, *8*, 710. [CrossRef]
21. Cantarella, H. Nitrogênio. In *Fertilidade do Solo*, 1st ed.; Novais, F.R., Alvarez, V.V.H., Barros, N.F., Fontes, R.L.F., Cantarutti, R.B., Neves, J.C.L., Eds.; Sociedade Brasileira de Ciência do Solo: Viçosa, Brazil, 2007; pp. 375–470.
22. Rao, B.R.R. Biomass and essential oil yields of rainfed palmarosa (*Cymbopogon martinii* (roxb.) Wats.Var. motia Burk.) supplied with different levels of organic manure and fertilizer nitrogen in semi-arid tropical climate. *Ind. Crops Prod.* **2001**, *14*, 171–178. [CrossRef]
23. Rodrigues, A.C.; Cavalcante, L.F.; Oliveira, A.P.; Sousa, J.T.; Mesquita, F.O. Produção e nutrição mineral do maracujazeiro-amarelo em solo com biofertilizante supermagro e potássio. *Rev. Bras. Eng. Agríc. Amb.* **2009**, *13*, 117–124. [CrossRef]
24. Silva, S.D.; Presotto, R.A.; Marota, H.B.; Zonta, E. Uso de torta de mamona como fertilizante orgânico. *Pesqu. Agropec. Trop.* **2012**, *42*, 19–27. [CrossRef]
25. Ming, L.C.; Ferreira, M.I.; Gonçalves, G.G. Pesquisas agronômicas das plantas medicinais da Mata Atlântica regulamentadas pela ANVISA. *Rev. Bras. Plantas Med.* **2012**, *14*, 131–137. [CrossRef]
26. Gitari, H.I.; Gachene, C.K.K.; Karanja, N.N.; Kamau, S.; Nyawade, S.; Schulte-Geldermann, E. Potato-legume intercropping on a sloping terrain and its effects on soil physico-chemical properties. *Plant Soil* **2019**, *438*, 447–460. [CrossRef]
27. Lima, R.L.S.; Severino, L.S.; Sampaio, L.R.; Sofiatti, V.; Gomes, J.A.; Beltrão, N.E.M. Blends of castor meal and castor husks for optimized use as organic fertilizer. *Ind. Crops Prod.* **2011**, *33*, 364–368. [CrossRef]
28. Silva, P.N.L.; Lanna, N.B.L.; Cardoso, A.I.I. Produção de beterraba em função de doses de torta de mamona em cobertura. *Hort. Bras.* **2016**, *34*, 416–421. [CrossRef]
29. Ramesha, Y.M.; Manjunatha, B.; Krishnamurthy, D. Effect of castor de-oiled cake and inorganic fertilizers on growth, yield and economics of rice (*Oryza Sativa* L.). *Acta Sci. Agric.* **2017**, *1*, 11–15.
30. Mello, G.A.B.; Carvalho, D.F.; Medici, L.O.; Silva, A.C.; Gomes, D.P.; Pinto, M.F. Organic cultivation of onion under castor cake fertilization and irrigation depths. *Acta Sci. Agron.* **2018**, *40*, e34993. [CrossRef]
31. Severino, L.S.; Auld, D.L.; Baldanzi, M.; Cândido, M.J.D.; Chen, G.; Crosby, W.; Tan, D.; He, X.; Lakshmamma, P.; Lavanya, C.; et al. A review on the challenges for increased production of castor. *Agron. J.* **2012**, *104*, 853–880. [CrossRef]
32. Zapata, N.; Vargas, M.; Reyes, J.F.; Belmar, G. Quality of biodiesel and press cake obtained from Euphorbia lathyris, Brassica napus and Ricinus communis. *Ind. Crops Prod.* **2012**, *38*, 1–5. [CrossRef]
33. Nogueira, T.A.R.; Miranda, B.G.; Jalal, A.; Lessa, L.G.F.; Filho, M.C.M.T.; Marcante, N.C.; Abreu-Junior, C.H.; Jani, A.D.; Capra, G.F.; Moreira, A.; et al. Nepheline Syenite and Phonolite as Alternative Potassium Sources for Maize. *Agronomy* **2021**, *11*, 1385. [CrossRef]
34. Soratto, R.P.; Crusciol, C.A.C.; Campos, M.; Costa, C.H.M.; Gilabel, A.P.; Castro, G.S.A.; Ferrari Neto, J. Silicate rocks as an alternative potassium fertilizer for upland rice and common bean crops. *Pesqu. Agropec. Bras.* **2021**, *56*, e01411. [CrossRef]
35. Soratto, R.P.; Crusciol, C.A.C.; Campos, M.; Gilabel, A.P.; Costa, C.H.M.; Castro, G.S.A.; Ferrari Neto, J. Efficiency and residual effect of alternative potassium sources in grain crops. *Pesqu. Agropec. Bras.* **2021**, *56*, e02686. [CrossRef]
36. Mancuso, M.A.C.; Soratto, R.P.; Crusciol, C.A.C.; Castro, G.S.A. Effect of potassium sources and rates on Arabica coffee yield, nutrition and macronutrient export. *Rev. Bras. Cienc. Solo* **2014**, *38*, 1448–1456. [CrossRef]
37. Martins, V.; Silva, D.R.G.; Marchi, G.; Leite, M.C.A.; Martins, É.S.; Gonçalves, A.S.F.; Guilherme, L.R.G. Effect of alternative multinutrient sources on soil chemical properties. *Rev. Bras. Cienc. Solo* **2015**, *39*, 194–204. [CrossRef]
38. Teixeira, A.M.S.; Garrido, F.M.S.; Medeiros, M.E.; Sampaio, J.A. Estudo do comportamento térmico da rocha fonolito com fins à produção de fertilizantes. *Holos* **2015**, *31*, 52–64. [CrossRef]
39. Teixeira, A.M.S.; Sampaio, J.A.; Garrido, F.M.S.; Medeiros, M.E. Avaliação da rocha fonolito como fertilizante alternativo de potássio. *Holos* **2012**, *28*, 21–33. [CrossRef]
40. Tavares, L.F.; Carvalho, A.M.X.; Camargo, L.G.B.; Pereira, S.G.F.; Cardoso, I.M. Nutrients release from powder phonolite mediated by bioweathering actions. *Inter. J. Recyc. Organic Waste Agric.* **2018**, *7*, 89–98. [CrossRef]
41. Ciceri, D.; Oliveira, M.; Allanore, A. Potassium fertilizer via hydrothermal alteration of K-feldspar ore. *Green Chem.* **2017**, *19*, 5187–5202. [CrossRef]

42. Okpara, D.A.; Njoku, J.C.; Asiegbu, J.E. Responses of two sweet potato varieties to four green manure sources and inorganic fertilizer in a humid tropical Ultisol. *Biol. Agric. Hortic.* **2004**, *22*, 81–90. [CrossRef]
43. Oliveira, A.P.; Santos, J.F.; Cavalcante, L.F.; Pereira, W.F.; Santos, M.C.C.; Oliveira, A.N.P.; Silva, N.V. Yield of sweet potato fertilized with cattle manure and biofertilizer. *Hort. Bras.* **2010**, *28*, 277–281. [CrossRef]
44. FAO—Food and Agriculture Organization of the United Nations. *World Reference Base for Soil Resources 2014: International Soil Classification System for Naming Soils and Creating Legends for Soil Maps*; FAO: Rome, Italy, 2015. Available online: http://www.fao.org/3/i3794en/I3794en.pdf (accessed on 24 September 2022).
45. van Raij, B.; Andrade, J.C.; Cantarella, H.; Quaggio, J.A. *Análise Química para Avaliação da Fertilidade de solos Tropicais*; Instituto Agronômico: Campinas, Brazil, 2001.
46. EMBRAPA—Empresa Brasileira de Pesquisa Agropecuária. *Manual de Métodos de Análise de Solo*; Embrapa-CNPS: Rio de Janeiro, Brasil, 1997. Available online: https://ainfo.cnptia.embrapa.br/digital/bitstream/item/169149/1/Manual-de-metodos-de-analise-de-solo-2-ed-1997.pdf (accessed on 24 September 2022).
47. Feltran, J.C.; Peresin, V.A.; Granja, N.P.; Silva Filho, H.M.; Lorenzi, J.O.; Fernandes, A.M.; Soratto, R.P.; Factor, T.L.; Rós, A.B.; Aguiar, E.B. Raízes e tubérculos. In *Boletim 100: Recomendações de Adubação e Calagem Para o Estado de São Paulo*; Cantarella, H., Quaggio, J.A., Mattos, D., Jr., Boaretto, R.M., van Raij, B., Eds.; Instituto Agronômico: Campinas, Brazil; pp. 314–338.
48. Malavolta, E.; Vitti, G.C.; Oliveira, S.A. *Avaliação do Estado Nutricional das Plantas: Princípios e Aplicações*, 2nd ed.; Associação Brasileira para Pesquisa da Potassa e do Fosfato: Piracicaba, Brazil, 1997.
49. Tabatabai, M.A. Soil Enzymes. In *METHODS of Soil analysis: Part 2: Microbiological and Biochemical Properties*; SSSA Book Series: 5; SSSA: Madison WI, USA, 1994; pp. 775–833.
50. Miranda, J.E.C.; França, F.H.; Carrijo, O.A.; Souza, A.F.; Pereira, W.; Lopes, C.A.; Silva, J.B.C. *A Cultura da Batata-Doce*; Coleção Plantar; EMBRAPA-CNPH: Brasília, Brazil, 1995.
51. Nelson, N. A photometric adaptation of the Somogyi method for the determination of glucose. *J. Biol. Chem.* **1944**, *153*, 375–390. [CrossRef]
52. AOAC—Association of Official Analytical Chemists. *Official Methods of Analysis of AOAC International*, 18th ed.; AOAC: Rockville, MD, USA, 2007.
53. Figueiredo, R.T.; Fernandes, A.M.; Garreto, F.G.S.; Silva, J.A.; Nunes, J.G.S.; Vargas, P.F. Sweetpotato responses to potassium rate and timing in tropical sandy soils. *Agron. J.* **2023**, *115*, 2044–2057. [CrossRef]
54. Yamaguchi, T.; Takei, T.; Yazawa, Y.; Wong, M.T.F.; Gilkes, R.J.; Swift, R.S. Effect of humic acid, sodium, and calcium additions on the formation of water-stable aggregates in Western Australian wheatbelt soils. *Aust. J. Soil Res.* **2004**, *42*, 435–439. [CrossRef]
55. Canellas, L.P.; Zandonadi, D.B.; Médici, L.O.; Peres, L.E.P.; Olivares, F.L. Bioatividade de substâncias húmicas—Ação sobre o desenvolvimento e metabolismo das plantas. In *Humosfera: Tratado Preliminar Sobre a Química das Substâncias Húmicas*; Canellas, L.P., Santos, G.A., Eds.; Ed. do Autor: Campos dos Goytacazes, Brazil, 2005; pp. 224–243.
56. Billingham, K.L. *Humic Products-Potential or Presumption for Agriculture? Can Humic Products Improve My Soil? 27th Annual Conference*; Grassland Society of NSW Inc.: Orange, NSW, USA, 2012.
57. Ampong, K.; Thilakaranthna, M.S.; Gorim, L.Y. Understanding the role of humic acids on crop performance and soil health. *Front. Agron.* **2022**, *4*, 848621. [CrossRef]
58. Canellas, L.P.; Olivares, F.L. Physiological responses to humic substances as plant growth promoter. *Chem. Biol. Techn. Agric.* **2014**, *1*, 1–11. [CrossRef]
59. Maji, D.; Misra, P.; Singh, S.; Kalra, A. Humic acid rich vermicompost promotes plant growth by improving microbial community structure of soil as well as root nodulation and mycorrhizal colonization in the roots of Pisum sativum. *App. Soil Ecol.* **2017**, *110*, 97–108. [CrossRef]
60. Li, Y.; Fang, F.; Wei, J.; Wu, X.; Cui, R.; Li, G.; Zheng, F.; Tan, D. Humic acid fertilizer improved soil properties and soil microbial diversity of continuous cropping peanut: A three-year experiment. *Sci. Rep.* **2019**, *9*, 12014. [CrossRef] [PubMed]
61. Fernandes, A.M.; Ribeiro, N.P.; Assunção, N.S.; Nunes, J.G.S.; Sorroche, C.P.; Leonel, M. Impact of nitrogen and green manure on yield and quality of sweet potato in sandy soil: A Brazilian case study. *J. Agric. Food Res.* **2021**, *4*, 100131. [CrossRef]
62. Okpara, D.A.; Okon, O.E.; Ekeleme, F. Optimizing nitrogen fertilization for production of white and orange-fleshed sweet potato in southeast Nigeria. *J. Plant Nutr.* **2009**, *32*, 878–891. [CrossRef]
63. Du, X.; Kong, L.; Xi, M.; Zhang, X. Split application improving sweetpotato yield by enhancing photosynthetic and sink capacity under reduced nitrogen condition. *Field Crops Res.* **2019**, *238*, 56–63.
64. Byju, G.; George, J. Potassium nutrition of sweet potato. Adv. *Hortic. Sci.* **2005**, *19*, 221–239.
65. Lv, Z.; Lu, G. New curve of critical leaf potassium concentration based on the maximum root dry matter for diagnosing k nutritional status of sweetpotato. *Front. Plant Sci.* **2021**, *12*, 714279. [CrossRef] [PubMed]
66. Oliveira, A.P.; Oliveira, M.R.T.; Barbosa, J.A.; Silva, G.G.; Nogueira, D.H.; Moura, M.F.; Braz, M.S.S. Yield and quality of sweet potato roots fertilized with urea. *Hort. Bras.* **2005**, *23*, 925–928, (In Portuguese, with English abstract). [CrossRef]
67. Santos Neto, A.R.; Silva, T.O.; Blank, A.F.; Silva, J.O.; Araújo Filho, R.N. Produtividade de clones de batata doce em função de doses de nitrogênio. *Hortic. Bras.* **2017**, *35*, 445–452. [CrossRef]
68. Fageria, N.K. Eficiência do uso de potássio pelos genótipos de arroz de terras altas. *Pesqu. Agropec. Bras.* **2000**, *35*, 2115–2120. [CrossRef]

69. Rietra, R.P.; Heinen, M.; Dimkpa, C.O.; Bindraban, P.S. Effects of nutrient antagonism and synergism on yield and fertilizer use efficiency. *Commun. Soil Sci. Plant Anal.* **2017**, *48*, 1895–1920. [CrossRef]
70. Baligar, V.C.; Fageria, N.K.; He, Z.L. Nutrient use efficiency in plants. *Comm. Soil Sci. Plant Anal.* **2001**, *32*, 921–950. [CrossRef]
71. Bourke, R.M. Influence of nitrogen and potassium fertilizer on growth of sweet potato (Ipomoea batatas) in Papua New Guinea. *Field Crops Res.* **1985**, *12*, 363–375. [CrossRef]
72. Mugo, N.J.; Karanja, N.N.; Gachene, C.K.; Dittert, K.; Ireri, G.H.; Schulte-Geldermann, E. Response of potato crop to selected nutrients in Central and Eastern highlands of Kenya. *Cogent Food Agric.* **2021**, *7*, 1898762. [CrossRef]
73. Lopes, A.A.C.; Sousa, D.M.G.; Chaer, G.M.; Reis Junior, F.B.; Goedert, W.J.; Mendes, I.C. Interpretation of microbial soil indicators as a function of crop yield and organic carbon. *Soil Sci. Soc. Am.* **2013**, *77*, 461–472. [CrossRef]
74. Lopes, A.A.C.; Sousa, D.M.G.; Reis Junior, F.B.; Figueiredo, C.C.; Malaquias, J.V.; Souza, L.M.; Mendes, I.C. Temporal variation and critical limits of microbial indicators in oxisols in the Cerrado, Brazil. *Geoderma Reg.* **2018**, *12*, 72–82. [CrossRef]
75. Ganeshamurthy, A.N.; Nielsen, N.E. Arylsulphatase and the biochemical mineralization of soil organic sulphur. *Soil Biol. Biochem.* **1990**, *22*, 1163–1165. [CrossRef]
76. Mendes, I.C.; Souza, L.M.D.; Sousa, D.M.G.; Lopes, A.A.D.C.; Reis Junior, F.B.D.; Lacerda, M.P.C.; Malaquias, J.V. Critical limits for microbial indicators in tropical Oxisols at post-harvest: The FERTBIO soil sample concept. *App. Soil Eco.* **2019**, *139*, 85–93. [CrossRef]
77. Nogueira, M.A.; Melo, W.J. Enxofre disponível para a soja e atividade de arilsulfatase em solo tratado com gesso agrícola. *Rev. Bras. Cienc. Solo* **2003**, *27*, 655–663. [CrossRef]
78. Nyawade, S.O.; Karanja, N.N.; Gachene, C.K.K.; Gitari, H.I.; Schulte-Geldermann, E.; Parker, M.L. Short-term dynamics of soil organic matter fractions and microbial activity in smallholder legume intercropping systems. *Appl. Soil Ecol.* **2019**, *142*, 123–135. [CrossRef]
79. Shao, Z.; Mwakidoshi, E.R.; Muindi, E.M.; Soratto, R.P.; Ranjan, S.; Padhan, S.R.; Wamukota, A.W.; Sow, S.; Wasonga, D.O.; Nasar, J.; et al. Synthetic fertilizer application coupled with bioslurry optimizes potato (*Solanum tuberosum*) growth and yield. *Agronomy* **2023**, *13*, 2162. [CrossRef]
80. Tabatabai, M.A.; Bremner, J.M. Factors affecting soil arylsulfatase activity. *Soil Sci. Soc. Am. Proc.* **1970**, *34*, 427–429. [CrossRef]

horticulturae

Article

Effect of Fertilization in Companion Cropping Systems of Andean Fruit Trees in the Municipality of Ipiales

Ovidio Javier Moran-Chamorro [1], Danita Andrade-Díaz [2,*], Juan Sebastian Chirivi-Salomon [2] and Pedro Alexander Velasquez-Vasconez [3]

1 Facultad de Ciencias Agrarias, Universidad de Nariño, Pasto 520001, Colombia; ovidiomoran302@gmail.com
2 Escuela de Ciencias Agrícolas, Pecuarias y del Medio Ambiente, Universidad Nacional Abierta y a Distancia, Bogota 111511, Colombia; juan.chirivi@unad.edu.co
3 Escuela de Ciencias Básicas, Tecnología e Ingeniería, Universidad Nacional Abierta y a Distancia, Pasto 520001, Colombia; pedro.velasquez@unad.edu.co
* Correspondence: danita.diaz@unad.edu.co; Tel.: +57-31-7886-3748

Abstract: Companion cropping offers a potential solution to the challenges of sustainable agriculture, such as optimizing resource use and reducing reliance on chemical inputs. The problem of achieving higher yields while maintaining environmental health remains critical. This practice enhances natural resource conservation, improves fertilization, and optimizes nutrient cycling through the balanced use of chemical and organic sources. Studies, such as those involving tree tomato and Hass avocado, have demonstrated a significant yield increase compared to monocultures, underscoring the viability of this practice. In addition to their environmental benefits, companion crops provide economic advantages by allowing producers to harvest multiple products simultaneously, thereby strengthening food security and the rural economy. This study evaluated three levels of fertilization and interactions between fruit trees at different altitudes, observing differential behavior in the variables evaluated. The combination of cape gooseberry and blackberry showed significantly positive results, with more leaves and fewer pests, demonstrating the benefits of companion plants. A trend towards the combined use of chemical and organic fertilizers was observed, a potential strategy to reduce costs and improve crop growth. The results indicated that the UF system (*P. peruviana* and *P. vulgaris*) had the highest plant height, while TF (tree tomato and bean) showed the best stem perimeter development. The incidence of pests was also significant, with *Trialeurodes vaporarioum* being most prevalent in the *P. peruviana* companion. These findings support companion cropping as a viable and promising strategy for more efficient and sustainable agriculture, offering both environmental and economic benefits.

Keywords: growth; development; pest incidence; disease incidence; technology transfer

Citation: Moran-Chamorro, O.J.; Andrade-Díaz, D.; Chirivi-Salomon, J.S.; Velasquez-Vasconez, P.A. Effect of Fertilization in Companion Cropping Systems of Andean Fruit Trees in the Municipality of Ipiales. *Horticulturae* **2024**, *10*, 1107. https://doi.org/10.3390/horticulturae10101107

Academic Editors: Francesco De Mastro, Gennaro Brunetti, Karam Farrag and Huadong Zang

Received: 12 August 2024
Revised: 11 September 2024
Accepted: 12 September 2024
Published: 18 October 2024

1. Introduction

Cultivated soil has experienced erosion due to practices such as conventional tillage and the excessive use of machinery, resulting in increased dependence on fertilizers for production [1]. However, accessing and acquiring fertilizers has become more difficult due to their high costs and the scarcity of raw materials needed for their manufacture [2,3]. This has raised production costs and made it difficult to obtain consistent profits in each production cycle [4]. In addition, their indiscriminate use is common, as it is usually not carried out considering soil analysis and without following technical criteria, which leads to contamination problems, especially by nitrogen fertilization, which can affect vulnerable populations [5].

The Department of Nariño is no stranger to this situation, and it is of great importance because it depends mainly on agriculture as an economic source, with fruit trees being the most representative sector, occupying more than 38% of the total area dedicated to agriculture in the region [6]. Specifically, in the townships of San Juan in the municipality of Ipiales, fruit trees have been adopted by producers and contribute significantly to their economy [7]. In this sense, Andean fruit trees, such as cape gooseberry, have experienced an increase in the region, with numerous farms dedicated to its production for fresh export [8].

Andean fruit trees, such as cape gooseberry (*Physalis peruviana*), tree tomato (*Solanum betaceum*), blackberry (*Rubus glaucus*), and passion fruit (*Passiflora edulis*), are native species to the mountainous ecosystems of the Andes. These species are characterized by their adaptability to different altitudes and microclimates, making them essential for agroforestry systems in the region. In addition to their nutritional value, these fruits have antioxidant properties and are highly appreciated in both local and international markets. The production of these fruit trees not only contributes to food security but also represents a significant source of income for rural families, diversifying their economies and strengthening sustainable rural development. Therefore, promoting sustainable agricultural practices, such as companion cropping, is crucial to maximize yields and conserve biodiversity in these systems [8].

To minimize the use of chemical fertilizers, which are not viable for small producers, research is being encouraged to develop agroecological strategies that are environmentally friendly and contribute to climate change adaptation [4].

Chemical fertilization can improve crop growth and yield, but productive, quality, and environmental benefits have also been reported with organic fertilization in Andean fruit trees, which requires further evaluation. Recent studies have highlighted that organic fertilization can enhance soil quality, improve microbial biodiversity, and contribute to sustainable agricultural practices [9,10].

In this context, companion cropping systems are presented as a sustainable alternative for the production of Andean fruit trees, as they contribute to reducing production costs and conserving soil resources [11,12]. Companion crops are complex cropping systems in which two or more species are planted in spatial proximity, which can result in competition or complementation between them, positively influencing their development and yield [13]. These diversified systems, such as companion cropping and agroforestry, are considered more sustainable and contribute to the conservation of natural resources [11,12]. These systems have been called the "new green revolution" due to their potential to increase soil productivity by taking advantage of complementarities between species and allowing intensive agriculture in small areas in a sustainable manner [9,10].

The interaction between species in companion crops allows for enhancing fertilization, promoting more sustainable organic fertilization and a balance between chemical and organic sources, since the soil depends on the biological component for nutrient cycling from organic to mineral forms available to plants [10]. These systems also benefit from nutrient cycling and organic matter availability, which improves the overall nutrition of the different companion crops [9]. These systems further benefit from nutrient cycling and organic matter availability, which improves the overall nutrition of the different companion crops [14]. Companion crops have proven to be viable in various species of economic importance, as was observed in an intercropping system of tree tomato and Hass avocado, where a positive influence was evidenced in 75% of the population in terms of size and yield, marking a significant difference compared to monoculture systems [15].

These companion cropping systems are not only more profitable for producers by obtaining multiple products in the same crop but also contribute to food security and improve the economy of low-income rural families who depend mainly on the production of a single crop or small self-consumption gardens [16]. Therefore, it is necessary to evaluate the interactions between plants and fertilization in each region, considering specific characteristics that can modify the results of possible combinations. By achieving a balance between chemical and organic fertilization in each companion system, a production alternative is presented that reduces the excessive use of chemical fertilizers, which could help mitigate production costs and generate greater benefits for producers, considering the high prices and global shortage of some chemical fertilizers, as well as their indiscriminate use [4].

The companion cropping systems proposed in this research focus on crops that have been widely accepted by producers in the area due to their market potential, profitability, and management knowledge. Therefore, the results of the different development variables in intercropping companion systems managed with three levels of fertilization for crops such as blackberry, purple passion fruit, cape gooseberry, and tree tomato complemented with bush beans should be demonstrated and shared for the benefit of producers [16,17].

Having said the above, this project aimed to evaluate the initial growth and determine the incidence of pests and diseases in five companion cropping systems with Andean fruit trees under three levels of fertilization in environmental conditions of the corregimiento of San Juan vereda Loma de Zuras. Additionally, the aim is to share the results obtained with producers in the area.

2. Materials and Methods

2.1. Location

This study was carried out between 2022 and 2023, in the municipality of Ipiales, specifically in the corregimiento of San Juan, located in the Department of Nariño. Evaluations were conducted in several townships, including Loma de Zuras, Camellones, Laguna de Vaca, Boquerón, and Guacan. The evaluated plots are located at altitudes ranging between 2575 and 2877 masl (Table 1; Figure 1). Ipiales, located in southwestern Colombia, has a cool Andean climate characterized by mild temperatures ranging from 11 °C to 13 °C (52 °F to 55 °F) year-round, with significant drops during the night. The area experiences moderate annual rainfall between 1000 and 1500 mm (39 to 59 inches), mainly concentrated in the rainy seasons from March to May and October to November, while the period from June to August is generally drier. The humidity is relatively high, averaging around 80%, with frequent cloud cover and strong winds, especially in the afternoons.

Table 1. Information and data of experimental plots.

Name—Plot	Block/System	X	Y	Area (m^2)	Masl
Tablon	1-low/1	947,277.0898	591,090.2	5072	2655
Cundala	1-low/2	947,119.3487	590,739.8	6839	2586
Chapicha	1-low/3	945,570.4425	590,613.2	5003	2676
Rancheria	1-low/4	947,074.3692	590,816.7	6418	2620
Santa Barbara	1-low/5	947,507.7808	591,300.7	5237	2576
Culacal	2-medium/1	944,307.8386	589,604.8	6358	2686
Cundala	2-medium/2	945,619.9373	589,926.4	5470	2742
Tuquer	2-medium/3	943,785.2733	589,759.5	5190	2695
Churumbuta	2-medium/4	946,619.0676	590,880.7	5087	2751
Yerba buena	2-medium/5	942,863.7872	588,375.7	6593	2736
Capuli	3-high/1	943,370.221	588,113.5	5178	2812
Churumbuta Laguna de vaca	3-high/2	946,547.2773	590,646.6	5120	2768
campanario	3-high/3	943,103.9658	588,347.8	5107	2800
Cundala	3-high/4	945,680.5052	589,754.3	8276	2770
Chuchala	3-high/5	943,173.5561	587,244.1	5140	2753

2.2. Planting Material

The 15 experimental plots with an area of 5000 m^2 were planted with cape gooseberry, blackberry, tomato, and purple passion fruit seedlings supplemented with beans, obtained from the BIOPASS nursery certified by the ICA by resolution number 065191 of 2020.

2.3. Experimental Design

According to the altitude data of each of the lots, stratification was made, in low, medium, and high taking the highest and lowest location, which corresponds to the repetitions or blocks (Table 1). In each stratum, one of the five proposed intercropping systems was randomly distributed (Figure 2) for a total of 15 trials. Within each of the intercropping systems, the companion models (Species 1 in monoculture and bush bean, Species 2 in monoculture and bush bean, Species 1 and 2 with bush bean) and the fertilization levels (fertilization level one (F1), 100% chemical fertilization; fertilization level two (F2), 50% chemical fertilization and 50% organic fertilization; and fertilization level three (F3), 100% organic fertilization) were distributed, for which a layout was made according to the experimental design of Divided Plots.

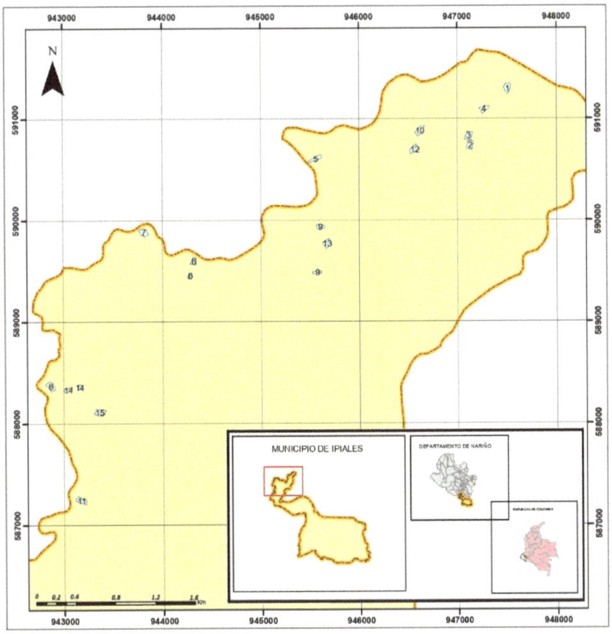

Figure 1. Location of the experimental sites: 1. Tablon; 2. Cundala; 3. Chapicha; 4. Rancheria; 5. Santa Barbara; 6. Culacal; 7. Cundala; 8. Tuquer; 9. Churumbuta; 10. Yerba Buena; 11. Capuli; 12. Churumbuta—Laguna de Vaca; 13. Campanario; 14. Cundala; and 15. Chuchala.

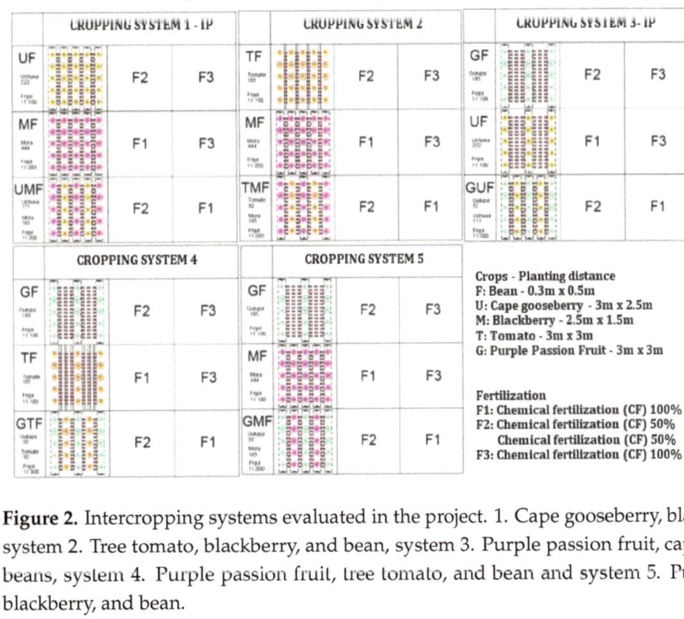

Figure 2. Intercropping systems evaluated in the project. 1. Cape gooseberry, blackberry, and bean, system 2. Tree tomato, blackberry, and bean, system 3. Purple passion fruit, cape gooseberry, and beans, system 4. Purple passion fruit, tree tomato, and bean and system 5. Purple passion fruit, blackberry, and bean.

The planting distances were as follows: cape gooseberry: 3 m between furrows and 3 m between plants; tree tomato: 3 m between furrows and 2.5 m between plants; purple passion fruit: 3 m between furrows and 3 m between plants; blackberry: 2.5 m between furrows and 1.5 m between plants. In the purple passion fruit trials, the distance between furrows in monoculture was 6 m. Beans were planted in two rows between the lanes of the fruit trees with distances between plants of 0.30 m and between furrows of 0.50 m.

The distribution of the factors within the strips was determined after conducting a soil fertility analysis, following the methodology proposed by AgroMel. This methodology is based on an integrated management model that allows for geo-localized collection, processing, and analysis of multiple agronomic variables on a detailed scale. It characterizes the different productive micro-environments of each block on the farms, comparing them with growth and vigor indexes obtained via satellite imagery. This approach enables the visualization of the soil's physical, chemical, and structural variables.

2.4. Variables Evaluated

Five evaluations were made every 45 days, on 2 plants per species in each treatment selected at random: total height (TH), taken from the base of the stem to the apex using a tape measure; stem perimeter (SP), measuring in cm the basal part of the stem at 10 cm from the ground; number of total leaves (NL) by direct counting; and total leaf area of the plant (LA). The leaf area index (LAI) was determined by the equation: LAI = ((leaf area) × (stocking density))/(planted area), and the pest and disease incidence (IPD), which records the presence of pests and diseases, was determined by the formula: IPD = (affected plants)/(total plants evaluated) × 100.

To measure the leaf area, the predictive models for leaf area obtained for these species by Velasquez and Andrade [18] were used. Once the average leaf area per leaf was calculated, it was multiplied by the total leaves per plant to obtain the leaf area of the plant.

2.5. Information Analysis

The information obtained for each of the variables was organized in an Excel spreadsheet to be analyzed using the methodology of a Functional Growth Analysis, which is used for measurements made at frequent time intervals (45, 90, 135, 180, and 225 dap) in each of the species that allows for determining the vigor in terms of the growth rate for the response variables. The data were transformed with the "log" function of the "R environment", version 4.3.1, to meet the linearity assumption and to obtain regression data using the linear regression model method [19].

The values of the regression coefficients (β) obtained in the Functional Analysis of Growth were analyzed based on the Analysis of Variance (ANDEVA) according to the split-plot design:

$$Yijk = \mu + Rk + Ai + (RA)ik + Bj + (RB)kj + (AB)ij + (RAB)kij.$$

where

Yijk = Response variable

μ = overall mean of the experiment

Rk = effect of the k-th block corresponding to the heights

Ai = effect of the factor associated with the i-th main plot corresponding to the companion crops (RA)ik = Error a of the main plot

Bj = effect of the factor associated with the j-th subplot corresponding to fertilization (RB)kj = error b associated with the subplot

(AB)ij = effect of the interaction between the main plot and subplot (i, j).

Based on the ANDEVA results, the hypotheses were either rejected or accepted. When a null hypothesis (Ho) was rejected, a comparison of means was made using the DUNKAN test with a significance level of 95% in order to determine the best treatments and interactions. Analyses were performed with the spltplot function [20] and plotted with the ggplot2 package [21] of the free "R 1.3.0" software. The pest and disease incidence variable

(IPD) was analyzed descriptively using graphs. From a contingency table with absolute frequency values of the incidence of pests and diseases, a chi-square test was performed to identify whether the companion and fertilization interaction was significant, to subsequently perform a simple correspondence analysis (ANACOR) using the FactoMineR package Husson [22]. Graphs were constructed with the ggplot2 package [21] in the free software "R".

2.6. Training with Producers and Dissemination of the Information Generated in This Project

An initial socialization was carried out with producers to make known the methodology to be used and the dynamics implemented. During the execution of the research, the partial and results of the research were shared by applying field school methodology (ECA).

3. Results

3.1. Functional Analysis of Growth and ANDEVA Analysis of Variance of the Regression Coefficients Obtained

For the variable plant height (TH) (Table 2), the results of the ANDEVA show that there is a significant difference in the source of companion variation; therefore, we can determine that the variation in growth was affected by the microclimate of the different blocks and the cultural management in each crop related to the formation pruning that affected the data collection and the growth register.

Table 2. ANDEVA mean squares for regression coefficients (β) under the split-strip design mode.

FV	G.L	TH	SP	NL	LA	LAI
Block	2	0.029 ns	0.023 ns	4989.3 *	0.135 ns	0.136 ns
Companion	8	0.034 *	1.621 **	16,039.7 **	0.517 *	0.525 *
Fertilization	2	0.001 ns	0.027 ns	41.4 ns	0.003 ns	0.417 ns
Companion × Fertilization	16	0.001 ns	0.018 ns	252.9 ns	0.006 ns	0.650 ns
Error a	15	0.029	0.07381	1203.9	176	0.173
Error b	34	0.003	0.01091	243.4	0.009	0.009
Mean		0.478	0.886	62.48	0.7	0.17
R^2		0.874	0.97	0.952	0.955	0.955
CV (%)		12.09	11.78	24.97	14.02	14.03

** Highly significant. * Significant. ns Not significant.

In all systems, a linear increase over time of the evaluated variable is observed (Figure 3A), and the differentiation in the behavior of this variable (Table 3) is evidenced by the fact that between the growth of some of the species, the gray shading does not overlap.

Table 3. Comparison of means for the regression coefficients (β) of the variables evaluated under the split-strip design model. The same letters correspond to results without statistical differences.

Treatment	TH	SP	NL	LA	LAI
UF	24.23 a	1.08 b	117.66 ab	0.93 ab	0.92 ab
MF	19.01 ab	0.50 c	134.47 a	0.96 ab	0.91 ab
MGF	18.19 ab	0.56 c	59.63 cd	1.02 a	1.36 a
TMF	15.93 ab	1.06 b	58.56 cd	0.69 ab	0.68 abc
GF	14.95 ab	0.45 c	34.89 de	0.64 adc	0.62 abc
UGF	14.74 ab	0.57 c	28.65 de	0.44 c	0.47 c
UMF	14.55 ab	0.71 c	84.31 bc	0.75 abc	0.72 abc
TGF	13.03 b	1.08 b	19.58 e	0.51 bc	0.49 c
TF	11.38 b	1.80 a	13.26 e	0.33 c	0.38 c
UF	24.23 a	1.08 b	117.66 ab	0.93 ab	0.92 ab

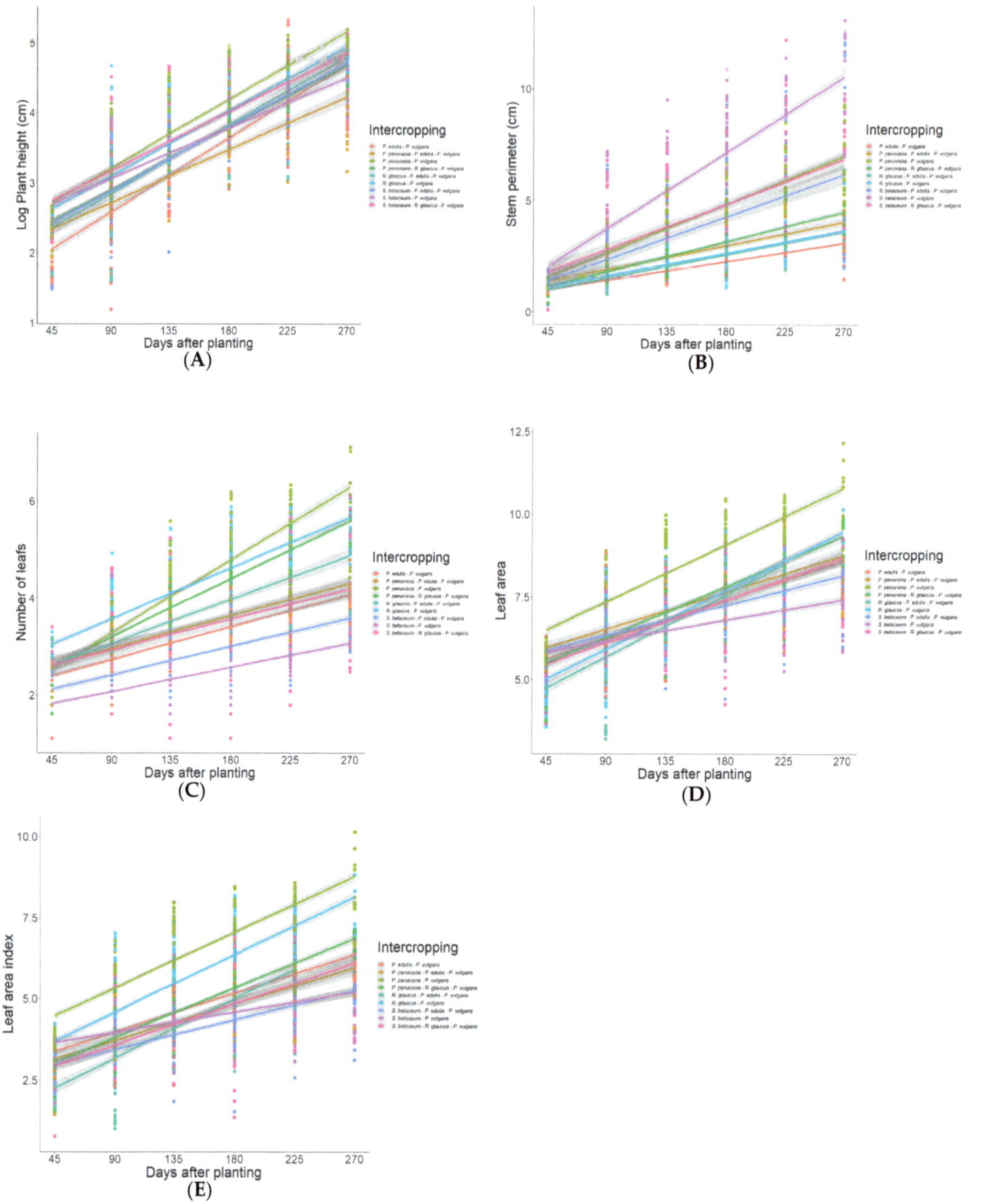

Figure 3. Integral analysis of growth variables with comparison of regression coefficients (β) under the split-strip design model. (**A**) Plant height, (**B**) stem perimeter, (**C**) number of leaves, (**D**) leaf area, and (**E**) leaf area index.

When analyzing the means of Table 3, we can highlight that the UF system (*P. peruviana* and *P. vulgaris*) outperforms the other fruit tree species with a minimum in growth.

According to the results of the chi-square test for the identification of the interaction between companion and fertilization, there was a highly significant interaction ($p = 0.01$).

The crops in companionship with TMF, UGF, GTF, MF, and UF presented greater severity in terms of diseases (Figure 3). Similarly, the presence of pests is significant in all the companions, with a greater amount of *Trialeurodes vaporarioum* in the UF companion.

3.2. Determination of the Incidence of Pests and Diseases

According to the results of the chi-square test for the identification of the interaction between companion and fertilization, there was a highly significant interaction ($p = -0.01$). The simple correlation analysis indicated that the crops in companionship with TMF, UGF, GTF, MF, and UF presented greater severity in terms of diseases (Figure 4). Similarly, the presence of pests was also significant in all the companions (Figure 5), with a greater amount of *Trialeurodes vaporarioum* in the UF companion. The beans in the companions were an important factor in the growth of this pest because of their ability to host it, according to observations and technical assistance carried out in the field.

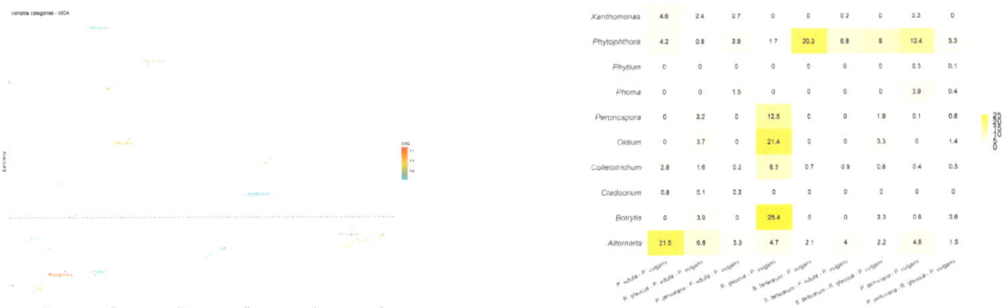

(A) (B)

Figure 4. Analysis of the presence of diseases in different companions—(**A**) heatmap and (**B**) simple correlation.

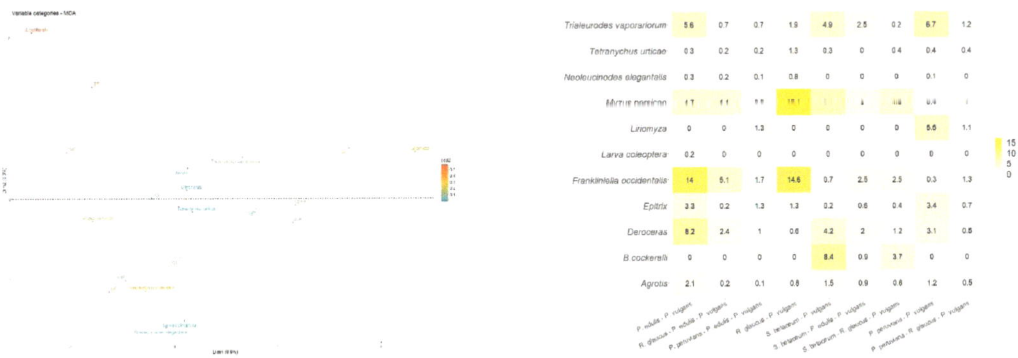

(A) (B)

Figure 5. Analysis of the presence of pests in different companions—(**A**) heatmap and (**B**) simple correlation.

3.3. Field Schools

Eight Farmer Field Schools (FFSs) were conducted during the period January to December 2022. These FFSs focused on topics such as protocol and biosecurity in food production and marketing; the development of methodologies to identify vulnerable areas; habitat description in terms of soil, topography, climate, and plants; climate risk management in terms of crop health; and the recognition of climatic and edaphic requirements of production systems. Box tests show a significant increase in participants' knowledge on each of these topics after completing the FFS, which demonstrates the success of the participatory "Learning by Doing" methodology used. Farmers showed a high level of interest and active participation in the FFSs, which contributed to the effective acquisition of new knowledge and skills.

4. Discussion

4.1. Functional Analysis of Growth and ANDEVA Analysis of Variance of the Regression Coefficients Obtained

The plant height variable is of great importance, since the greater the height, the lower is the risk of diseases affecting the fruit due to contact with the soil, and the greater the height, the greater are the number of fruits with better quality and the possibility of storing reserves for times of high requirements [23].

The TF-TGF-UF-TMF companions showed better development in terms of stem perimeter (Figure 3B). TF was the companion with the best performance for this variable. TF is among the top three companions, with no noticeable difference among the three types of fertilization. As for TMF, a slight improvement was noted in the interaction with organic fertilizers, with a better development of stem perimeter. On the contrary, the TGF companion is better favored by the F1 fertilization (100% chemical). The results can be attributed to a combination of specific plant interactions and varied responses to fertilization. In these companion cropping systems, the complementarity between species can lead to a more efficient utilization of resources, such as light, water, and nutrients, thus improving stem growth. In addition, the different species in the companionships can positively influence the soil structure and soil health, which, together with the use of different types of fertilizers, can optimize the availability and uptake of nutrients needed for robust stem development. These plant–plant and plant–soil interactions, as well as the adaptive response to fertilization, can result in improved stem growth in these specific companion systems [24].

With respect to leaf production in UF-MF-UMF (Figure 3C), in the interactions between plant age after planting, significant statistical differences were observed when comparing age after planting days. The development of the UF and MF companions could be influenced by the nutrient content of the dry matter input to the soil. Thus, leaf production is expected to increase at an early age, where synthetic fertilizers have been found to play a better role in this variable [25].

Regarding the leaf area and leaf area index, a consistent upward trend can be observed in all intercropping combinations, which indicates that FA and LAI tend to increase over time, particularly in those companions represented by the highest lines and steeper slopes; these are the treatments that show faster growth, indicating greater efficiency in light capture, better utilization of available resources, greater leaf density, and possibly a more efficient canopy for photosynthesis, as presented in the MGF system that over-emitted for both variables and presented no difference with UF and MF (Figure 3D,E).

The rate of leaf emission in the different plots of the municipality showed a constant growth. Leaf development is related to solar brightness and photosynthetically active radiation, since there is evidence that leaf formation is constant, with different degrees of shade generating a greater or lesser number [26].

Similarly, it can be affirmed that by obtaining significant differences in the blocks and these being distributed at different altitudes, we can say that this factor directly affected the leaf production of these crops in the aforementioned companions and also gives us an understanding that these crops share a similar potential at a certain altitude point with respect to this variable.

The leaf area is related to the photosynthetic rate, evapotranspiration, and vegetative development, as well as water and nutrient uptake. This is why we highlight the importance of this variable along with leaf number (Figure 3D).

Temperatures between 15 and 22 °C offer an exponential growth of the leaf; on the other hand, if the temperature reaches more than 29 °C, a longitudinal growth of very high branches originates, with a large number of nodes, but it in turn retracts the growth of leaves [27]. Therefore, according to the results obtained in the foliar part and the significant differences found between the blocks, we can say that the leaf area was affected by the temperature of the different altitudes in each block and the adaptations of each companion to these.

The continuous application of organic fertilizers is critical for sustainable crop management. While organic fertilizers improve soil structure, microbial biodiversity, and long-term soil health, they may also pose challenges to efficiency and crop yield when used continuously without complementary measures.

It has been reported that continuous use can lead to nutrient imbalances, negatively impacting crop yields [28]. Similarly, potential issues with the carbon/nitrogen ratio have been noted, affecting nutrient availability [29]. Conversely, other research shows that continuous application could improve water retention and crop yields under stress conditions [30]. However, the positive effects depend on the crop type, local climate, and fertilizer composition.

To mitigate potential yield declines, integrating organic with mineral fertilizers or complementary practices such as crop rotation is recommended [31,32]. Monitoring soil nutrient levels and adjusting practices accordingly is essential.

4.2. Determination of the Incidence of Pests and Diseases

The high severity of diseases observed in crops under companionship with the TMF, UGF, GTF, MF, and UF systems may primarily result from the widespread dissemination of fungi among these companion crops. This dissemination can be attributed to several factors, including poor management practices by producers, such as inadequate disinfection of tools, insufficient crop hygiene, and improper handling of infected plant material. These practices can facilitate the spread of fungal pathogens that thrive in environments with high moisture, shade, and poor air circulation [1]. Recent studies have demonstrated that dense cropping systems, where plants are closely spaced, can create a favorable microenvironment for fungal growth, leading to a higher disease incidence and severity [1].

Similarly, the significant presence of pests across all companion systems, particularly the increased levels of *Trialeurodes vaporariorum* (whitefly) in the UF companion, underscores the susceptibility of these systems to pest infestations. The whitefly is a well-known pest in Colombia, notable for its rapid reproduction rate, broad host range, and resistance to various chemical control methods. Its population growth is exacerbated by environmental conditions, such as prolonged dry seasons, which favor its life cycle and reduce the effectiveness of natural predators [33]. Additionally, the presence of beans in the companion systems contributes to the proliferation of whiteflies, as beans serve as a preferred host that supports their reproduction and acts as a reservoir for spreading infestations to other crops [34]. This finding highlights the importance of carefully selecting companion crops to manage pest populations effectively.

The growth and spread of pests in companion cropping systems are also significantly influenced by the proximity of neighboring plots or crop systems. When host plants are removed or controlled in one plot, adult pests frequently migrate to adjacent plots, increasing the risk of infestation and damage. This phenomenon, known as "pest spillover," has been widely documented in mixed or fragmented cropping systems, where pests easily move between different plant species [35]. In this study, pest spillover was evident in the GMF, GTF, UGF, and UF companion systems, where pests such as aphids, leaf-miner flies (Agromyzidae), and thrips were prevalent. The purple passion fruit, which was most affected by pests under conditions of free exposure, further supports the conclusion that the spatial arrangement of crops significantly impacts pest dynamics [36]. Moreover, the severe impact of slugs from the genus Deroceras on purple passion fruit around 45 days after planting (dap) delayed crop growth in the GF companion system. However, the reduced presence of this pest in the GMF and GTF companion systems suggests that certain companion crops may offer protective effects, possibly through physical barriers, chemical deterrents, or by providing habitats for natural predators [37].

The choice of companion crops is crucial in determining pest and disease outcomes in intercropping systems. While some companion plants may enhance pest control by attracting beneficial insects or repelling harmful ones, others may inadvertently serve as hosts or attractants for pests, as seen with beans and whiteflies. Agroecological principles, such as increasing plant diversity and promoting beneficial ecological interactions, are essential in designing effective companion cropping strategies [38]. Incorporating plants with pest-repellent properties or those that attract beneficial insects can help reduce pest pressure and minimize reliance on chemical controls [39]. Additionally, companion plants that improve soil health or provide structural support for the main crops contribute to the overall resilience of the cropping system, making it less susceptible to both pests and diseases [40].

Moreover, the findings emphasize the need for improved management practices, such as regular monitoring, enhanced sanitation protocols, and the use of integrated pest management (IPM) strategies, to reduce the incidence of pests and diseases in companion cropping systems. IPM approaches that combine biological, cultural, mechanical, and chemical controls provide a comprehensive pest management strategy that promotes environmental sustainability. For example, incorporating natural enemies such as parasitoids and predators, creating habitats for beneficial organisms, and using organic mulches can help suppress pest populations and improve crop health, Future research should focus on identifying the most effective combinations of companion crops, environmental conditions, and management practices to optimize pest and disease control in diverse agricultural settings.

Furthermore, understanding the specific interactions between companion plants, pests, and diseases, as well as the role of environmental factors such as temperature, humidity, and soil health, is vital for developing more targeted and effective management practices. For instance, research on the microclimatic effects of different companion crops could provide new insights into how these interactions influence pest and disease dynamics. Integrating this knowledge into decision-making frameworks can help farmers tailor their cropping systems to local conditions, reducing pest pressures and improving overall crop productivity and resilience [12].

4.3. Field Schools

The success of the Farmer Field Schools (FFSs) conducted between January and December 2022 reflects the effectiveness of the participatory "Learning by Doing" methodology in enhancing farmers' knowledge and skills in key areas such as biosecurity, climate risk management, and sustainable production practices. This approach aligns with recent studies emphasizing the importance of experiential learning in agricultural education. For example, [41] found that participatory learning methods, such as those used in FFSs, significantly improve farmers' capacity to adopt new technologies and practices by fostering a deeper understanding of local ecological conditions and adaptive management strategies. By directly engaging farmers in the learning process, FFS programs help build a sense of ownership and confidence in applying new knowledge, which is crucial for the long-term sustainability of agricultural innovations.

Moreover, the increase in participants' knowledge demonstrated by the Box tests suggests that FFS programs can effectively address gaps in technical knowledge related to food production, marketing, habitat management, and climate risk adaptation. Recent literature supports the idea that FFSs can play a critical role in building farmers' resilience to climate change. A study by Davis [42] highlights that farmers who participate in FFS programs are better equipped to understand and manage climate risks, such as erratic rainfall and temperature fluctuations, by applying context-specific knowledge and practices. The focus on topics such as soil, topography, and climate within the FFS curriculum is particularly relevant given the growing need for site-specific management practices that consider the unique climatic and edaphic conditions of different agricultural systems. For example, the authors of [43] suggest that climate adaptation in African agriculture must be based on an approach that considers local variations in soils and climates, supporting resilient agricultural practices.

The high level of interest and active participation observed among farmers in the FFS sessions further underscores the value of this approach. Research by Murphy [44] suggests that the participatory nature of FFSs not only enhances knowledge acquisition but also strengthens social capital within farming communities. This social dimension is crucial for fostering collaboration and collective action among farmers, which are essential components for successful adaptation to environmental changes and market dynamics. Additionally, the interactive format of the FFSs encourages peer-to-peer learning and the exchange of local knowledge, which can be more effective than conventional top-down extension methods in promoting the adoption of sustainable practices. This conclusion is supported by studies demonstrating that participatory extension approaches, such as farmer-to-farmer training programs, can significantly improve the dissemination of agricultural innovations in rural contexts [45].

The focus on biosecurity protocols and methodologies to identify vulnerable areas within FFS curricula is particularly timely in light of recent global challenges, such as the COVID-19 pandemic and increasing biosecurity threats from transboundary pests and diseases. According to [46], integrating biosecurity training into agricultural extension programs such as the FFSs can significantly improve farmers' capacity to detect and respond to pest and disease outbreaks, thereby safeguarding local and regional food security. This integration of biosecurity measures within the FFSs is consistent with broader efforts to enhance food system resilience through a more holistic approach that combines ecological, social, and economic dimensions. The authors of [47], for example, highlight the importance of integrated approaches to addressing food security in contexts of high vulnerability.

Finally, the FFS methodology's emphasis on recognizing climatic and edaphic requirements of production systems aligns with the broader shift toward agroecological approaches in agriculture. Recent studies have shown that such approaches, which integrate ecological principles into agricultural production, can enhance ecosystem services, improve crop yields, and promote sustainable land management practices [38]. By equipping farmers with the knowledge and tools to manage their production systems more effectively within their specific environmental contexts, FFS programs contribute to the development of more resilient agricultural landscapes and communities.

5. Conclusions

This study addressed key challenges in agricultural practices, particularly the management of pests, diseases, and crop productivity in companion cropping systems. The initial issues identified included high disease severity and pest incidence, which were linked to factors such as poor management practices, suboptimal plant spacing, and the selection of companion crops that inadvertently served as hosts for pests. Additionally, climate risks and the need for improved biosecurity and sustainable practices were highlighted as critical areas for intervention.

The implementation of Farmer Field Schools (FFSs) demonstrated significant benefits by improving farmers' knowledge and skills in various aspects of agricultural management, including biosecurity, climate risk adaptation, and sustainable production methods. The participatory "Learning by Doing" methodology effectively increased participants' understanding and application of new techniques, fostering greater resilience in agricultural practices.

The achievements of this study include identifying the most effective companion cropping combinations that enhance pest and disease management and promote crop growth. The findings emphasize the importance of selecting appropriate companion crops and implementing integrated pest management (IPM) strategies to optimize productivity. Furthermore, this study demonstrates that targeted educational initiatives, such as FFSs, can play a vital role in empowering farmers, building community resilience, and supporting sustainable agricultural development.

Author Contributions: Conceptualization, O.J.M.-C., J.S.C.-S., P.A.V.-V. and D.A.-D.; methodology, D.A.-D.; software, O.J.M.-C., J.S.C.-S., P.A.V.-V. and D.A.-D.; validation, D.A.-D. and J.S.C.-S.; formal analysis, D.A.-D. and P.A.V.-V.; investigation, O.J.M.-C., J.S.C.-S. and D.A.-D.; resources, D.A.-D.; data curation, O.J.M.-C.; writing—original draft preparation, O.J.M.-C. and D.A.-D.; writing—review and editing, J.S.C.-S. and D.A.-D.; visualization, O.J.M.-C., P.A.V.-V. and D.A.-D.; supervision, J.S.C.-S.; project administration, J.S.C.-S.; funding acquisition, J.S.C.-S. and D.A.-D. All authors have read and agreed to the published version of the manuscript.

Funding: The study was funded by the Sistema General de Regalias (SGR) of the Ministerio de Ciencia Tecnología e Innovación, Colombia (MINCIENCIAS). Research project: "Estudio de sistemas de cultivo asociados a los frutales andinos estrategia innovadora para la reactivación económica de los municipios de Sandoná, La Florida, Arboleda, Providencia y El Peñol" with BPIN number 2020000100677.

Data Availability Statement: Data are contained within the article. For any additional information, contact the author by correspondence.

Acknowledgments: The authors would like to extend their heartfelt gratitude to Johana M. Belalcazar, Jenifer B. Vargas, Laura M. Pantoja, Luisa F. Vallejo, Javier M. Chamorro, Carlos Charfuelan, and Tania M. Pantoja for their invaluable support and dedication in collecting and organizing the data for this study; their contributions are deeply appreciated. The authors are deeply grateful to Martha I. Cabrera Otalora for their invaluable support and guidance.

Conflicts of Interest: The authors declare no conflicts of interest.

References

1. Moreno, M.A.M. *Efectos de la Labranza Convencional y Labranza de Conservación en la Producción Agrícola: Revisión de Literatura*; Escuela Agrícola Panamericana, Zamorano: Tegucigalpa, Honduras, 2021; 43p, Available online: https://bdigital.zamorano.edu/server/api/core/bitstreams/a8681a48-2637-44e2-9867-5b12a8ee8a51/content (accessed on 12 May 2024).
2. Benavides, M.A. *Alternativas Para la Comercialización Internacional de Fertilizantes Agrícolas de la Empresa Agrochuldey el Fomento a Las Exportaciones*; UPEC: Paris, France, 2022.
3. Higuera, J.M.I.; Villa, C.U. Estudio de Factibilidad Para el Proyecto de Fabricación de Módulos de Acuaponía y Prestación de Servicios Afines Para el Área Rural de Uribia. Manaure y Maicao en La Guajira de la Fundación Aquaponics Relief. 2020. Available online: https://bibliotecadigital.udea.edu.co/handle/10495/16651 (accessed on 12 May 2024).
4. FAO. *El Mercado Mundial de Fertilizantes: Balance de la Situación de un Mercado en Dificultades*; Organanización de La Naciones Unidad Para La Alimentación y La Agricultura: Rome, Italy, 2022; Available online: https://investing.com (accessed on 12 May 2024).
5. Rivas, A.C.; Melendez, J.P.; Mayorquín, N. Alternativas microbiológicas para la remediación de suelos y aguas contaminados con fertilizantes nitrogenados Microbiological alternatives for the remediation of soils and water contaminated with nitrogen fertilizers. *Sci. Tech. Año XXV* **2020**, *25*, 172–183.
6. Ministerio de Agricultura. *Plan Departamental de Extension Agropecuaria del Departamento de Nariño PDEA—Nariño*; Gobernacion de Nariño: Nariño, Colombia, 2019; 222p. Available online: https://www.minagricultura.gov.co/ministerio/direcciones/Documents/PDEA'sAprobados/PDEA%20Nari%C3%B1o.pdf%0A (accessed on 12 May 2024).
7. Alcaldía Municipal de Ipiales. *Plan de Desarrollo 2020–2023 Hablemos con hechos. Turismo Activo y Progresivo*; Fundación Empresarios por la Educación: Bogota, Columbia, 2024; p. 141. Available online: https://www.obsgestioneducativa.com/download/plan-de-desarrollo-municipal-ipiales-2020-2023/ (accessed on 12 May 2024).
8. Instituto Colombiano Agropecuario—ICA. *En Ipiales, Productores de Uchuva se Blindan Contra las Moscas de la Fruta*; Ministerio de Agricultura y Desarrollo Rural: Bogotá, Columbia, 2019. Available online: https://www.ica.gov.co/noticias/ica-productores-uchuva-blindan-moscas-fruta-narino (accessed on 12 May 2024).
9. Martínez-Camacho, Y.D.; Negrete-Yankelevich, S.; Maldonado-Mendoza, I.E.; la Mora, A.N.-D.; Amescua-Villela, G. Agroecological Management with Intra- and Interspecific Diversification as an Alternative to Conventional Soil Nutrient Management in Family Maize Farming. *Agroecol. Sustain. Food Syst.* **2022**, *46*, 364–391. [CrossRef]
10. Morales-Belpaire, I.; Alfaro-Flores, A.; Ramos, K.L.; Velarde, O.P.; Ordonez, P.A.; Armengot, L. Soil Quality Indicators under Five Different Cacao Production Systems and Fallow in Alto Beni, Bolivia. *Agrofor. Syst.* **2024**, *2024*, 1–16. [CrossRef]
11. Tinoco-Jaramillo, L.; Vargas-Tierras, Y.; Habibi, N. Agroforestry Systems of Cocoa (*Theobroma cacao* L.) in the Ecuadorian Amazon. *Forests* **2024**, *15*, 195. [CrossRef]

12. Prabhu, R.; Barrios, E.; Bayala, J.; Diby, L.; Donovan, J., Gyau, A.; Graudal, L.; Jamnadass, R.; Kahia, J.; Kehlenbeck, K.; et al. Agroforestry: Realising the Promise of an Agroecological Approach. In *Agroecology for Food Security and Nutrition: Proceedings of the FAO International Symposium, Rome, Italy, 18–19 September 2014*; Food and Agriculture Organization of the United Nations: Rome, Italy, 2015; pp. 201–224.
13. Aguirre Cuellar, B.; Lobo, J.C. Shading in Family Coffee Farms as an Environmental Incentive Promoter for Ecosystem Services in Tolima, Colombia. *Int. J. Agric. Sustain.* **2022**, *20*, 1015–1027. [CrossRef]
14. Alvez, N.V.; Alayon Luaces, P. Evaluación de policultivos frutihortícolas agroecológicos del Nordeste Argentino. *Bol. Soc. Argent. Bot.* **2020**, *55*, 273–284. [CrossRef]
15. Gómez Herrera, D.A.; Muñoz Monsalve, C. Análisis del Comportamiento Morfológico de las Variables Altura y Diámetro de Tallo, en los Sistemas de Siembra Mono Cultivo y Asocio, de las Especies Aguacate (Persea americana) Variedad Hass y Tomate de Árbol (Solanum betaceum), Durante 12 Meses, en el. 2020. Available online: https://repository.unad.edu.co/bitstream/handle/10596/39011/dagomezhe.pdf?sequence=1&isAllowed=y (accessed on 12 May 2024).
16. Rosero Alpala, M.G.; Tapie, W.A.; Rosero Alpala, D.A. Diversidad fenotípica de papas nativas en las comunidades indígenas de la etnia de los Pastos (Nariño, Colombia): Agricultura ecológica para la seguridad alimentaria y el desarrollo rural. *Rev. Peru. Biol.* **2020**, *27*, 509–516. [CrossRef]
17. Tumbani, Z. *Evaluacion de la Asociacion con dos Especies Fabaceas (*Crotalaria júncea *y* Lupinus mutabilis*) con el Cultivo de Cilantro (*Coriandrum sativum *L.) en Invernadero de la Comunicad de la Comunidad Sullkataca baja del Municipio de Laja*; Universidad Mayor de San Andrés: San Andrés, Bolivia, 2018; pp. 50–107. Available online: http://dspace.ups.edu.ec/handle/123456789/17965 (accessed on 12 May 2024).
18. Velasquez-Vasconez, P.A.; Andrade Díaz, D. LeafArea Package: A Tool for Estimating Leaf Area in Andean Fruit Species. *Int. J. Plant Biol.* **2024**, *15*, 102–109. [CrossRef]
19. Gardner, E.S. Exponential smoothing: The state of the art. *J. Forecast.* **1985**, *4*, 1–28. [CrossRef]
20. Popat, R.; Banakara, K. Doebioresearch: Analysis of Design of Experiments for Biological Research. 2020. Available online: https://CRAN.R-project.org/package=doebioresearch (accessed on 28 May 2022).
21. Wickham, H. *ggplot2: Elegant Graphics for Data Analysis*; Springer: New York, NY, USA, 2016.
22. Sébastien, L.; Julie, J.; François, H. FactoMineR: An R Package for Multivariate Analysis. *J. Stat. Softw. Found. Open Access Stat.* **2008**, *25*, 1–18.
23. López Cerda, I.G. *Análisis de Cuatro Cultivos Precolombinos en Asocio, Establecidos en dos Métodos de Labranza: Camellones Prehispánicos y Labranza Convencional. Diriamba, Carazo 2018*; Universidad Nacional Agraria: Managua, Nicaragua, 2018; Available online: https://repositorio.una.edu.ni/4307/ (accessed on 12 May 2024).
24. Paola De Sá, F.; Cerioni Belniaki, A.; Panobianco, M.; Gabira, M.M.; Kratz, D.; Lima, E.A.D.; Wendling, I.; Luiz Esteves Magalhães, W. Peach palm residue compost as substrate for *Bactris gasipaes* self-sustaining seedlings production. *Int. J. Recycl. Org. Waste Agric.* **2024**, *9*, 183–192. [CrossRef]
25. Rebolledo-Martínez, L.; Megchún-García, J.V.; Rebolledo-Martínez, A.; Orozco-Corona, D.M. Association of Persian Lime (*Citrus latifolia*) and Coconut Palm (*Cocos nucifera* L.) Trees with the Contribution of Dry Matter by Annual Crops. *Ibero-Am. J. Bioecon. Clim. Change* **2019**, *5*, 1248–1266. [CrossRef]
26. Larramendi, L.A.R.; Hernández, F.G.; Castro, H.G.; Flores, M.F.; Castañeda, J.C.G.; Ruiz, R.P. Leaf Anatomy Related to the Photosynthetic Pathway in Coffee Trees (*Coffea arabica* L.; var. *Caturra rojo*) Exposed to Different Levels of Solar Radiation in the Sierra Maestra, Granma, Cuba. *Acta Agron.* **2019**, *65*, 248–254. [CrossRef]
27. Fischer, G.; Melgarejo, L.M. The ecophysiology of cape gooseberry (*Physalis peruviana* L.)—An Andean fruit crop. A review. *Rev. Colomb. Cienc. Hortíc.* **2020**, *14*, 76–89. [CrossRef]
28. Liu, Y.; Ge, T.; Wang, J. Long-term effects of organic and inorganic fertilizers on soil organic matter and microbial community structure in a tropical agricultural ecosystem. *J. Clean. Prod.* **2020**, *246*, 118993.
29. Goss, M.J.; Tubeileh, A.; Goorahoo, D. A Review of the Use of Organic Amendments and the Risk to Human Health. *Agric. Ecosyst. Environ.* **2019**, *272*, 224–239.
30. Singh, B.P.; Setia, R.; Marschner, P. Organic amendments enhance soil functioning and crop performance in various agroecosystems: A review. *Soil Use Manag.* **2018**, *34*, 592–602.
31. Chivenge, P.; Vanlauwe, B.; Six, J. Does the combined application of organic and mineral nutrient sources influence maize productivity? A meta-analysis. *Agron. Sustain. Dev.* **2011**, *31*, 143–156. [CrossRef]
32. Möller, K. Assessment of alternative farm practices in organic farming using life cycle assessment methodology: A review. *Agron. Sustain. Dev.* **2018**, *38*, 23.
33. Czarnecka, D.; Czubacka, A.; Agacka-Mołdoch, M. The occurrence of fungal diseases in maize in organic farming versus an integrated management system. *Agronomy* **2022**, *12*, 558. [CrossRef]
34. Amata, R.L.; Otipa, M.J.; Waiganjo, M.; Wasilwa, L.A. Management strategies for fungal diseases in passion fruit production systems in Kenya. *Acta Hortic.* **2009**, *911*, 207–213. [CrossRef]
35. Fones, H.N.; Fisher, M.C.; Gurr, S.J. Emerging Fungal Threats to Plants and Animals Challenge Agriculture and Ecosystem Resilience. In *The Fungal Kingdom*; Wiley: Hoboken, NJ, USA, 2017. [CrossRef]
36. Jabbour, R.; Barbercheck, M.E. Soil management effects on entomopathogenic fungi during the transition to organic agriculture in a feed grain. *Biol. Control.* **2009**, *51*, 435–443. [CrossRef]

37. Fisher, M.C.; Gurr, S.J.; Cuomo, C.A.; Blehert, D.S.; Jin, H. Threats posed by the fungal kingdom to humans, wildlife, and agriculture. *MBio* **2020**, *11*, 10–1128. [CrossRef]
38. Altieri, M.A.; Nicholls, C.I. Agroecology: Principles for designing sustainable farming systems. *Agron. J.* **2020**, *112*, 5031–5043. [CrossRef]
39. Gurr, G.M.; Wratten, S.D.; Luna, J.M.; Snyder, W.E. Integrated pest management for sustainable agriculture. *Agric. Ecosyst. Environ.* **2019**, *274*, 67–73.
40. Wezel, A.; Bellon, S.; Doré, T.; Francis, C.; Vallod, D.; David, C. Agroecology as a science, a movement, and a practice: A review. *Agron. Sustain. Dev.* **2021**, *41*, 38. [CrossRef]
41. Friis-Hansen, E.; Duveskog, D.; Taylor, E.W. Farmer field schools: A transformative learning approach in rural Africa. *J. Agric. Educ. Ext.* **2020**, *26*, 29–44.
42. Davis, K.; Lion, K.A.; Ramasamy, S. Agricultural extension and rural advisory services: Understanding the capacities needed to adapt to climate change. *Glob. Food Secur.* **2021**, *28*, 100498.
43. Mbow, C.; Van Noordwijk, M.; Luedeling, E.; Neufeldt, H.; Minang, P.A.; Kowero, G. Agroforestry solutions to address food security and climate change challenges in Africa. *Curr. Opin. Environ. Sustain.* **2019**, *6*, 61–67. [CrossRef]
44. Murphy, J.; Lyon, S.; Dalzell, B. Building community resilience through farmer field schools: A review of evidence from African case studies. *J. Rural. Stud.* **2019**, *72*, 110–121.
45. Franzel, S.; Sinja, J.; Simpson, B. Farmer-to-farmer extension in Kenya: The perspectives of volunteer farmer trainers on their motivations, challenges, and experiences. *Dev. Pract.* **2014**, *24*, 403–414.
46. Lee, J.; Settle, Q. Integrating biosecurity into agricultural extension programs: Lessons from COVID-19. *J. Agric. Educ.* **2021**, *62*, 95–109.
47. Tomich, T.P.; Brodt, S.; Dahlgren, R.; Scow, K.; Stephens, S. Agroecology for food security and resilience in the face of climate change. *Glob. Food Secur.* **2019**, *4*, 11–18.

horticulturae

MDPI

Article

Effect of Application of Nitrogen Fertilizer, Microbial and Humic Substance-Based Biostimulants on Soil Microbiological Properties During Strawberry (*Fragaria* × *ananassa* Duch.) Cultivation

Jana Maková [1], Renata Artimová [2], Soňa Javoreková [1], Samuel Adamec [3,4], Oleg Paulen [4], Alena Andrejiová [4], Ladislav Ducsay [5] and Juraj Medo [1,*]

1 Institute of Biotechnology, Slovak University of Agriculture in Nitra, Tr. A. Hlinku 2, 94976 Nitra, Slovakia; jana.makova@uniag.sk (J.M.); sona.javorekova@uniag.sk (S.J.)
2 Institute of Forest Ecology, Slovak Academy of Sciences, Ľ. Štúra 2, 96001 Zvolen, Slovakia; artimova@ife.sk
3 Organix, s.r.o., Rastislavova 1067/323, 95141 Lužianky, Slovakia; samo.joker@gmail.com
4 Institute of Horticulture, Slovak University of Agriculture in Nitra, Tr. A. Hlinku 2, 94976 Nitra, Slovakia; oleg.paulen@uniag.sk (O.P.); alena.andrejiova@uniag.sk (A.A.)
5 Institute of Agrochemistry and Soil Science, Slovak University of Agriculture in Nitra, Tr. A. Hlinku 2, 94976 Nitra, Slovakia; ladislav.ducsay@uniag.sk
* Correspondence: juraj.medo@uniag.sk

Abstract: Plant biostimulants have been the subject of intense interest in recent years. The aim of this study was to assess, during the years 2021–2022, the effect of mineral nitrogen (N) fertilizer, experimental (PGPB) and commercial (G) microbial biostimulants, and humic substance product (A) on the soil microbial communities, and yield of strawberries, under field conditions. Dehydrogenase activity was significantly affected by nitrogen fertilization, but an increase occurred in the treatment N+G. The treatments N+G, N+G+A, and N+PGPB+A increased FDA hydrolysis, and phosphatase activity. All plant biostimulants increased basal as well as substrate-induced respiration. Culturable bacteria (total counts, dormant forms, actinomycetes) were not clearly affected by treatment. Based on 16S rRNA analysis, bacterial community composition was different in N+PGPB+A and N+G+A treatments. The number of cultivable fungi was significantly lower in N+PGPB and N+PGPB+A treatments. The genus of fungi *Pilidium*, a potential phytopathogen of strawberries, was present in the second year, but in these treatments, it was absent. In the second year, strawberry yield was shown to be 95% higher in the N+PGPB+A treatment than in the control. Microbial biostimulants in combination with humic substances represent a potential solution in increasing strawberry production.

Keywords: soil microbiome; plant biostimulants; mineral fertilizer; soil enzymatic activity; *Pilidium*

Academic Editors: Francesco De Mastro, Gennaro Brunetti, Karam Farrag and Huadong Zang

Received: 20 December 2024
Revised: 20 January 2025
Accepted: 20 January 2025
Published: 22 January 2025

Citation: Maková, J.; Artimová, R.; Javoreková, S.; Adamec, S.; Paulen, O.; Andrejiová, A.; Ducsay, L.; Medo, J. Effect of Application of Nitrogen Fertilizer, Microbial and Humic Substance-Based Biostimulants on Soil Microbiological Properties During Strawberry (*Fragaria* × *ananassa* Duch.) Cultivation. *Horticulturae* **2025**, *11*, 119. https://doi.org/10.3390/horticulturae11020119

1. Introduction

The strawberry (*Fragaria* × *ananassa* Duch.) is one of the most popular horticulture crops belonging to the *Rosaceae* family, subfamily *Rosoideae* [1]. Strawberries are a rich source of a wide variety of nutritive compounds such as sugars, vitamins (especially vitamin C), and minerals (K, P, Ca, Fe), as well as non-nutritive, bioactive compounds such as flavonoids, anthocyanins, and phenolic acids, which contribute to their nutritional and health benefits [2,3]. The overall demand for strawberries has been steadily increasing, which is associated with a 20% increase in the area under cultivation worldwide over the last decade [4]. According to the Food and Agriculture Organization of the United Nations, the largest strawberry producers in the world in 2022 include China (148,015 ha), Poland (31,300 ha), Turkey (22,272 ha), USA (22,784 ha), and Egypt (19,236 ha) [5]. In Slovakia, strawberries are grown on about 240 ha producing 1240 tons per year [5]. The increased consumption of strawberries leads to the need to develop and implement interventions that will result in higher plant productivity while maintaining plant health in a sustainable manner.

Nutrient management is a critical factor influencing strawberry yield and quality. To improve soil quality and optimize plant growth, various organic fertilizers such as manures, organic mulches, composts, vermicomposts, and plant biostimulants are applied in horticulture [6]. These organic inputs have been shown to positively affect plant growth, root development, flowering, fruit set, yield, and quality of strawberry [3]. The use of organic fertilizers reduces the chemical inputs to the environment while enhancing soil and environmental health status [7].

Plant biostimulant is generally referred to as any substance or microorganism that, when applied to a plant or soil, stimulates biological and chemical processes in the plant and/or associated microorganisms. European Union regulation 2019/1009 [8] defines biostimulants as products that, independent of the nutrient content of the product, are able to improve one or more of the following properties of the plant or plant rhizosphere: (i) nutrient-use efficiency, (ii) tolerance to abiotic stress, (iii) quality characteristics, and (iv) availability of limited nutrients in the soil and rhizosphere [9]. Biostimulants are classified into several groups, including humic substances (e.g., humic and fulvic acids), protein hydrolysates, seaweed and plant extracts, chitosan and other biopolymers, beneficial elements (e.g., silicon, selenium), beneficial fungi (e.g., arbuscular mycorrhizal fungi, *Trichoderma*), and beneficial bacteria (e.g., *Azotobacter, Azospirillum, Rhizobium*) [10]. Humic substances are derived from the microbial decomposition and chemical degradation of organic matter in the soil [11]. These carbon-rich amendments, commonly derived from manure, compost, vermicompost, lignite, coal, and peat, are widely used in agriculture [12]. The main stimulatory effects associated with the promotion of strawberry growth by humic substances are increased nutrient utilization [13,14] and reduced incidence of plant diseases [15]. Beneficial microorganisms, such as plant growth-promoting rhizobacteria, enhance plant growth through mechanisms such as biological nitrogen fixation, phytohormone synthesis, and nutrient solubilization (e.g., phosphorus and potassium) [16]. Initially used in organic farming, plant biostimulants are now widely used in conventional and integrated farming systems, as well as in field and greenhouse cultivation. Their use covers a wide range of crops, including vegetables, fruit trees, berries, vines, ornamentals, cereals, and turf [10].

Recent studies have reported positive effects of biostimulants on plant growth, yield crop quality parameters, and stress tolerance [17–19]. However, limited information is available on their effects on soil quality, especially on soil microbiological parameters from the rhizosphere of strawberries. Soil microbiological properties, such as microbial biomass, respiration, enzymatic activity, and microbial community structure, are important indicators of soil fertility [20].

The aim of this study was to evaluate the effects of mineral nitrogen fertilizers and plant biostimulants based on plant growth-promoting microorganisms and humic substances on the abundance, activity, and structure of microbial communities as well as on the yield of strawberries grown in humic soils.

2. Materials and Methods

2.1. Experiment Design and Treatments

The field trials were carried out on the experimental site of the Slovak University of Agriculture (SUA) in Nitra (48°18′53″ N, 18°5′15″ E) during two growing seasons from 2021 to 2022. Soil type on the site is fluvisol with a high clay content (particles < 0.01 mm 65%) and a pH value of 6.83. According to an agrochemical analysis of the soil performed before the trial was established, the soil properties were the following: soil organic carbon 1.84%, humus 3.17%, total nitrogen 0.25%, inorganic nitrogen 18.1 mg/kg, total phosphorus 85 mg/kg.

The trial was set up in a randomized block design with three replications (plots) per treatment. The area of each plot was 3.24 m^2 with 9 strawberry plants in a 0.6 × 0.6 m pattern. Experimental plots were arranged in a 7 × 3 layout where a single treatment appeared only once in either columns or rows. Frigo seedlings of strawberries (variety JOLY) were transplanted to soil in March 2021 and grown for two seasons.

Seven experimental treatments were analyzed. A control treatment (C) without any fertilization was prepared to evaluate the fertilization effect. Other treatments included the application of mineral nitrogen fertilizer solely (N) or in combination with commercial microbial biostimulant Groundfix (N+G), experimental microbial biostimulant (N+PGPB), humic substance-based biostimulant Agriful (N+A), and microbial and humic biostimulants together (N+G+A, N+PGPB+A). Timeplan of treatments and soil sampling is summarized in Table 1.

Table 1. Timetable of interventions during strawberry cultivation in each treatment.

Activity/Application	Date	C	N	N+G	N+PGPB	N+A	N+G+A	N+PGPB+A
40 kg N	19 March 2021		•	•	•	•	•	•
Transplantation	26 March 2021	•	•	•	•	•	•	•
Agriful	26 March 2021					•	•	•
PGPB	26 March 2021				•			•
Grounfix	26 March 2021			•			•	
Soil sampling	12 April 2021	•	•	•	•	•	•	•
Agriful	12 April 2021					•	•	•
40 kg N	30 April 2021		•	•	•	•	•	•
Agriful	5 May 2021					•	•	•
PGPB	5 May 2021				•			•
Grounfix	5 May 2021			•			•	
Agriful	28 May 2021					•	•	•
Harvesting	4–18 June 2021	•	•	•	•	•	•	•
Soil sampling	8 July 2021	•	•	•	•	•	•	•
40 kg N	16 March 2022		•	•	•	•	•	•
Agriful	23 March 2022					•	•	•
PGPB	23 March 2022				•			•
Grounfix	23 March 2022			•			•	
Soil sampling	8 April 2022	•	•	•	•	•	•	•
Agriful	6 April 2022					•	•	•
40 kg N	18 April 2022		•	•	•	•	•	•
Agriful	21 April 2022					•	•	•
PGPB	21 April 2022				•			•
Grounfix	21 April 2022			•			•	
Agriful	4 May 2022					•	•	•
Harvesting	23 May–17 June 2022	•	•	•	•	•	•	•
Soil sampling	2 July 2022	•	•	•	•	•	•	•

• indicates that activity occurred on a selected date for a particular treatment.

Basic nitrogen fertilization was applied at a dose of 80 kg N per hectare per year in the form of DASA fertilizer divided into two applications. DASA is a granular fertilizer containing 26% N in the form of an ammonium sulfate and an ammonium nitrate. Groundfix is a commercial microbial biostimulant containing *Bacillus subtilis*, *Bacillus megatherium* var. *phosphaticum*, *Azotobacter chroococcum*, *Enterobacter* sp. and *Paenibacillus polymyxa*. The total number of living microorganisms is 1.0 × 10^9 CFU/mL. This biostimulant also contains other beneficial microorganisms (Lactobacilli, enzyme-producing bacteria), vitamins, phytohormones, amino acids, and other physiologically active substances. The experimental microbial biostimulant PGPB was designed at the Institute of Biotechnology, SUA, in Nitra. It contains five bacterial strains with previously tested PGP properties [17,21] in a concentration of 1×10^9 cells or spores/mL. The used microbial strains were identified as *Bacillus pumilus*, *Bacillus arybhattai*, *Streptomyces puniceus*, *Streptomyces olivochromogenes*, and *Azotobacter* sp. Agriful is a humic substance product containing humic acids 25% and fulvic acids 25%. The nitrogen content is 4.5%, phosphorus content (P_2O_5) 1.0%, potassium content 1.0% (K_2O) 1.0%, and organic matter 45.0%. Each of the biostimulants was suspended solely or in combination in a concentration of 1% in irrigation water. Then, 1 L of suspension per plant was applied directly to each strawberry plant crown as irrigation. On each application date, plants were irrigated with either clean irrigation water or water containing biostimulants or their mixture to ensure the same condition.

Soil cultivation and weed control were carried out by manual hoeing. During the vegetation, additional sprinkler irrigation was applied as required, based on climatic conditions and soil moisture status. Supplementary Table S1 shows the total rainfall and average air temperature on the experimental site in the years 2021 and 2022.

2.2. Soil Sampling and Analyses

Soil samples for analysis of microbiological parameters were collected on four dates April 2021, July 2021, April 2022, and July 2022 (Table 1). Soil was collected with a sterile core sampler from a depth of 0.00–0.15 m in the area of the root system of the strawberry plants. Five sub-samples were collected from a single plot and then mixed together. Soil from each experimental plot was analyzed separately; this means that all analyses were performed in three biological replications. The samples were stored at 4 °C until analysis. The actual soil moisture and maximum water capacity were determined by the method of Cassel and Nielson [22].

The plate dilution method was used for evaluation of microbial counts using various media. Plate count agar was used for total bacterial count and dormant forms (inoculum was heated at 80 °C for 10 min). Pochon's medium was used for cultivation of actinomycetes and malt extract agar for microscopic filamentous fungi. Plates were cultivated at 30 °C for 72 h in case of total counts and dormant forms. Actinomycetes and microscopic fungi were cultivated at 25 °C for 7 days.

Three enzymatic activities (dehydrogenase activity, phosphatase activity, and hydrolysis of fluorescein diacetate) were measured in soil samples. Dehydrogenase activity (DHA) was determined using the assay of Casida Jr. et al. [23] as the reduction of triphenyltetrazolium chloride (TTC) to triphenylphormazan (TPF). The concentration of TPF was calculated according to the calibration curve and expressed as µg TPF per 1 g dry soil matter. Fluorescein diacetate hydrolysis (FDA) was measured as described by [24]. The concentration of released fluorescein was calculated according to the standard curve and expressed as µg FDA per 1 g dry soil matter per 3 h. Phosphatase activity (PA) was determined according to the Tabatabai and Bremner [25] method.

Microbial biomass carbon (Cmic) was determined following the fumigation extraction (FE) method [26]. Microbial respiration (basal and potential) was measured according to Alef and Nannipieri [27] as CO_2 sorption in a closed system on the 1 L jar for 24 h. Potential respiration was measured in the presence of glucose (2 g/kg soil). Results of basal and potential respiration were expressed as µg CO_2-C per g dry soil per hour. The metabolic quotient (qCO_2) was calculated as the amount of CO_2 mineralized per unit of microbial biomass carbon per hour (µg CO_2-C/mg Cmic/h) [28].

DNA was extracted from 0.25 g of soil using DNeasy® PowerSoil® kit (Qiagen, Hilden, Germany). The prokaryotic microbial community was analyzed using the V4 region of the 16S rRNA gene while ITS2 was used for assessing fungal diversity. Primer pairs 515F and 806R [29] and gITS7 and ITS4 [30] were used for prokariotic and fungi, respectively. PCR amplification was conducted using Q5 polymerase (New England Biolabs, Ipswich,

MA, USA) according to manufacturer instructions. PCR products were controlled by gel electrophoresis, purified using AMPure XP reagent (Beckman Coulter, Brea, CA, USA), quantified by qubit HS DNA assay (Invitrogen, Waltham, MA, USA), and pooled together in an equimolar ratio. TruSeq LT PCR free kit (Illumina, San Diego, CA, USA) was used for sequencing library preparation. Sequencing was performed using MiSeq Reagent Kit v3 (600-cycle)on Illumina MiSeq sequencer (Illumina, San Diego, CA, USA).

2.3. *Strawberry Yield*

Strawberry fruits were collected five times during the harvesting season in 2021, and nine times in 2022. The total yield and average weight of a single fruit were calculated for each experimental plot.

2.4. *Bioinformatics and Statistical Analysis*

SEED2 environment ver. 2.1.4 [31] was used for demultiplexing of acquired sequences. Then, QIIME 2 pipeline version 2023.9 [32] using DADA2 algorithm [33] was used for denoising and ASV calling. Prokaryotic ASVs were identified using RDP classifier ver. 2.13 [34]. Fungal ITS2 region was extracted from ASVs by ITSextractor ver. 1.1.3 [35] and identified using the GenBank ITS fungal reference database. Then, diversity indices were calculated after rarefaction to the lowest sequence counts. ASVs were aligned using MAFFT ver. 7.487 [36] and weighted Unifrac ver. 1.8 [37] distances between samples were calculated. Beta diversity analysis including permutational analysis of variance PERMANOVA and non-metric multidimensional scaling (NMDS) was performed using Vegan package ver. 2.6-4 [38] in R environment, version 4.2.2 [39]. Before PERMANOVA, homogeneity was tested using betadisper in Vegan package. Differential abundance was assessed by LefSe ver. 1.16.0 [40]. For redundancy analysis (RDA), concatenated total sum scaled ASV tables of both prokaryotic and fungal communities were paired with all other measured variables, and significant variables were selected using backward selection.

Analysis of variance (ANOVA) followed by the Tukey test (significance level 0.05) was used to assess the effect of treatment and sampling on soil microorganism counts, enzymatic activities, respiration activities, and diversity indices strawberry yields. All statistical analyses were conducted in an R environment. Data used for ANOVA were checked for equivalence of variance using the Lavene test. ANOVA residuals were checked for normality using Shapiro–Wilk test. Prior ANOVA microbial counts were logarithmically transformed to ensure normal distribution.

3. Results

3.1. *Soil Enzymatic Activity*

Analysis of all three evaluated enzymatic activities (Table 2) indicated that the treatment factors had a significant impact on the measured values. ($p < 0.05$). The addition of N-mineral fertilizer had a significant negative impact on dehydrogenase activity in all treatments in both experimental years with a 36 to 52% reduction in N treatments compared to the control ($p < 0.05$). ..

The addition of nitrogen-mineral fertilizer was found to have a significant ($p < 0.05$) negative impact on dehydrogenase activity with a reduction from 4.78 to 2.66 µg TPF/g soil/h in the N treatment in comparison to the control. The detrimental impact of nitrogen was similar when plant biostimulants, namely, PGPB and Agriful, were applied ($p < 0.05$). In April 2021, the effect was observed to be mitigated in the N+G (3.51) and N+G+A treatments (2.96). The positive impact of these treatments was observed to be consistent across all sampling periods. Moreover, the DHA levels were found to be even higher in the N+G treatment (8.35) than in the control (5.83) while the N+G+A treatment (5.96) was equivalent to the control in July 2021 ($p < 0.05$).

Table 2. Effect of nitrogen fertilization and plant biostimulants application during strawberry cultivation on soil enzymatic activities.

Treatment	April 2021	July 2021	April 2022	July 2022	Average
		Dehydrogenase activity in µg TPF/g soil/h			
C	3.82 ± 0.07 [1] a C [2]	5.83 ± 0.22 c D	4.34 ± 0.41 a C	5.11 ± 0.07 b E	4.78 ± 0.82 C
N	1.82 ± 0.17 a A	3.75 ± 0.08 c B	2.08 ± 0.08 a A	2.98 ± 0.12 b AB	2.66 ± 0.80 A
N+G	3.51 ± 0.05 a C	8.35 ± 0.31 c E	3.30 ± 0.17 a B	4.07 ± 0.21 b D	4.81 ± 2.16 C
N+PGPB	2.00 ± 0.13 a A	4.96 ± 0.32 c C	2.58 ± 0.60 a AB	3.52 ± 0.09 b BC	3.27 ± 1.21 AB
N+A	1.76 ± 0.14 a A	3.51 ± 0.16 d B	2.43 ± 0.04 b A	2.84 ± 0.18 c A	2.64 ± 0.68 A
N+G+A	2.96 ± 0.23 a B	5.96 ± 0.15 c D	3.34 ± 0.18 ab B	3.86 ± 0.34 b CD	4.03 ± 1.23 BC
N+PGPB+A	2.14 ± 0.09 a A	2.72 ± 0.11 b A	2.37 ± 0.01 a A	3.47 ± 0.21 c BC	2.68 ± 0.54 A
Average	2.57 ± 0.82 a	5.01 ± 1.82 c	2.92 ± 0.78 a	3.69 ± 0.74 b	
		FDA hydrolysis in mg FDA/g soil/3 h			
C	1.56 ± 0.16 a AB	1.52 ± 0.16 a AB	1.78 ± 0.07 a A	1.48 ± 0.10 a AB	1.59 ± 0.16 AB
N	1.32 ± 0.08 a A	1.25 ± 0.08 a A	1.69 ± 0.22 b A	1.29 ± 0.04 a A	1.39 ± 0.21 A
N+G	2.13 ± 0.09 c D	1.91 ± 0.08 ab B	2.03 ± 0.11 c AB	1.73 ± 0.12 a BC	1.95 ± 0.18 C
N+PGPB	1.42 ± 0.10 a AB	1.55 ± 0.12 a AB	1.56 ± 0.23 a A	1.29 ± 0.17 a A	1.46 ± 0.18 A
N+A	1.40 ± 0.08 a AB	1.36 ± 0.05 a A	1.59 ± 0.18 a A	1.30 ± 0.12 a A	1.42 ± 0.15 A
N+G+A	2.05 ± 0.23 a CD	1.96 ± 0.11 a B	1.93 ± 0.25 a AB	1.72 ± 0.02 a BC	1.92 ± 0.20 C
N+PGPB+A	1.76 ± 0.14 a BC	1.60 ± 0.34 a AB	2.36 ± 0.10 b B	1.84 ± 0.12 a C	1.89 ± 0.34 BC
Average	1.66 ± 0.33 ab	1.59 ± 0.28 ab	1.85 ± 0.31 b	1.52 ± 0.24 a	
		Phosphatase activityµg PNF/g soil/h			
C	12.46 ± 0.45 c B	7.77 ± 0.35 a C	9.36 ± 0.46 b A	6.94 ± 0.24 a A	9.13 ± 2.23 AB
N	12.11 ± 0.19 d B	4.84 ± 0.60 a AB	8.52 ± 0.17 c A	7.30 ± 0.34 b A	8.20 ± 2.75 A
N+G	14.18 ± 0.42 c C	10.73 ± 1.05 b D	10.51 ± 0.21 b B	8.66 ± 0.20 a B	11.02 ± 2.14 C
N+PGPB	12.23 ± 0.67 c B	6.32 ± 0.09 a BC	8.96 ± 0.25 b A	6.77 ± 0.54 a A	8.57 ± 2.47 A
N+A	12.09 ± 0.20 c B	4.70 ± 0.29 a A	8.91 ± 0.64 b A	7.81 ± 0.69 b AB	8.38 ± 2.79 A
N+G+A	14.26 ± 0.34 c C	9.75 ± 0.56 a D	11.20 ± 0.44 b BC	8.83 ± 0.63 a B	11.01 ± 2.19 C
N+PGPB+A	10.15 ± 0.43 b A	9.70 ± 0.38 b D	11.74 ± 0.22 c C	8.77 ± 0.18 a B	10.09 ± 1.16 BC
Average	12.50 ± 1.38 c	7.69 ± 2.38 a	9.89 ± 1.23 b	7.87 ± 0.92 a	

[1] Data are presented as mean ± standard deviation. [2] Means followed by the same letters are not statistically different (ANOVA, Tukey test $\alpha = 0.05$). Lowercase letters indicate differences between sampling dates (rows), and uppercase letters indicate statistically significant differences between treatments (columns).

The addition of nitrogen in the form of mineral fertilizer did not affect FDA hydrolysis. The application of plant biostimulants was observed to have a positive effect on FDA hydrolysis, with the most pronounced results observed in April 2021 in the N+G (2.13 mg FDA/g soil/3 h) and N+G+A treatments (2.05) in comparison to control (1.56) or N (1.32) treatment ($p < 0.05$), as well as in the second cultivation season in the N+PGPB+A treatment (2.36 in April 2022 and 1.84 in July 2022).

The application of nitrogen fertilizer resulted in a negative impact ($p < 0.05$) on phosphatase activity in the N treatment (4.84 µg PNF/g soil/h) compared to the control (7.77), which was observed exclusively in July 2021. Once again, the N+G and N+G+A treatments were found to have a statistically significant ($p < 0.05$) effect on the activity. Furthermore, an increase was observed for the N+PGPB+A treatment from July 2021 to the end of the experiment.

All enzyme activities were affected by the soil sampling date, although the trend was not consistent across all activities. DHA was demonstrably lower for all treatments at the April sampling than at the post-harvest sampling. FDA hydrolysis was the least affected by sampling dates among all enzymatic activities. Significant differences ($p < 0.05$) between sampling dates were only found for N treatment and N+G treatment. The most pronounced effect was a demonstrable decrease in phosphatase activity after harvest.

3.2. Soil Respiration Activities

As illustrated in Table 3, the addition of nitrogen and biostimulants led to an enhancement in the basal and potential respiration of microorganisms. It was observed that the application of nitrogen (N) without biostimulants resulted in a significant ($p < 0.05$) increase in basal respiration values (2.23 ± 0.12 µg CO_2-C/g soil/h), reaching 32% higher value than the control (1.69) in the first experimental year (based on two sampling periods). The application of nitrogen in conjunction with all plant biostimulants resulted in a significant ($p < 0.05$) enhanced respiration rate in comparison to the control (8 to 57%). The greatest increase was observed in both experimental years with the N+PGPB+A combination. In the second experimental year, the basal respiration rate was observed to be higher in all treatments (more than twice in treatments C, N+G, N+A, N+G+A) compared to the first year ($p < 0.05$).

Table 3. Effect of nitrogen fertilization and plant biostimulants application during strawberry cultivation on soil respiration activity and metabolic quotient.

Treatment	July 2021	July 2022	Average
	Basal respiration in µg CO_2-C/g soil/h		
C	1.69 ± 0.08 [1] a A [2]	3.05 ± 0.04 b A	2.37 ± 0.75 A
N	2.23 ± 0.12 a B	3.19 ± 0.03 b AB	2.71 ± 0.53 B
N+G	2.28 ± 0.06 a B	3.57 ± 0.08 b D	2.93 ± 0.71 BC
N+PGPB	2.09 ± 0.11 a B	3.38 ± 0.04 b C	2.74 ± 0.71 B
N+A	2.57 ± 0.17 a C	3.29 ± 0.12 b BC	2.93 ± 0.42 BC
N+G+A	2.04 ± 0.05 a B	3.74 ± 0.03 b DE	2.89 ± 0.93 B
N+PGPB+A	2.66 ± 0.03 a C	3.78 ± 0.05 b E	3.22 ± 0.61 C
Average	2.22 ± 0.32 a	3.43 ± 0.27 b	
	Potential respiration in µg CO_2-C/g soil/h		
C	13.84 ± 0.27 a A	17.50 ± 0.14 b A	15.67 ± 2.01 A
N	18.61 ± 0.02 a D	20.36 ± 0.05 b CD	19.49 ± 0.96 C
N+G	14.91 ± 0.03 a B	20.17 ± 0.12 b BC	17.54 ± 2.88 B
N+PGPB	17.72 ± 0.71 a C	20.75 ± 0.07 b E	19.24 ± 1.72 C
N+A	18.91 ± 0.05 a D	20.50 ± 0.09 b D	19.70 ± 0.87 C
N+G+A	17.55 ± 0.19 a C	20.01 ± 0.06 b B	18.78 ± 1.35 BC
N+PGPB+A	16.99 ± 0.06 a C	21.60 ± 0.04 b F	19.29 ± 2.53 C
Average	16.93 ± 1.81 a	20.13 ± 1.21 b	
	Metabolic quotient in µg CO_2-C/mg Cmic/h		
C	2.35 ± 0.09 a A	3.45 ± 0.13 b AB	2.90 ± 0.61 A
N	4.33 ± 0.15 a AB	3.94 ± 0.11 a CD	4.14 ± 0.36 AB
N+G	4.72 ± 0.81 a AB	3.82 ± 0.28 a BC	4.27 ± 0.73 AB
N+PGPB	7.76 ± 1.60 b CD	3.28 ± 0.14 a A	5.52 ± 2.66 BC
N+A	8.94 ± 0.03 b D	4.30 ± 0.13 a D	6.62 ± 2.54 BC
N+G+A	5.42 ± 0.53 a BC	4.79 ± 0.00 a E	5.11 ± 0.48 ABC
N+PGPB+A	9.03 ± 1.63 b D	4.85 ± 0.15 a E	6.94 ± 2.51 C
Average	6.08 ± 2.55 b	4.06 ± 0.60 a	

[1] Data are presented as mean ± standard deviation. [2] Means followed by the same letters are not statistically different (ANOVA, Tukey test $\alpha = 0.05$). Lowercase letters indicate differences between sampling dates (rows) and uppercase letters indicate statistically significant differences between treatments (columns).

Potential respiration reflects the capacity of microorganisms to degrade glucose. Application of nitrogen resulted in a demonstrable 34% increase in potential respiration, similar to basal respiration, compared to the control in the first experimental year. Of all the biostimulants, N+A treatment had the greatest effect on this parameter in the first experimental year with 18.91 ± 0.05 µg CO_2-C/g soil/h (up 37% compared to control), and N+PGPB+A in the second experimental year with 21.60 ± 0.04 µg CO_2-C/g soil/h (up 23% compared to control). Similarly, potential respiration was demonstrably higher in the second experimental year than in the first one ($p < 0.05$).

The soil of the experimental plot exhibited a high biomass of microorganisms (775 µg Cmic/g) at the beginning of the experiment. The addition of N and biostimulants resulted in a statistically demonstrable increase in qCO_2 compared to the control in treatments to which humic substance (N+A) was added and also in the combination of N+PGPB+A and N+G+A ($p < 0.05$). These results were confirmed in both experimental years ($p < 0.05$).

3.3. Numbers of Cultivable Microorganisms in Soil

The evaluated groups of cultivable microorganisms exhibited no clear influence from soil treatment; however, the effect of soil sampling was more prominent (Table 4). In April 2021, no difference in total bacterial counts was observed between the treatments under investigation. The values ranged from 6.05 ± 0.05 log CFU/g d. m. soil in control to 6.13 ± 0.03 log CFU/g d. m. soil in N treatment. However, following the strawberry harvest (July 2021), the N+G and N+G+A treatments exhibited the lowest counts (5.93 ± 0.05 and 5.72 ± 0.03 log CFU/g d.m. soil, respectively) in comparison to the other treatments ($p < 0.05$). In contrast, in the second year, the N+G 6.23 ± 0.13 log CFU/g d. m. soil in July) and N+G+A (5.92 ± 0.06 log CFU/g d. m. soil in April) treatments exhibited statistically significantly higher ($p < 0.05$) total bacterial counts compared to the control (5.43 ± 0.12 log CFU/g d. m. soil in April 2022 and 5.69 ± 0.07 log CFU/g d. m. soil in July 2022).

Table 4. Effect of nitrogen fertilization and plant biostimulants application during strawberry cultivation on the numbers of cultivable microorganisms in log CFU/g dry weight soil.

Treatment	April 2021	July 2021	April 2022	July 2022	Average
			Total microbial count		
C	6.05 ± 0.05 [1] c A [2]	6.23 ± 0.02 c B	5.43 ± 0.12 a AB	5.69 ± 0.07 b AB	5.85 ± 0.33 AB
N	6.13 ± 0.03 b A	7.30 ± 0.05 c C	5.43 ± 0.19 a AB	5.72 ± 0.15 a AB	6.15 ± 0.75 B
N+G	6.12 ± 0.07 bc A	5.93 ± 0.05 ab A	5.69 ± 0.13 a BC	6.23 ± 0.13 c C	5.99 ± 0.23 AB
N+PGPB	6.08 ± 0.06 bc A	6.35 ± 0.17 c B	5.65 ± 0.11 a BC	5.86 ± 0.10 ab ABC	5.99 ± 0.29 AB
N+A	6.06 ± 0.10 b A	6.21 ± 0.05 b B	5.26 ± 0.02 a A	5.58 ± 0.31 a A	5.78 ± 0.42 A
N+G+A	6.10 ± 0.08 b A	5.72 ± 0.03 a A	5.92 ± 0.06 b C	5.94 ± 0.10 b ABC	5.92 ± 0.15 AB
N+PGPB+A	6.07 ± 0.06 b A	6.15 ± 0.05 b B	5.34 ± 0.16 a AB	6.03 ± 0.11 b BC	5.90 ± 0.35 AB
Average	6.09 ± 0.06 bc	6.27 ± 0.48 c	5.53 ± 0.24 a	5.86 ± 0.25 b	
			Dormant forms count		
C	6.09 ± 0.08 c AB	5.31 ± 0.09 a A	5.54 ± 0.05 b ABC	5.44 ± 0.07 ab AB	5.60 ± 0.32 A
N	6.26 ± 0.04 b C	5.63 ± 0.05 a ABC	5.60 ± 0.19 a BC	5.49 ± 0.15 a ABC	5.75 ± 0.33 ABC
N+G	6.36 ± 0.03 c C	5.85 ± 0.04 b BC	5.49 ± 0.15 a ABC	5.96 ± 0.11 b D	5.92 ± 0.33 C
N+PGPB	6.23 ± 0.05 c BC	6.05 ± 0.06 c C	5.35 ± 0.13 a AB	5.72 ± 0.10 b BCD	5.84 ± 0.36 BC
N+A	6.28 ± 0.04 c C	5.59 ± 0.24 b AB	5.21 ± 0.09 a A	5.43 ± 0.05 ab AB	5.63 ± 0.43 AB
N+G+A	6.04 ± 0.06 a A	5.72 ± 0.29 a ABC	5.72 ± 0.04 a C	5.80 ± 0.15 a CD	5.82 ± 0.20 BC
N+PGPB+A	6.06 ± 0.04 c A	5.71 ± 0.18 b ABC	5.31 ± 0.10 a AB	5.31 ± 0.14 a A	5.60 ± 0.35 A
Average	6.19 ± 0.12 c	5.70 ± 0.26 b	5.46 ± 0.20 a	5.59 ± 0.24 ab	
			Actinomycetes		
C	4.57 ± 0.04 c B	4.35 ± 0.10 ab C	4.23 ± 0.12 a B	4.52 ± 0.05 bc A	4.42 ± 0.16 B
N	4.57 ± 0.09 b B	3.84 ± 0.18 a ABC	4.00 ± 0.17 a AB	4.80 ± 0.04 b C	4.30 ± 0.43 AB
N+G	4.72 ± 0.10 b B	3.46 ± 0.62 a AB	4.24 ± 0.06 ab B	4.61 ± 0.14 b ABC	4.26 ± 0.59 AB
N+PGPB	4.57 ± 0.07 b B	4.28 ± 0.08 a C	4.16 ± 0.06 a AB	4.57 ± 0.06 b AB	4.39 ± 0.19 B
N+A	4.32 ± 0.10 bc A	3.14 ± 0.17 a A	4.07 ± 0.12 b AB	4.67 ± 0.14 c ABC	4.05 ± 0.60 A
N+G+A	4.62 ± 0.07 b B	3.65 ± 0.27 a ABC	4.04 ± 0.14 a AB	4.49 ± 0.03 b A	4.20 ± 0.42 AB
N+PGPB+A	4.57 ± 0.06 c B	4.06 ± 0.00 b BC	3.87 ± 0.08 a A	4.77 ± 0.02 d BC	4.32 ± 0.39 AB
Average	4.56 ± 0.13 c	3.83 ± 0.48 a	4.09 ± 0.16 b	4.63 ± 0.13 c	
			Microscopic fungi		
C	4.43 ± 0.05 ab A	4.48 ± 0.16 b BC	4.23 ± 0.07 a A	4.48 ± 0.02 b C	4.41 ± 0.13 A
N	4.02 ± 0.41 a A	4.21 ± 0.10 a ABC	5.73 ± 0.22 b B	4.20 ± 0.02 a A	4.54 ± 0.75 A
N+G	4.43 ± 0.09 ab A	4.57 ± 0.10 b C	4.16 ± 0.20 a A	4.67 ± 0.06 b D	4.46 ± 0.23 A
N+PGPB	4.50 ± 0.12 b A	3.87 ± 0.10 a A	4.42 ± 0.35 b A	4.29 ± 0.02 ab AB	4.27 ± 0.30 A
N+A	4.40 ± 0.14 ab A	4.07 ± 0.11 a A	4.33 ± 0.13 ab A	4.41 ± 0.13 b BC	4.30 ± 0.18 A
N+G+A	4.44 ± 0.13 a A	4.16 ± 0.21 a AB	4.45 ± 0.13 a A	4.51 ± 0.05 a CD	4.39 ± 0.19 A
N+PGPB+A	4.05 ± 0.36 a A	4.19 ± 0.13 a AB	4.33 ± 0.05 a A	4.20 ± 0.06 a A	4.19 ± 0.20 A
Average	4.32 ± 0.27 ab	4.22 ± 0.25 a	4.52 ± 0.54 b	4.39 ± 0.17 ab	

[1] Data are presented as mean ± standard deviation. [2] Means followed by the same letters are not statistically different (ANOVA, Tukey test $\alpha = 0.05$). Lowercase letters indicate differences between sampling dates (rows) and uppercase letters indicate statistically significant differences between treatments (columns).

The impact of the treatments on dormant bacterial forms exhibited variability across treatments. The N+PGPB+A treatment demonstrated the lowest total counts, statistically equivalent to the control, while the N+G treatment exhibited the statistically highest counts (except for April 2022. The highest number of dormant forms was identified in the initial sampling (April 2021), in comparison to subsequent samplings ($p < 0.05$).

The abundance of actinomycetes showed no clear effect of nitrogen and biostimulant treatments. In the first experimental year, the abundance of actinomycetes was significantly lower in the N+A treatment compared to the control ($p < 0.05$), but the difference evened out in the following year.

The microscopic filamentous fungi exhibited a relatively consistent population across the sampling period, with a slight increase observed in April 2022. The significantly highest fungal counts in this sampling were recorded in the N treatment (5.73 ± 0.22 log CFU/g d. m. soil). In July 2022, the lowest fungal counts were observed in the N (4.20 ± 0.02 log CFU/g d. m. soil), N+PGPB (4.29 ± 0.02 log CFU/g d. m. soil), and N+PGPB+A (4.20 ± 0.06 log CFU/g d. m. soil) treatments, while the remaining treatments demonstrated an increase in fungal abundance.

3.4. Microbial Communities in Soil

A total of 2,284,033 sequences of the 16S rRNA gene were collected for 84 samples. Following denoising, 3353 ASVs were identified. For fungi, a total of 2,159,936 sequences of the fungal ITS2 region were collected, comprising 2800 amplicon sequence variants (ASVs). The Shannon's index of alpha diversity (Table 5) for bacteria exhibited a range of 6.3 to 8.1, indicating significant alterations between sampling points and treatments. The lowest diversity was observed in the N+PGPB+A and N+G+A treatments. The Shannon's index of fungal community demonstrated considerably higher variability, as illustrated in Table 5. The range remained stable at 5.36–5.78, with no discernible impact of treatment in 2021. A significant decline was observed in the 2022 samples, wherein the impact of the treatments became discernible. In April 2022, only the PGPB-containing treatments (N+PGPB and N+PGPB+A) demonstrated diversity values above four. In July 2022, the N+G treatment also exhibited values that were equivalent to those observed previously.

Table 5. Effect of nitrogen fertilization and plant biostimulants application during strawberry cultivation on Shannon's diversity index (mean ± standard deviation).

Treatment	April 2021	July 2021	April 2022	July 2022	Average
			Prokaryotic community		
C	7.87 ± 0.19 [1] bc B [2]	6.95 ± 0.11 a B	8.06 ± 0.12 c C	7.33 ± 0.33 ab AB	7.55 ± 0.49 BC
N	8.05 ± 0.1 c B	7.57 ± 0.09 b CD	7.42 ± 0.18 b B	6.7 ± 0.12 a AB	7.44 ± 0.52 BC
N+G	7.68 ± 0.34 a AB	7.5 ± 0.12 a C	7.48 ± 0.07 a B	7.13 ± 0.89 a AB	7.45 ± 0.46 BC
N+PGPB	8.1 ± 0.11 b B	7.76 ± 0.19 b CD	8.06 ± 0.06 b C	6.9 ± 0.43 a AB	7.71 ± 0.55 C
N+A	7.59 ± 0.17 a AB	7.84 ± 0.05 ab D	8.03 ± 0.06 b C	7.46 ± 0.23 a B	7.73 ± 0.26 C
N+G+A	7.68 ± 0.22 b AB	6.39 ± 0.12 a A	6.58 ± 0.26 a A	6.29 ± 0.15 a A	6.74 ± 0.61 A
N+PGPB+A	7.15 ± 0.13 a A	7.58 ± 0.05 a CD	7.17 ± 0.27 a B	7.23 ± 0.15 a AB	7.28 ± 0.23 B
Average	7.73 ± 0.35 c	7.37 ± 0.50 b	7.54 ± 0.55 bc	7 ± 0.52 a	
			Fungal community		
C	5.6 ± 0.51 c A	5.76 ± 0.37 c A	2.75 ± 0.49 b AB	1.07 ± 0.19 a AB	3.8 ± 1.84 A
N	5.63 ± 0.1 b A	5.65 ± 0.44 b A	2.76 ± 0.43 a A	2.22 ± 0.29 a AB	4.06 ± 1.59 A
N+G	5.53 ± 0.4 b A	5.73 ± 0.57 b A	3.55 ± 0.59 a B	4.24 ± 0.13 a C	4.76 ± 0.89 B
N+PGPB	5.41 ± 0.14 b A	5.78 ± 0.37 b A	4.89 ± 0.47 b C	4.32 ± 0.44 a C	5.1 ± 0.81 B
N+A	5.56 ± 1.14 ab A	5.76 ± 0.09 b A	3.37 ± 0.08 a AB	2.61 ± 0.12 a BC	4.33 ± 1.24 A
N+G+A	5.52 ± 0.15 b A	5.36 ± 0.10 b A	2.81 ± 0.12 a A	3.1 ± 0.07 a BC	4.2 ± 1.40 A
N+PGPB+A	5.46 ± 0.15 b A	5.58 ± 0.37 b A	5.64 ± 0.04 a C	4.48 ± 0.19 b C	5.29 ± 0.73 B
Average	5.53 ± 0.48 c	5.66 ± 0.2 d	3.68 ± 1.12 b	3.15 ± 10 a	

[1] Data are presented as mean ± standard deviation. [2] Means followed by the same letters are not statistically different (ANOVA, Tukey test $\alpha = 0.05$). Lowercase letters indicate differences between sampling dates (rows) and uppercase letters indicate statistically significant differences between treatments (columns).

The results of the beta diversity analysis confirmed that both the bacterial and fungal communities were strongly dependent on the date of sampling. The NMDS scatterplots (Figure 1) illustrate that each sampling date is represented by a distinct cluster. The PERMANOVA analysis of the prokaryotic community confirmed that there were highly significant (p = 0.001) differences between the clusters, i.e., sampling dates, which collectively explained 27% of the total variability in Unifrac distances. The impact of the treatments was less pronounced (R^2 = 0.113), although still statistically significant (p = 0.001) with each treatment cluster within larger sampling dates clusters. Nevertheless, no significant differences were observed between the treatments in some pairs (see Supplementary Table S2). The most distant clusters were formed by the N+PGPB+A, which was significantly different from all other treatments (p values in the range of 0.001 to 0.039). Treatments N+G and N+G+A were different in comparison to control (p = 0.008 and 0.015, respectively).

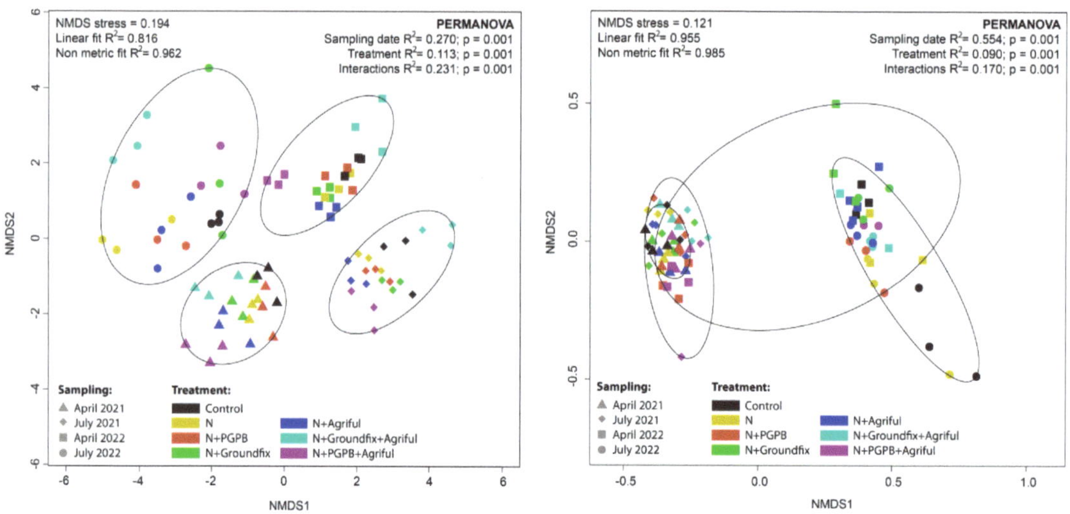

Figure 1. The NMDS scatterplots based on Unifrac weighted distances for prokaryotic (**a**) and fungal (**b**) communities in soil treated by nitrogen fertilization and plant biostimulants application during strawberry cultivation.

Even though beta diversity analysis demonstrated the considerable impact of the sampling date on the fungal community, the 2021 samples formed clusters that are overlaid. The April 2022 samples were found to cluster according to the treatment, with the N+PGPB and N+PGPB+A treatments maintaining a fungal community similar to that observed in 2022, while the other treatments formed a separate cluster in the NMDS scatterplot. This caused longer distances between samples in the April 2022 sampling, and the homogeneity of sampling date clusters was thus significantly different (April 2022 vs. April 2021 and July 2021, p = 0.033 and 0.023, respectively). The effect of sampling explained 55% of the observed variability (p = 0.001), while only 9% of the variation could be attributed to the treatments (p = 0.001) with significant interactions (R^2= 0.17; p = 0.001). However, it should be noted that the effect of sampling may be overestimated due to the significant difference in homogeneity as PERMANOVA may report increased differences in case of significant differences in homogeneity. The significant effect of the treatments was confirmed by the pairwise PERMANOVA comparison only between C and N treatment (p = 0.041), N and N+G (p = 0.044), N+G and N+PGPB (p = 0.043) N, and N+PGPB+A (p = 0.043) treatments, but the R2 values were low (see Supplementary Table S3).

The distribution of bacterial phyla is shown in the bar chart (Figure 2a), which demonstrates the strong effect of the sampling date. The most prevalent phylum, Actinomycetota, exhibited a significant ($p < 0.001$) decline in abundance, with a nearly threefold reduction observed between July 2022 and April 2022. Conversely, Bacteroidota and Planctomycetota exhibited significantly ($p < 0.001$) higher abundance in July 2022. The impact of the treatments is most evident in April 2022. The most prevalent changes are observed in the N+PGPB+A and N+G+A treatments.

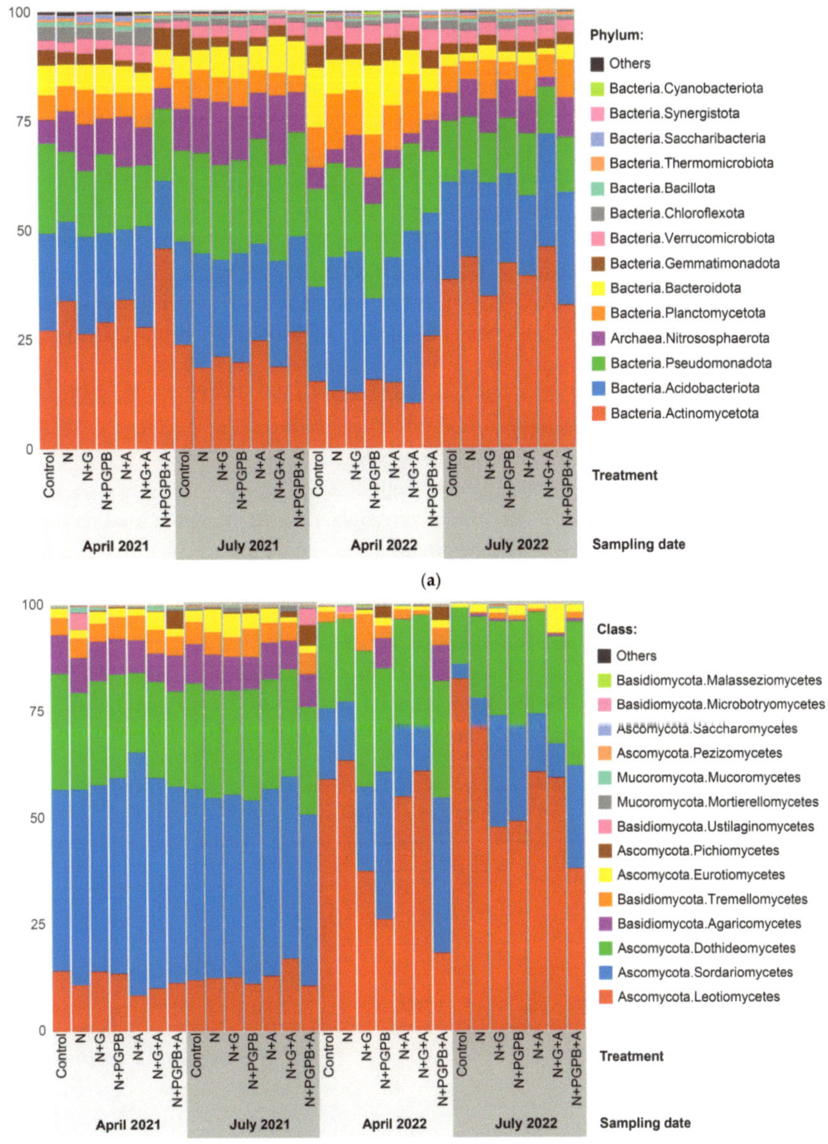

Figure 2. Distribution of bacterial phyla (**a**) and fungal classes (**b**) in soil treated by nitrogen fertilization and plant biostimulants application during strawberry cultivation.

The bar chart of fungal classes (Figure 2b) illustrates a notable increase in the prevalence of Leotiomycetes from 12–13% in 2021 to 46–58% in 2022 and an accompanied decline in the occurrence of Sordariomycetes from 43–27% in 2021 to 12–21% in 2022), Agaricomycetes (from 8% in 2021 to 2% in April 2022 and 0.004% in July), and Tremellomycetes (from 5% in 2021 to 3.2% in April 2022 and 0.007% in July).

Heatmap (Figure 3a) of the most common prokaryotic genera (or lowest possible taxonomic levels) shows the most prevalent groups were *Acidobacteria* GP6 (varied from 8 to 28% according to sample date and treatment), *Nitrososphaera* (2–16%), *Gaiella* (3–12%), and *Tepidisphaera* (3–13%). Once again, pronounced clustering according to the sampling date is evident on the heatmap. Variability in the abundance of genera between sampling dates and separately among treatments in each sampling date is shown in Supplementary Figures S3–S7. Among the other above-mentioned genera are those significantly affected by treatments according to their high linear discriminant analysis (LDA) score.

A heatmap of the most common fungal genera (Figure 3b) illustrates the rapid decline of fungal diversity in 2022. The principal reason for this discrepancy is the high prevalence of the *Pilidium* genus in 2022. In 2021, the abundance of *Pilidium* was below 5%, while other genera, including *Fusarium* (12–36%), *Cladosporium* (6–12%), *Chaetoscypha* (3–9%), and *Alternaria* (3–7%), were more prevalent. In 2022, however, the most common ASV of *Pilidium* constituted over 70% of the fungal community in some samples. The *Pilidium* genus is represented at a significantly lower level in the N+PGPB (18%) and N+PGPB+A (8%) treatments during the April 2022 sampling period. This trend persists in the July 2022 sampling (35 and 34%, respectively), where the *Pilidium* genus is less dominant in both treatments and the N+G treatment (35%). Additionally, an increased abundance in samples was recorded for *Didimella* (3–13% in 2022 and 2–3% in 2021) and *Paraphoma* (rise from less than 1% to 1–16% in 2022). Lefse LDA score dot plots for each sampling date are in Supplementary Figures S8–S12.

Redundancy analysis (RDA) of constrained data (Figure 4) shows basal respiration, potential respiration, and yield linked to microbiome composition in both years. In the year 2021, Dehydrogenase activity and FDA hydrolysis were also significant.

The strawberry yield (Table 6) in the experimental plot, expressed in t/ha, did not differ between treatments. In the second year of cultivation, the yield in the N+PGPB+A treatment was demonstrably higher than the yield in the control (95% higher) as well as in the N-only treatment (72% higher). The weight per fruit was not affected by the application of N and biostimulants.

Table 6. Effect of nitrogen fertilization and plant biostimulants application on the strawberry yield and the weight of a single fruit.

Treatment	2021	2022	Average
	Total yield in t/ha		
C	0.90 ± 0.19 [1] a A [2]	2.25 ± 0.17 b A	1.58 ± 0.76 A
N	0.86 ± 0.09 a A	2.56 ± 0.51 b A	1.71 ± 0.98 A
N+G	1.00 ± 0.10 a A	2.98 ± 0.46 b AB	1.99 ± 1.12 AB
N+PGPB	0.78 ± 0.13 a A	3.39 ± 0.46 b AB	2.09 ± 1.46 AB
N+A	0.90 ± 0.16 a A	3.07 ± 1.12 b AB	1.98 ± 1.39 AB
N+G+A	1.05 ± 0.13 a A	3.56 ± 0.54 b AB	2.31 ± 1.42 AB
N+PGPB+A	1.15 ± 0.42 a A	4.38 ± 0.66 b B	2.77 ± 1.83 B
Average	0.95 ± 0.21 a	3.17 ± 0.84 b	
	Average weight of a single fruit in g		
C	8.81 ± 1.68 a A	7.68 ± 0.38 a A	8.25 ± 1.30 A
N	9.53 ± 1.70 a A	8.62 ± 0.74 a A	9.07 ± 1.28 A
N+G	10.20 ± 0.58 b A	8.12 ± 0.66 a A	9.16 ± 1.27 A
N+PGPB	8.08 ± 0.52 a A	8.93 ± 0.47 a A	8.50 ± 0.64 A
N+A	10.33 ± 0.77 a A	9.69 ± 1.97 a A	10.01 ± 1.38 A
N+G+A	10.05 ± 0.67 a A	10.26 ± 1.71 a A	10.16 ± 1.17 A
N+PGPB+A	8.70 ± 1.15 a A	9.47 ± 0.65 a A	9.09 ± 0.94 A
Average	9.39 ± 1.24 a	8.97 ± 1.26 a	

[1] Data are presented as mean ± standard deviation. [2] Means followed by the same letters are not statistically different (ANOVA, Tukey test α = 0.05). Lowercase letters indicate differences between sampling dates (rows) and uppercase letters indicate statistically significant differences between treatments (columns).

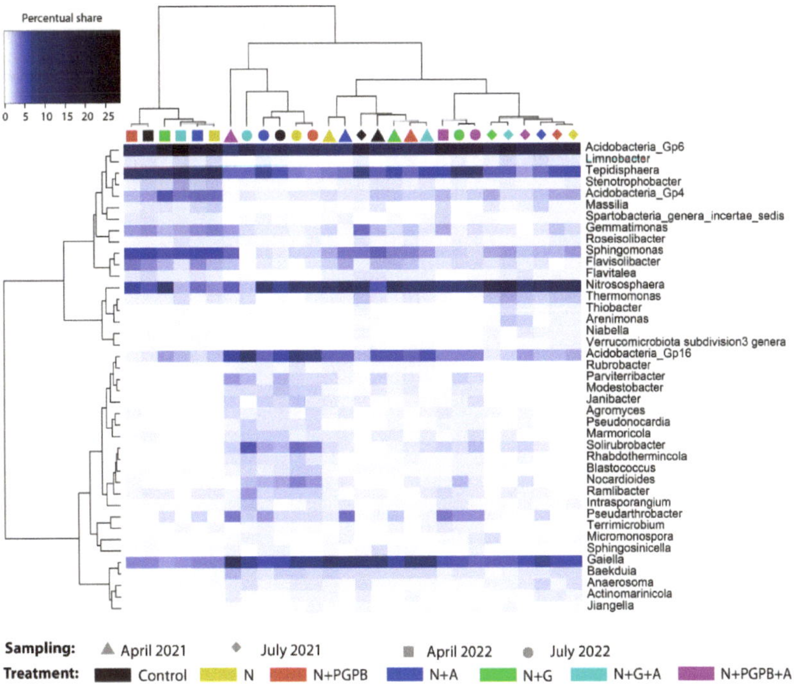

(a)

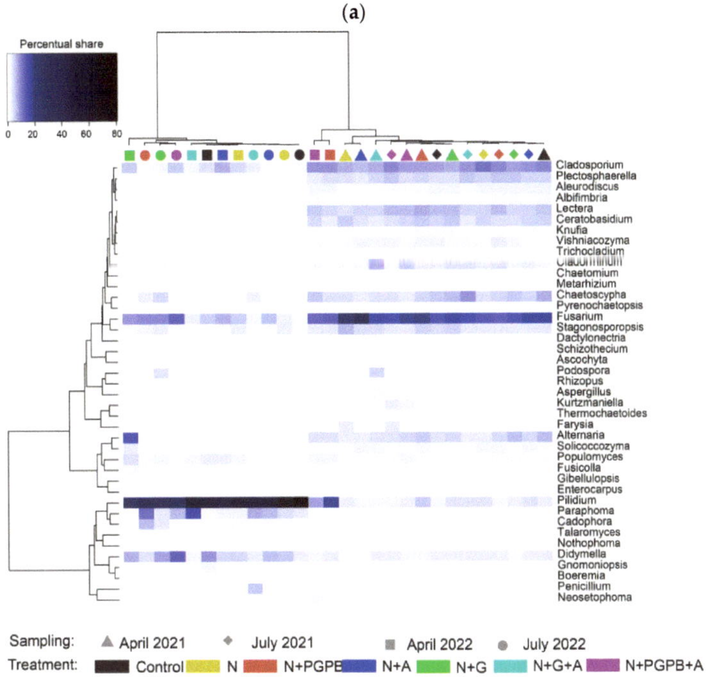

(b)

Figure 3. Heatmaps of 40 most common prokaryotic (**a**) and fungal (**b**) genera in soil treated by nitrogen fertilization and plant biostimulants application during strawberry cultivation. Dendrograms were constructed using the Bray–Curtis distance and complete clustering method.

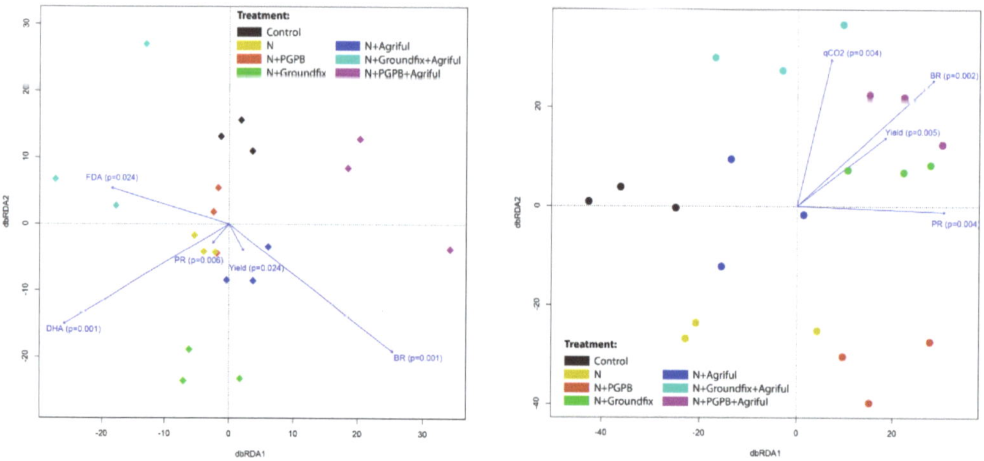

(a) (b)

Figure 4. The RDA biplots for July 2021 (**a**) and July 2022 (**b**) microbiome data, microbial activities, and strawberry yield after nitrogen fertilization and plant biostimulants application during strawberry cultivation.

4. Discussion

Soil microorganisms play a crucial role in the functioning of ecosystems, influencing biogeochemical cycles and, consequently, the availability of nutrients and soil fertility [41]. They exhibit high responsiveness and sensitivity to management practices, including mineral fertilization, pesticide application, and microbial products [42]. The activity and population dynamics of microorganisms in soil are also dependent on environmental factors and are highly variable over time. In our experiment, this variability was reflected in the different compositions of the microbial community and its activity at four dates. As a result, it was challenging to assess the effect of individual treatments, with the possibility that the effects of some treatments may have been masked by this variability.

The soil of our experimental plot was characterized by high soil organic carbon (SOC), humus, total nitrogen, and high activity of microorganisms. Higher SOC and humus content may provide sufficient substrate to support higher microbial biomass and thus higher enzyme production [43], which was confirmed in our experiment. Dehydrogenases play a significant role in the biological oxidation of soil organic matter by transferring hydrogen from organic substrates to inorganic acceptors [44]. Pukalchik et al. [45] reported that the activity of oxidative enzymes, which include DHA, usually decreases after the addition of humic products, which is also in agreement with our results. Latkovic et al. [46] also described the inhibition of dehydrogenase and proteinase activity as well as microbial biomass carbon in soil under wheat and maize in treatments with high rates of NPK nutrients. They also describe the stimulation of DHA and proteinase activity by low rates of NPK and bacterial inoculants (*Enterobacter* spp. strains and *Klebsiella planticola*). Also in our experiment, we confirmed that the negative effect of N on DHA was partially mitigated by the addition of a commercial biostimulant, even in combination with a humic substance. The reversal of the negative effect of N treatment can be explained by the fact that inoculation with microbial biostimulants will increase the colonization of biostimulant-derived microorganisms mainly to the rhizosphere region, leading to higher accumulation of available nutrients as a substrate for dehydrogenase activity [47]. FDA hydrolysis measures the activity of nonspecific esterases, proteases, and lipases involved in the decomposition of soil organic matter. Since more than 90% of the energy flux in the soil ecosystem is passed by microbial decomposers and is dominated by heterotrophic microorganisms, it is assumed that FDA hydrolysis reflects the overall microbial activity of the soil [20]. In our experiment, the application of a commercial stimulant alone, as well as with a humic substance, positively influenced FDA hydrolysis. The positive effect of biostimulant treatment on FDA hydrolysis in rice cultivation was reported by Buragohain et al. [48]. On the other hand, Soltaniband et al. [49] found that there was no increase in FDA hydrolysis when various algae, fungi, and bacteria-based plant biostimulants were applied in strawberry cultivation compared to control. Both plant biostimulants of microbial origin when combined with humic substances were shown to enhance phosphatase activity (treatments N+G+A and N+PGPB+A). In blueberry cultivation, a 43% increase in phosphatase activity occurred with the treatment of the microbial consortium Oiko bac 174 containing beneficial bacteria and fungi, and Biosolve (Oikos Chile Ltd.) containing humic substances [50].

The determination of CO_2 production is one of the effective methods used for assessing the overall microbial activity in soil [51], which records the catabolic degradation of native organic substances under aerobic conditions [52]. In general, basal respiration is high in the presence of an active microbial community and an abundance of readily mineralizable organic matter. The soil of the experimental trial met these requirements, and the limiting element for increased mineralization of organic matter was N. Therefore, the application of N alone but also N with all plant biostimulants promoted both basal and potential respiration compared to the control. The microbial metabolic quotient qCO_2 used as a so-called eco-physiological indicator provides information on the catabolic requirements of the microbial community for a certain short period [53]. In soils with a neutral soil reaction, qCO_2 values range from 0.5 to 2.0 μgCO_2-C/mg Cmic/h, according to Anderson and Domsch [54]. All of our determined qCO_2 data, for soils with the same pH, had qCO_2 values higher than those reported by these authors. The lowest mean qCO_2 value in the control was 2.90 ± 0.61 μgCO_2-C/mg Cmic/h. Anderson [55] points out that when qCO_2 values are higher than two, it indicates a less energy-efficient use of organic substrates by microbial biomass. In our experiment, the addition of N and biostimulants resulted in even further statistically demonstrable increases in qCO_2 to values greater than 4 and 9 in 2022 and 2021, respectively. N+A, N+G+A, and N+PGPB+A treatments contributed to the observed increase in both experimental years. The results showed that the added humic substances were not utilized by the microorganisms to support their biomass increase but for energy acquired by respiration. As the microorganisms were not exposed to substrate limitation their carbon utilization efficiency decreased [56].

Although the accuracy of culture methods in estimating microbial community size is affected by the relatively large fraction of nonculturable microorganisms, it represents a simple and comparable method of estimation [57].

Soils with a high stock of soil organic matter provide a favorable environment for the presence of an abundant microorganism community. Soils with a high organic matter content have sufficient nutrients and a naturally rich microbiome. In our opinion, this is the reason why the total number of microorganisms was not significantly affected by the addition of nitrogen, humic substances, and microbial biostimulant treatments compared to the control. In contrast to our results, the experiment by Hawrot-Paw et al. [58] confirms higher or equal bacterial counts than the control (135% of control values during strawberry fruiting toward the end of the research) when soil was inoculated with Rhizocell C-containing *Bacillus amyloliquefaciens* IT45 biofertilizer during strawberry fruiting.

The survival strategy of soil microorganisms is based on a dormant population, expending energy to maintain a metabolic alertness for the immediate use of any substrate that enters the soil [59]. We found a slightly increased abundance of dormant forms of bacteria in treatments with the application of microbial stimulants in the second year of cultivation. Both stimulants also contained bacteria belonging to the genus *Bacillus*, which are able to go dormant and necrotic in the case of fresh organic substrates [53] and transform into the active form.

Soil actinomycetes are producers of many secondary metabolites with antibiotic effects and some also significantly promote plant growth by phytostimulation through direct or indirect mechanisms [60]. In our experiment, we did not observe changes in actinomycete numbers caused by fertilizer or biostimulant application, even after the application of a formulation containing two different strains of actinomycetes. Hawrot-Paw et al. [58] described a decrease in actinobacteria after application of the Rhizocell C biopreparation compared to the control. In contrast, Bezuglova et al. [61] reported an increase in the number of ammonifying bacteria, cellulose-degrading fungi, and actinobacteria upon application of the humic preparation BIO-Don (2.24 g/l humic acid) in winter wheat cultivation.

For microscopic filamentous fungi, we found the smallest treatment-induced differences. Only in July 2022, compared to the control, was the count lower in the treatments N+PGPB and N+PGPB+A. The experimental microbial biostimulant contains microbial strains with proven antifungal activities; *Bacillus arybhattai* with antifungal activity against *Leptosphaeria maculans* [21], as well as *Streptomyces puniceus* and *Streptomyces olivochromogenes* with antifungal activity against *Penicillium expansum*, *Alternaria tennuissima*, and *Leptosphaeria maculans* [17], which may have reflected.

The experimental microbial biostimulant contains microbial strains with proven antifungal activity [17,21], which could have been reflected.

High-throughput sequencing helps to overcome problems with assessing non-cultivable microorganisms in soil. We used the most common metabarcoding method based on PCR amplification of phylogenetic markers [62]. Both prokaryotic and fungal communities changed between sampling dates. Temporal variability [63], together with the plant development stage [64], is a well-known factor affecting microbiome compositions. Fertilizing by N [65] or also S [66] was reported to change the soil and rhizosphere microbiome. However, we found fewer changes between the control and N treatment than after the application of biostimulants. Biostimulants based on plant growth-promoting microbial strains significantly affected the soil microbiome in sugarcane [64].

Overgrowth of potential phytopathogen causes a strong decrease in microbiome diversity in plants [67] but also in plant-associated soil. *Pilidium concavum* causes brown spot in strawberry leaves and fruits, although it can also live endophytically and saprophytically without directly damaging fruits. Its occurrence has been reported in Poland, Iran, China, the USA, and other countries [68]. We did not find apparent fruit decay caused by this potential phytopathogen. However, as it primarily develops on older leaves its proliferation may have happened in late 2021. Its abundance was significantly reduced in treatments where an experimental PGPB mixture was applied. Thảo et al. [69] analyzed the antifungal activity of bacterial strains against the fungus of the genus *Pilidium*, with the strongest activity confirmed for *Bacillus subtilis*. Our experimental PGPB mixture used *Bacillus arybhattai* with a previously confirmed ability to suppress fungal growth. PGPB possesses biocontrol mechanisms against phytopathogens such as the production of antibiotics, lytic enzymes (chitinases, lipases, proteases, etc.) attacking cell walls, and Fe^{3+} sequestering siderophores, and induces plant defenses by systemic resistance [16].

Mineral fertilizers affect yield directly by supplying nutrients to crop plants, whereas microbial plant biostimulants affect yield indirectly by improving soil biological properties, plant nutrient intake, or health status [19]. Despite there not being differences in strawberry fruit size due to the effect of N fertilizer and plant biostimulants applied, a positive effect on total yield was confirmed when the N+PGPB+A combination was applied. In comparison to the control, the yield increased by 95%. It is in congruence with Silva et al. [70] who found better plant growth properties and yield after inoculation by PGP bacterial mixture. The strawberry yield in our experiment suggests that the use of plant biostimulants is a suitable alternative to chemical fertilization, can improve the efficiency of mineral fertilizer, and ultimately increase strawberry production.

5. Conclusions

This study demonstrated that the combined application of mineral nitrogen fertilizers and microbial biostimulants, particularly in conjunction with humic substances, significantly influenced soil microbial activity and community composition during strawberry cultivation. However, the application of biostimulants did not have a consistent effect on all measured properties of soil microbiome during the whole experimental period. The effect of a particular treatment may be reduced by the high content of organic matter in the soil. The N+PGPB+A treatment enhanced enzymatic activities, such as phosphatase activity, and FDA hydrolysis, and probably reduced the prevalence of potentially phytopathogenic fungus *Pilidium*. Furthermore, this treatment led to the highest fruit yield in the second experimental year, outperforming other combinations and the control group. These findings highlight the potential of integrated biostimulant applications to sustainably improve strawberry production by fostering a healthier soil microbial ecosystem and the possible mitigation of phytopathogenic fungi. Future research should explore the long-term effects of these biostimulants under varying environmental conditions.

Supplementary Materials: The following supporting information can be downloaded at: https://www.mdpi.com/article/10.3390/horticulturae11020119/s1, Table S1. Sums of rainfall (mm), average air temperature (°C) during 2021-2022, and evaluation according to climatology normal 1990-2020. Table S2. PERMANOVA pairwise comparison of prokaryotic community composition based on Unifrac Distances P values indicating difference are in upper triangle while corresponding R2 values are in lower triangle. Table S3. PERMANOVA pairwise comparison of fungal community composition based on Unifrac Distances P values indicating difference are in upper triangle while corresponding R2 values are in lower triangle. Figure S1. LefSe LDA score dotplot of prokaryotic biomarker phyla for each sampling date in soil treated by nitrogen fertilization and plant biostimulants application during strawberry cultivation. Figure S2. LefSe LDA score dotplot of fungal biomarker phyla for each sampling date in soil treated by nitrogen fertilization and plant biostimulants application during strawberry cultivation. Figure S3. LefSe LDA score dotplot of prokaryoticbiomarker genera for each sampling date in soil treated by nitrogen fertilization and plant biostimulants application during strawberry cultivation. Figure S4. LefSe LDA score dotplot of prokaryoticbiomarker genera for nitrogen fertilization and plant biostimulants treatments in the first sampling date. Figure S5. LefSe LDA score dotplot of prokaryoticbiomarker genera for nitrogen fertilization and plant biostimulants treatments in the second sampling date. Figure S6. LefSe LDA score dotplot of prokaryoticbiomarker genera for nitrogen fertilization and plant biostimulants treatments in the third sampling date. Figure S7. LefSe LDA score dotplot of prokaryoticbiomarker genera for nitrogen fertilization and plant biostimulants treatments in the fourth sampling date. Figure S8. LefSe LDA score dotplot of fungal biomarker genera for each sampling date in soil treated by nitrogen fertilization and plant biostimulants application during strawberry cultivation. Figure S9. LefSe LDA score dotplot of fungal biomarker genera for nitrogen fertilization and plant biostimulants treatments in the first sampling date. Figure S10. LefSe LDA score dotplot of fungal biomarker genera for nitrogen fertilization and plant biostimulants treatments in the second sampling date. Figure S11. LefSe LDA score dotplot of fungal biomarker genera for nitrogen fertilization and plant biostimulants treatments in the third sampling date. Figure S12. LefSe LDA score dotplot of fungal biomarker genera for nitrogen fertilization and plant biostimulants treatments in the fourth sampling date.

Author Contributions: Conceptualization, J.M. (Jana Maková), S.J., R.A., S.A., O.P., A.A., L.D. and J.M. (Juraj Medo); methodology, J.M. (Jana Maková), S.J., R.A., S.A., O.P., A.A., L.D. and J.M. (Juraj

Medo); software, J.M. (Juraj Medo) and J.M. (Jana Maková); validation, R.A., S.J., J.M. (Jana Maková) and J.M. (Juraj Medo); formal analysis, J.M. (Jana Maková), J.M. (Juraj Medo) and R.A.; investigation R.A., S.A., J.M. (Jana Maková) and J.M. (Juraj Medo); resources, R.A., J.M. (Jana Maková) and J.M. (Juraj Medo); data curation, J.M. (Jana Maková) and J.M. (Juraj Medo); writing—original draft preparation; J.M. (Jana Maková) and J.M. (Juraj Medo); writing—review and editing, all authors; project administration, J.M. (Juraj Medo). All authors have read and agreed to the published version of the manuscript.

Funding: This publication was supported by the Operational Program Integrated Infrastructure within the project: Demand-driven research for sustainable and innovative food, Drive4SIFood 313011V336, co-financed by the European Regional Development Fund and project VEGA 1/0573/23 Compost microbiome and its role in improving soil quality and crop production funded by Ministry of Education, Research, Development, and Youth.

Data Availability Statement: All data are available upon request of the corresponding author. Sequence data are accessible under NCBI bioproject PRJNA1200229.

Acknowledgments: The authors would thank to the technical staff of the Institute of Horticulture for maintaining field experiments and Jana Petrová, Henrieta Blaškovičová, and Daniela Košťálová for their assistance in the laboratory.

Conflicts of Interest: Author Samuel Adamec was employed by the company Organix. The remaining authors declare that the research was conducted in the absence of any commercial or financial relationships that could be construed as a potential conflict of interest.

References

1. Liu, Z.; Ma, H.; Jung, S.; Main, D.; Guo, L. Developmental mechanisms of fleshy fruit diversity in Rosaceae. *Annu. Rev. Plant Biol.* **2020**, *71*, 547–573. [CrossRef] [PubMed]
2. Giampieri, F.; Forbes-Hernandez, T.Y.; Gasparrini, M.; Alvarez-Suarez, J.M.; Afrin, S.; Bompadre, S.; Quiles, J.L.; Mezzetti, B.; Battino, M. Strawberry as a health promoter: An evidence based review. *Food Funct.* **2015**, *6*, 1386–1398. [CrossRef] [PubMed]
3. Kumar, R.; Bakshi, P.; Singh, M.; Singh, A.; Vikas, V.; Srivatava, J.; Kumar, V.; Gupta, V. Organic production of strawberry: A review. *Int. J. Chem. Stud.* **2018**, *6*, 1231–1236.
4. Sangiorgio, D.; Cellini, A.; Spinelli, F.; Donati, I. Promoting strawberry (*Fragaria* × *ananassa*) stress resistance, growth, and yield using native bacterial biostimulants. *Agronomy* **2023**, *13*, 529. [CrossRef]
5. FAOSTAT. Crops and Livestock Products: Strawberries. 2022. Available online: https://www.fao.org/faostat/en/#data/QCL (accessed on 22 November 2024).
6. Negi, Y.K.; Sajwan, P.; Uniyal, S.; Mishra, A. Enhancement in yield and nutritive qualities of strawberry fruits by the application of organic manures and biofertilizers. *Sci. Hortic.* **2021**, *283*, 110038. [CrossRef]
7. Hernández-Martínez, N.R.; Blanchard, C.; Wells, D.; Salazar-Gutiérrez, M.R. Current state and future perspectives of commercial strawberry production: A review. *Sci. Hortic.* **2023**, *312*, 111893. [CrossRef]
8. EU. *Regulation (EU) 2019/1009 of the European Parliament and of the Council of 5 June 2019 Laying Down Rules on the Making Available on the Market of EU Fertilising Products and Amending Regulations (EC) No 1069/2009 and (EC) No 1107/2009 and Repealing Regulation (EC) No 2003/2003*; EU: Luxembourg, 2019.
9. Soppelsa, S.; Kelderer, M.; Casera, C.; Bassi, M.; Robatscher, P.; Matteazzi, A.; Andreotti, C. Foliar applications of biostimulants promote growth, yield and fruit quality of strawberry plants grown under nutrient limitation. *Agronomy* **2019**, *9*, 483. [CrossRef]
10. Rouphael, Y.; Kyriacou, M.C.; Petropoulos, S.A.; De Pascale, S.; Colla, G. Improving vegetable quality in controlled environments. *Sci. Hortic.* **2018**, *234*, 275–289. [CrossRef]
11. Calvo, P.; Nelson, L.; Kloepper, J.W. Agricultural uses of plant biostimulants. *Plant Soil* **2014**, *383*, 3–41. [CrossRef]
12. Lumactud, R.A.; Gorim, L.Y.; Thilakarathna, M.S. Impacts of humic-based products on the microbial community structure and functions toward sustainable agriculture. *Front. Sustain. Food Syst.* **2022**, *6*, 977121. [CrossRef]
13. Lodolini, E.; Savini, G.; Sabbatini, P.; Bonanomi, G.; Zucconi, F.; Neri, D. Foliar application of humic acids on strawberry (cv Onda). *Int. Symp. Foliar Nutr. Perenn. Fruit Plants* **2001**, *594*, 297–302.
14. Ameri, A.; Tehranifar, A. Effect of humic acid on nutrient uptake and physiological characteristic *Fragaria ananassa* var: *camarosa*. *J. Biol. Environ. Sci.* **2012**, *6*, 77–79.
15. Singh, R.; Gupta, R.; Patil, R.; Sharma, R.; Asrey, R.; Kumar, A.; Jangra, K. Sequential foliar application of vermicompost leachates improves marketable fruit yield and quality of strawberry (*Fragaria* × *ananassa* Duch.). *Sci. Hortic.* **2010**, *124*, 34–39. [CrossRef]

16. Gupta, G.; Parihar, S.S.; Ahirwar, N.K.; Snehi, S.K.; Singh, V. Plant growth promoting rhizobacteria (PGPR): Current and future prospects for development of sustainable agriculture. *J. Microb. Biochem. Technol.* **2015**, *7*, 096–102.
17. Cinkocki, R.; Lipková, N.; Javoreková, S.; Petrová, J.; Maková, J.; Medo, J.; Ducsay, L. The impact of growth-promoting streptomycetes isolated from rhizosphere and bulk soil on oilseed rape (*Brassica napus* L.) growth parameters. *Sustainability* **2021**, *13*, 5704. [CrossRef]
18. Andrejiová, A.; Adamec, S.; Hegedűsová, A.; Hegedűs, O.; Rosa, R. Verification of the humic substances and PGPB biostimulants beneficial effects on the potato yield and bioactive substances content. *Slovak J. Food Sci.* **2023**, *17*, 1–15. [CrossRef]
19. Wadduwage, J.; Egidi, E.; Singh, B.K.; Macdonald, C.A. Impacts of biostimulants on crop yield and biological activity under drought conditions. *J. Sustain. Agric. Environ.* **2024**, *3*, e12093. [CrossRef]
20. Nannipieri, P.; Ascher, J.; Ceccherini, M.; Landi, L.; Pietramellara, G.; Renella, G. Microbial diversity and soil functions. *Eur. J. Soil Sci.* **2017**, *68*, 12–26. [CrossRef]
21. Lipková, N.; Medo, J.; Artimová, R.; Maková, J.; Petrová, J.; Javoreková, S.; Michalko, J. Growth promotion of rapeseed (*Brassica napus* L.) and blackleg disease (*Leptosphaeria maculans*) suppression mediated by endophytic bacteria. *Agronomy* **2021**, *11*, 1966. [CrossRef]
22. Cassel, D.; Nielsen, D. Field capacity and available water capacity. In *Methods of Soil Analysis: Part 1 Physical and Mineralogical Methods*; ASA-SSSA Inc.: Madison, WI, USA, 1986; Volume 5, pp. 901–926.
23. Casida, L., Jr.; Klein, D.A.; Santoro, T. Soil dehydrogenase activity. *Soil Sci.* **1964**, *98*, 371–376. [CrossRef]
24. Green, V.S.; Stott, D.E.; Diack, M. Assay for fluorescein diacetate hydrolytic activity: Optimization for soil samples. *Soil Biol. Biochem.* **2006**, *38*, 693–701. [CrossRef]
25. Tabatabai, M.A.; Bremner, J.M. Use of p-nitrophenyl phosphate for assay of soil phosphatase activity. *Soil Biol. Biochem.* **1969**, *1*, 301–307. [CrossRef]
26. Vance, E.D.; Brookes, P.C.; Jenkinson, D.S. An extraction method for measuring soil microbial biomass C. *Soil Biol. Biochem.* **1987**, *19*, 703–707. [CrossRef]
27. Alef, K. Soil respiration. In *Methods in Applied Soil Microbiology and Biochemistry*; Alef, K., Nannipieri, P., Eds.; Academic Press: London, UK, 1995; pp. 214–218.
28. Dilly, O.; Munch, J.-C. Ratios between estimates of microbial biomass content and microbial activity in soils. *Biol. Fertil. Soils* **1998**, *27*, 374–379. [CrossRef]
29. Caporaso, J.G.; Lauber, C.L.; Walters, W.A.; Berg-Lyons, D.; Lozupone, C.A.; Turnbaugh, P.J.; Fierer, N.; Knight, R. Global patterns of 16S rRNA diversity at a depth of millions of sequences per sample. *Proc. Natl. Acad. Sci. USA* **2011**, *108*, 4516–4522. [CrossRef]
30. Ihrmark, K.; Bödeker, I.T.; Cruz-Martinez, K.; Friberg, H.; Kubartova, A.; Schenck, J.; Strid, Y.; Stenlid, J.; Brandström-Durling, M.; Clemmensen, K.E. New primers to amplify the fungal ITS2 region–evaluation by 454-sequencing of artificial and natural communities. *FEMS Microbiol. Ecol.* **2012**, *82*, 666–677. [CrossRef]
31. Větrovský, T.; Baldrian, P.; Morais, D. SEED 2: A user-friendly platform for amplicon high-throughput sequencing data analyses. *Bioinformatics* **2018**, *34*, 2292–2294. [CrossRef]
32. Bolyen, E.; Rideout, J.R.; Dillon, M.R.; Bokulich, N.A.; Abnet, C.C.; Al-Ghalith, G.A.; Alexander, H.; Alm, E.J.; Arumugam, M.; Asnicar, F. Reproducible, interactive, scalable and extensible microbiome data science using QIIME 2. *Nat. Biotechnol.* **2019**, *37*, 852–857. [CrossRef]
33. Callahan, B.J.; McMurdie, P.J.; Rosen, M.J.; Han, A.W.; Johnson, A.J.A.; Holmes, S.P. DADA2: High-resolution sample inference from Illumina amplicon data. *Nat. Methods* **2016**, *13*, 581–583. [CrossRef]
34. Wang, Q.; Garrity, G.M.; Tiedje, J.M.; Cole, J.R. Naive Bayesian classifier for rapid assignment of rRNA sequences into the new bacterial taxonomy. *Appl. Environ. Microbiol.* **2007**, *73*, 5261–5267. [CrossRef]
35. Bengtsson-Palme, J.; Ryberg, M.; Hartmann, M.; Branco, S.; Wang, Z.; Godhe, A.; De Wit, P.; Sánchez-García, M.; Ebersberger, I.; de Sousa, F. Improved software detection and extraction of ITS1 and ITS 2 from ribosomal ITS sequences of fungi and other eukaryotes for analysis of environmental sequencing data. *Methods Ecol. Evol.* **2013**, *4*, 914–919. [CrossRef]
36. Katoh, K.; Standley, D.M. MAFFT multiple sequence alignment software version 7: Improvements in performance and usability. *Mol. Biol. Evol.* **2013**, *30*, 772–780. [CrossRef] [PubMed]
37. Lozupone, C.; Lladser, M.E.; Knights, D.; Stombaugh, J.; Knight, R. UniFrac: An effective distance metric for microbial community comparison. *ISME J.* **2011**, *5*, 169–172. [CrossRef]
38. Oksanen, J.; Blanchet, F.G.; Kindt, R.; Legendre, P.; Minchin, P.R.; O'hara, R.; Simpson, G.L.; Solymos, P.; Stevens, M.H.H.; Wagner, H. Package 'vegan'. *Community Ecol. Package Version* **2013**, *2*, 1–295.
39. R_Core_team. *R: A Language and Environment for Statistical Computing*; R Foundation for Statistical Computing: Vienna, Austria, 2022.
40. Segata, N.; Izard, J.; Waldron, L.; Gevers, D.; Miropolsky, L.; Garrett, W.S.; Huttenhower, C. Metagenomic biomarker discovery and explanation. *Genome Biol.* **2011**, *12*, R60. [CrossRef]
41. Gałązka, A.; Gawryjołek, K.; Frąc, J.; Księżak, J. Microbial community diversity and the interaction of soil under maize growth in different cultivation techniques. *Plant Soil Environ.* **2017**, *63*, 264–270. [CrossRef]

42. Luo, P.; Han, X.; Wang, Y.; Han, M.; Shi, H.; Liu, N.; Bai, H. Influence of long-term fertilization on soil microbial biomass, dehydrogenase activity, and bacterial and fungal community structure in a brown soil of northeast China. *Ann. Microbiol.* **2015**, *65*, 533–542. [CrossRef]
43. Mierzwa-Hersztek, M.; Wolny-Koładka, K.; Gondek, K.; Gałązka, A.; Gawryjołek, K. Effect of coapplication of biochar and nutrients on microbiocenotic composition, dehydrogenase activity index and chemical properties of sandy soil. *Waste Biomass Valorization* **2020**, *11*, 3911–3923. [CrossRef]
44. Campos, J.A.; Peco, J.D.; Garcia-Noguero, E. Antigerminative comparison between naturally occurring naphthoquinones and commercial pesticides. Soil dehydrogenase activity used as bioindicator to test soil toxicity. *Sci. Total Environ.* **2019**, *694*, 133672. [CrossRef]
45. Pukalchik, M.; Kydralieva, K.; Yakimenko, O.; Fedoseeva, E.; Terekhova, V. Outlining the potential role of humic products in modifying biological properties of the soil—A review. *Front. Environ. Sci.* **2019**, *7*, 80. [CrossRef]
46. Latkovic, D.; Maksimovic, J.; Dinic, Z.; Pivic, R.; Stanojkovic, A.; Stanojkovic-Sebic, A. Case study upon foliar application of biofertilizers affecting microbial biomass and enzyme activity in soil and yield related properties of maize and wheat grains. *Biology* **2020**, *9*, 452. [CrossRef] [PubMed]
47. Hassan, M. Application of biofertilization and biological control for cowpea production. *Ann. Agric. Sci. Moshtohor* **2017**, *55*, 271–286. [CrossRef]
48. Buragohain, S.; Sarma, B.; Nath, D.J.; Gogoi, N.; Meena, R.S.; Lal, R. Effect of 10 years of biofertiliser use on soil quality and rice yield on an Inceptisol in Assam, India. *Soil Res.* **2017**, *56*, 49–58. [CrossRef]
49. Soltaniband, V.; Brégard, A.; Gaudreau, L.; Dorais, M. Biostimulants promote plant development, crop productivity, and fruit quality of protected strawberries. *Agronomy* **2022**, *12*, 1684. [CrossRef]
50. Schoebitz, M.; López, M.D.; Serri, H.; Aravena, V.; Zagal, E.; Roldán, A. Characterization of bioactive compounds in blueberry and their impact on soil properties in response to plant biostimulants. *Commun. Soil Sci. Plant Anal.* **2019**, *50*, 2482–2494. [CrossRef]
51. Makova, J.; Javoreková, S.; Medo, J.; Majerčíková, K. Characteristics of microbial biomass carbon and respiration activities in arable soil and pasture grassland soil. *J. Cent. Eur. Agric.* **2011**, *12*, 745–758. [CrossRef]
52. Bolat, İ. Microbial biomass, basal respiration, and microbial indices of soil in diverse croplands in a region of northwestern Turkey (Bartın). *Environ. Monit. Assess.* **2019**, *191*, 695. [CrossRef]
53. Joergensen, R.G.; Wichern, F. Alive and kicking: Why dormant soil microorganisms matter. *Soil Biol. Biochem.* **2018**, *116*, 419–430. [CrossRef]
54. Anderson, T.-H.; Domsch, K.H. Soil microbial biomass: The eco-physiological approach. *Soil Biol. Biochem.* **2010**, *42*, 2039–2043. [CrossRef]
55. Anderson, T.-H. Microbial eco-physiological indicators to asses soil quality. *Agric. Ecosyst. Environ.* **2003**, *98*, 285–293. [CrossRef]
56. Zhu, Z.; Ge, T.; Luo, Y.; Liu, S.; Xu, X.; Tong, C.; Shibistova, O.; Guggenberger, G.; Wu, J. Microbial stoichiometric flexibility regulates rice straw mineralization and its priming effect in paddy soil. *Soil Biol. Biochem.* **2018**, *121*, 67–76. [CrossRef]
57. Zhang, L.; Xu, Z. Assessing bacterial diversity in soil: A brief review. *J. Soils Sediments* **2008**, *8*, 379–388. [CrossRef]
58. Hawrot-Paw, M.; Mikiciuk, G.; Mikiciuk, M.; Izwikow, M.; Kiniorska, J. Number and Activity of Microorganisms in Soil Inoculated with the Rhizocell C Biopreparation In Strawberry Cultivation (*Fragaria* × *ananasa* Duch.). *J. Ecol. Eng.* **2016**, *17*, 135–140. [CrossRef] [PubMed]
59. De Nobili, M.; Contin, M.; Mondini, C.; Brookes, P. Soil microbial biomass is triggered into activity by trace amounts of substrate. *Soil Biol. Biochem.* **2001**, *33*, 1163–1170. [CrossRef]
60. Amaresan, N.; Kumar, M.S.; Annapurna, K.; Kumar, K.; Sankaranarayanan, A. *Beneficial Microbes in Agro-Ecology: Bacteria and Fungi*; Academic Press: Cambridge, MA, USA, 2020.
61. Bezuglova, O.; Polienko, E.; Gorovtsov, A.; Lyhman, V.; Pavlov, P. The effect of humic substances on winter wheat yield and fertility of ordinary chernozem. *Ann. Agrar. Sci.* **2017**, *15*, 239–242. [CrossRef]
62. Semenov, M. Metabarcoding and metagenomics in soil ecology research: Achievements, challenges, and prospects. *Biol. Bull. Rev.* **2021**, *11*, 40–53. [CrossRef]
63. Chernov, T.; Zhelezova, A. The dynamics of soil microbial communities on different timescales: A review. *Eurasian Soil Sci.* **2020**, *53*, 643–652. [CrossRef]
64. Wei, J.-L.; Qin, Y.; Khan, Q.; Liang, W.-T.; He, W.-L.; Dong, D.-F.; Xing, Y.-X.; Li, Y.-R. Different responses of sugarcane and rhizosphere soil microorganisms to single or mixture application of PGPB. *Res. Square* **2024**. [CrossRef]
65. Beltran-Garcia, M.J.; Martínez-Rodríguez, A.; Olmos-Arriaga, I.; Valdes-Salas, B.; Di Mascio, P.; White, J.F. Nitrogen fertilization and stress factors drive shifts in microbial diversity in soils and plants. *Symbiosis* **2021**, *84*, 379–390. [CrossRef]
66. Damo, J.L.C.; Shimizu, T.; Sugiura, H.; Yamamoto, S.; Agake, S.-i.; Anarna, J.; Tanaka, H.; Sugihara, S.; Okazaki, S.; Yokoyama, T. The Application of Sulfur Influences Microbiome of Soybean Rhizosphere and Nutrient-Mobilizing Bacteria in Andosol. *Microorganisms* **2023**, *11*, 1193. [CrossRef]

67. Olimi, E.; Kusstatscher, P.; Wicaksono, W.A.; Abdelfattah, A.; Cernava, T.; Berg, G. Insights into the microbiome assembly during different growth stages and storage of strawberry plants. *Environ. Microbiome* **2022**, *17*, 21. [CrossRef] [PubMed]
68. Ayoubi, N.; Soleimani, M.; Zare, R. Pilidium concavum, causing tan-brown rot on strawberry in Iran. *J. Plant Pathol.* **2016**, *98*, 667–669.
69. Thảo, H.H.M.; Lê Thị, V.H.; Minh, Đ.H.N.; Quốc, N.V.; Việt, T.Q.; Duyên, N.N.; Tâm, B.N.; Ẩn, N.N.; Việt, P.T.; Hạnh, N.T.D. Screening of antagonistic bacteria strains against Pilidium sp. MD1 causing strawberry (*Fragaria ananassa*) leaf spot. *J. Sci. Technol.-IUH* **2022**, *55*, 134–141.
70. Silva, L.I.d.; Oliveira, I.P.d.; Jesus, E.d.C.; Pereira, M.C.; Pasqual, M.; Araújo, R.C.d.; Dória, J. Fertilizer of the future: Beneficial bacteria promote strawberry growth and yield and may reduce the need for chemical fertilizer. *Agronomy* **2022**, *12*, 2465. [CrossRef]

MDPI AG
Grosspeteranlage 5
4052 Basel
Switzerland
Tel.: +41 61 683 77 34

Horticulturae Editorial Office
E-mail: horticulturae@mdpi.com
www.mdpi.com/journal/horticulturae